Conference Proceedings of the Society for Experimental Mechanics Series

Series Editor

Kristin B. Zimmerman, Ph.D.
Society for Experimental Mechanics, Inc.,
Bethel, CT, USA

More information about this series at http://www.springer.com/series/8922

Luciano Lamberti • Ming-Tzer Lin • Cosme Furlong • Cesar Sciammarella
Phillip L. Reu • Michael A Sutton
Editors

Advancement of Optical Methods & Digital Image Correlation in Experimental Mechanics, Volume 3

Proceedings of the 2018 Annual Conference on Experimental and Applied Mechanics

Springer

Editors
Luciano Lamberti
Dipartimento Meccanica
Matematica e Management
Politechnico di Bari
Bari, Italy

Cosme Furlong
Mechanical Engineering Department
Worcester Polytechnic Institute
Worcester, MA, USA

Phillip L. Reu
Sandia National Laboratory
Albuquerque, NM, USA

Ming-Tzer Lin
Graduate Institute of Precision Engineering
National Chung Hsing University
Taichung, Taiwan

Cesar Sciammarella
Department of Mechanical, Materials
and Aerospace Engineering
Illinois Institute of Technology
Chicago, IL, USA

Department of Mechanical Engineering
Northern Illinois University
DeKalb, IL, USA

Michael A Sutton
Department of Mechanical Engineering
University of South Carolina
Columbia, SC, USA

ISSN 2191-5644 ISSN 2191-5652 (electronic)
Conference Proceedings of the Society for Experimental Mechanics Series
ISBN 978-3-030-07362-6 ISBN 978-3-319-97481-1 (eBook)
https://doi.org/10.1007/978-3-319-97481-1

This Springer imprint is published by the registered company Springer Nature Switzerland AG
The registered company address is: Gewerbestrasse 11, 6330 Cham, Switzerland

Preface

Advancement of Optical Methods & Digital Image Correlation in Experimental Mechanics represents one of the eight volumes of technical papers presented at the 2018 SEM Annual Conference & Exposition on Experimental and Applied Mechanics organized by the Society for Experimental Mechanics and held in Greenville, SC, June 4–7, 2018. The complete proceedings also include volumes on *Dynamic Behavior of Materials*; *Challenges in Mechanics of Time-Dependent Materials*; *Mechanics of Biological Systems & Micro-and Nanomechanics*; *Mechanics of Composite, Hybrid & Multifunctional Materials*; *Fracture, Fatigue, Failure and Damage Evolution*; *Residual Stress, Thermomechanics & Infrared Imaging, Hybrid Techniques and Inverse Problems*; and *Mechanics of Additive and Advanced Manufacturing.*

Each collection presents early findings from experimental and computational investigations on an important area within experimental mechanics, optical methods, and DIC being important areas.

With the advancement in imaging instrumentation, lighting resources, computational power, and data storage, optical methods have gained wide applications across the experimental mechanics society during the past decades. These methods have been applied for measurements over a wide range of spatial domain and temporal resolution. Optical methods have utilized a full-range of wavelengths from X-ray to visible lights and infrared. They have been developed not only to make two-dimensional and three-dimensional deformation measurements on the surface but also to make volumetric measurements throughout the interior of a material body.

Besides regular papers, the volume includes invited papers presented in a symposium on optical methods celebrating the 75th anniversary of the Society for Experimental Mechanics.

The area of digital image correlation has been an integral track within the SEM Annual Conference spearheaded by Professor Michael Sutton from the University of South Carolina. The contributed papers within this section of the volume span technical aspects of DIC.

The conference organizers thank the authors, presenters, and session chairs for their participation, support, and contribution to this very exciting area of experimental mechanics.

Bari, Italy — Luciano Lamberti
Taichung, Taiwan — Ming-Tzer Lin
Worcester, MA, USA — Cosme Furlong
Chicago, IL, USA — Cesar Sciammarella
Albuquerque, NM, USA — Phillip L. Reu
Columbia, SC, USA — Michael A Sutton

Contents

Contributors

T. Archer LMT (ENS Cachan/CNRS/University Paris-Saclay), Cachan, France

SAFRAN, Safran Ceramics, Le Haillan, France

ONERA, The French Aerospace Lab, Palaiseau, France

R. Balcaen KU Leuven, Ghent, Belgium

C. Barile Dipartimento di Meccanica, Matematica e Management (DMMM), Politecnico di Bari, Bari, Italy

P. Beauchêne ONERA, The French Aerospace Lab, Palaiseau, France

M. Berny LMT (ENS Cachan/CNRS/University Paris-Saclay), Cachan, France

SAFRAN, Safran Ceramics, Le Haillan, France

Kaushik Bhattacharya Division of Engineering and Applied Science, California Institute of Technology, Pasadena, CA, USA

V. Bianco NEAPoLIS Numerical and Experimental Advanced Program on Liquids and Interface Systems, Naples, Italy

Dipartimento di Ingegneria Chimica, dei Materiali e della Produzione Industriale—DICMaPI, University of Naples Federico II, Naples, Italy

B. Blaysat Université Clermont Auvergne, SIGMA, Institut Pascal, UMR CNRS 6602, Clermont-Ferrand, France

H. Bruck University of Maryland, College Park, MD, USA

Frédéric Canévet LC-DRIME, Joint Research Laboratory, Cooper Standard—Institut de Physique UMR 6251, Rennes Cedex, France

Cooper Standard France, Rennes, France

C. Casavola Dipartimento di Meccanica, Matematica e Management (DMMM), Politecnico di Bari, Bari, Italy

Terry Yuan-Fang Chen Department of Mechanical Engineering, National Cheng Kung University, Tainan, Taiwan

Yung-Hsiang Chen ITRC, NARL, Hsinchu, Taiwan, Republic of China

Fa-Yen Cheng Department of Mechanical Engineering, National Cheng Kung University, Tainan, Taiwan

Jeffrey Tao Cheng Eaton-Peabody Laboratory, Massachusetts Eye and Ear (MEE), Boston, MA, USA

Department of Otology and Laryngology, Harvard Medical School, Boston, MA, USA

Fu-Pen Chiang Department of Mechanical Engineering, State University of New York at Stony Brook, Stony Brook, NY, USA

Gary L. Cloud Michigan State University, Composite Vehicle Research Center, Lansing, MI, USA

Kevin B. Connolly Southern Research Institute, Birmingham, AL, USA

James Dawson Medtronic, Inc., Minneapolis, MN, USA

E. Di Maio NEAPoLIS Numerical and Experimental Advanced Program on Liquids and Interface Systems, Naples, Italy

Dipartimento di Ingegneria Chimica, dei Materiali e della Produzione Industriale—DICMaPI, University of Naples Federico II, Naples, Italy

John F. Eberth Department of Cell Biology and Anatomy, University of South Carolina, Columbia, SC, USA

Horacio D. Espinosa Department of Mechanical Engineering, Northwestern University Technological Institute, Evanston, IL, USA

P. Ferraro CNR-ISASI, Istituto di Scienze Applicate e Sistemi Intelligenti "E. Caianiello", Pozzuoli NA, Italy

NEAPoLIS Numerical and Experimental Advanced Program on Liquids and Interface Systems, Naples, Italy

V. Ferraro NEAPoLIS Numerical and Experimental Advanced Program on Liquids and Interface Systems, Naples, Italy

Dipartimento di Ingegneria Chimica, dei Materiali e della Produzione Industriale—DICMaPI, University of Naples Federico II, Naples, Italy

Cosme Furlong Center for Holographic Studies and Laser micro-mechaTronics (CHSLT), Worcester Polytechnic Institute, Worcester, MA, USA

Mechanical Engineering Department, Worcester Polytechnic Institute, Worcester, MA, USA

Eaton-Peabody Laboratory, Massachusetts Eye and Ear (MEE), Boston, MA, USA

Department of Otology and Laryngology, Harvard Medical School, Boston, MA, USA

Amos Gilat Department of Mechanical Engineering, The Ohio State University, Scott Laboratory, Columbus, OH, USA

Austin Giordano Department of Mechanical Engineering, State University of New York at Stony Brook, Stony Brook, NY, USA

M. Grédiac Université Clermont Auvergne, SIGMA, Institut Pascal, UMR CNRS 6602, Clermont-Ferrand, France

M. Grewer LaVision GmbH, Goettingen, Germany

S. Grilli CNR-ISASI, Istituto di Scienze Applicate e Sistemi Intelligenti "E. Caianiello", Pozzuoli NA, Italy

NEAPoLIS Numerical and Experimental Advanced Program on Liquids and Interface Systems, Naples, Italy

V. Herb SAFRAN, Safran Ceramics, Le Haillan, France

F. Hild LMT (ENS Cachan/CNRS/University Paris-Saclay), Cachan, France

Matthew S. Hoehler National Institute of Standards and Technology, Gaithersburg, MD, USA

Kevin Huang Department of Mechanical Engineering, University of South Carolina, Columbia, SC, USA

Chi-Hung Hwang ITRC, NARL, Hsinchu, Taiwan, Republic of China

M. Iadicola National Institute of Standards, Gaithersburg, MD, USA

Takanori Igarashi Skincare Products Research, Kao Corporation, Tokyo, Japan

Isabelle Jeanneau LC-DRIME, Joint Research Laboratory, Cooper Standard—Institut de Physique UMR 6251, Rennes Cedex, France

Cooper Standard France, Rennes, France

E. M. C. Jones Sandia National Laboratories, Albuquerque, NM, USA

R. Kitey Department of Aerospace Engineering, Indian Institute of Technology, Kanpur, India

B. Lacombe SAFRAN, Safran Ceramics, Le Haillan, France

L. Lamberti Dipartimento Meccanica, Matematica e Management, Politecnico di Bari, Bari, Italy

P. Lava MatchID, Ghent, Belgium

Jean-Benoit Le Cam Univ Rennes, CNRS, IPR (Institute de Physique de Rennes)–UMR 6251, Rennes, France

LC-DRIME, Joint Research Laboratory, Cooper Standard—Institut de Physique UMR 6251, Rennes Cedex, France

Susan M. Lessner Department of Cell Biology and Anatomy, University of South Carolina, Columbia, SC, USA

Ning Li Department of Mechanical Engineering, University of South Carolina, Columbia, SC, USA

Xin Li Collaborative Innovation Center of Advanced Nuclear Energy Technology, The Key Laboratory of Advanced Reactor Engineering and Safety of MOE, Institute of Nuclear and New Energy Technology of Tsinghua University, Beijing, China

Ming-Tzer Lin Graduate Institute of Precision Engineering, National Chung Hsing University, Taichung, Taiwan

Didier Loison Univ Rennes, CNRS, IPR (Institute de Physique de Rennes)–UMR 6251, Rennes, France

P. L. Maffettone NEAPoLIS Numerical and Experimental Advanced Program on Liquids and Interface Systems, Naples, Italy

Dipartimento di Ingegneria Chimica, dei Materiali e della Produzione Industriale—DICMaPI, University of Naples Federico II, Naples, Italy

Nima Maftoon Eaton-Peabody Laboratory, Massachusetts Eye and Ear (MEE), Boston, MA, USA

Department of Otology and Laryngology, Harvard Medical School, Boston, MA, USA

B. Mandracchia CNR-ISASI, Istituto di Scienze Applicate e Sistemi Intelligenti "E. Caianiello", Pozzuoli NA, Italy

NEAPoLIS Numerical and Experimental Advanced Program on Liquids and Interface Systems, Naples, Italy

Gérard Mauvoisin Laboratoire de Génie Civil et Génie Mécanique EA 3913, IUT-Université de Rennes 1, Rennes, France

A. Mavel ONERA, The French Aerospace Lab, Palaiseau, France

P. Memmolo CNR-ISASI, Istituto di Scienze Applicate e Sistemi Intelligenti "E. Caianiello", Pozzuoli NA, Italy

NEAPoLIS Numerical and Experimental Advanced Program on Liquids and Interface Systems, Naples, Italy

F. Merola CNR-ISASI, Istituto di Scienze Applicate e Sistemi Intelligenti "E. Caianiello", Pozzuoli NA, Italy

NEAPoLIS Numerical and Experimental Advanced Program on Liquids and Interface Systems, Naples, Italy

L. Miccio CNR-ISASI, Istituto di Scienze Applicate e Sistemi Intelligenti "E. Caianiello", Pozzuoli NA, Italy

NEAPoLIS Numerical and Experimental Advanced Program on Liquids and Interface Systems, Naples, Italy

Yukihiro Miyazaki Makeup Products Research, Kao Corporation, Odawara, Japan

Yoshiharu Morimoto Wakayama University, Wakayama, Japan

4D Sensor Inc., Wakayama, Japan

Paul Moy U.S. Army Research Laboratory, Weapons and Materials Research Directorate, Adelphi, MD, USA

M. Mugnano CNR-ISASI, Istituto di Scienze Applicate e Sistemi Intelligenti "E. Caianiello", Pozzuoli NA, Italy

NEAPoLIS Numerical and Experimental Advanced Program on Liquids and Interface Systems, Naples, Italy

Yuko Nagai Makeup Products Research, Kao Corporation, Odawara, Japan

Rupesh Nasre Department of Engineering Design, Indian Institute of Technology, Madras, Chennai, India

V. Pagliarulo CNR-ISASI, Istituto di Scienze Applicate e Sistemi Intelligenti "E. Caianiello", Pozzuoli NA, Italy

NEAPoLIS Numerical and Experimental Advanced Program on Liquids and Interface Systems, Naples, Italy

Wen Chieh Pan Graduate Institute of Precision Engineering, National Chung Hsing University, Taichung, Taiwan

Ashutosh Pandey Department of Applied Mechanics, Indian Institute of Technology Madras, Chennai, India

G. Pappalettera Dipartimento di Meccanica, Matematica e Management (DMMM), Politecnico di Bari, Bari, Italy

C. Pappalettere Dipartimento di Meccanica, Matematica e Management, Politecnico di Bari, Bari, Italy

F. Pierron Engineering and the Environment, University of Southampton, Southampton, UK

Lionel Pirsig 4D Sensor Inc., Wakayama, Japan

K. Pooladvand Center for Holographic Studies and Laser micro-mechaTronics (CHSLT), Worcester, MA, USA

Mechanical Engineering Department, Worcester Polytechnic Institute, Worcester, MA, USA

W. Carter Ralph Southern Research Institute, Birmingham, AL, USA

K. Ramesh Department of Applied Mechanics, Indian Institute of Technology Madras, Chennai, India

Payam Razavi Center for Holographic Studies and Laser micro-mechaTronics (CHSLT), Worcester Polytechnic Institute, Worcester, MA, USA

Mechanical Engineering Department, Worcester Polytechnic Institute, Worcester, MA, USA

P. L. Reu Sandia National Laboratories, Albuquerque, NM, USA

Eric Robin Univ Rennes, CNRS, IPR (Institute de Physique de Rennes)–UMR 6251, Rennes, France

LC-DRIME, Joint Research Laboratory, Cooper Standard—Institut de Physique UMR 6251, Rennes Cedex, France

John J. Rosowski Eaton-Peabody Laboratory, Massachusetts Eye and Ear (MEE), Boston, MA, USA

Department of Otology and Laryngology, Harvard Medical School, Boston, MA, USA

S. Roux LMT, ENS Paris Saclay/CNRS UMR 8535/University Paris Saclay, Cachan, France

Benoit Ruellan Univ Rennes, CNRS, IPR (Institute de Physique de Rennes)–UMR 6251, Rennes, France

LC-DRIME, Joint Research Laboratory, Cooper Standard—Institut de Physique UMR 6251, Rennes Cedex, France

Cooper Standard France, Rennes, France

Kasumi Sakai Department of Mechanical Engineering, Aoyama Gakuin University, Sagamihara, Japan

Cesar A. Sciammarella Department of Mechanical, Materials and Aerospace Engineering, Illinois Institute of Technology, Chicago, IL, USA

Department of Mechanical Engineering, Northern Illinois University, DeKalb, IL, USA

Jeremy D. Seidt Department of Mechanical Engineering, The Ohio State University, Scott Laboratory, Columbus, OH, USA

Jianguang Shuai Department of Mechanical Engineering of Tsinghua University, Beijing, China

T. Siebert Dantec Dynamics GmbH, Ulm, Germany

M. Simonsen Correlated Solutions Inc., Columbia, SC, USA

Sarthak S. Singh Department of Aerospace Engineering, Indian Institute of Technology, Kanpur, India

Christopher M. Smith Berkshire Hathaway Specialty Insurance, Boston, MA, USA

Jarrod L. Smith Department of Mechanical Engineering, The Ohio State University, Scott Laboratory, Columbus, OH, USA

Karl A. Stetson Karl Stetson Associates, LLC, Coventry, CT, USA

Sankara J. Subramanian Department of Engineering Design, Indian Institute of Technology, Madras, Chennai, India

F. Sur Université de Lorraine, CNRS, INRIA projet Magrit, Vandoeuvre-lès-Nancy Cedex, France

Y. Surrel Engineering and the Environment, University of Southampton, Southampton, UK

Yves Surrel Expertise and Consultancy, St-Étienne, France

Michael A. Sutton Department of Mechanical Engineering, University of South Carolina, Columbia, SC, USA

Akifumi Takagi 4D Sensor Inc., Wakayama, Japan

Haimi Tang Center for Holographic Studies and Laser micro-mechaTronics (CHSLT), Worcester Polytechnic Institute, Worcester, MA, USA

Mechanical Engineering Department, Worcester Polytechnic Institute, Worcester, MA, USA

Mullai Thiagu Department of Engineering Design, Indian Institute of Technology, Madras, Chennai, India

E. Toussaint Université Clermont Auvergne, Clermont-Ferrand, France

Ang-Ting Tsai Graduate Institute of Precision Engineering, National Chung Hsing University, Taichung, Taiwan

D. Turner Sandia National Laboratories, Albuquerque, NM, USA

Masaki Ueki 4D Sensor Inc., Wakayama, Japan

M. Villone NEAPoLIS Numerical and Experimental Advanced Program on Liquids and Interface Systems, Naples, Italy

Dipartimento di Ingegneria Chimica, dei Materiali e della Produzione Industriale—DICMaPI, University of Naples Federico II, Naples, Italy

Timothy Walter U.S. Army Research Laboratory, Weapons and Materials Research Directorate, Adelphi, MD, USA

Wei-Chung Wang Department of Power Mechanical Engineering, National Tsing Hua University, Hsinchu, Taiwan, Republic of China

Z. Wang CNR-ISASI, Istituto di Scienze Applicate e Sistemi Intelligenti "E. Caianiello", Pozzuoli NA, Italy

NEAPoLIS Numerical and Experimental Advanced Program on Liquids and Interface Systems, Naples, Italy

Charles Wilson Medtronic, Inc., Minneapolis, MN, USA

Xinxin Wu Collaborative Innovation Center of Advanced Nuclear Energy Technology, The Key Laboratory of Advanced Reactor Engineering and Safety of MOE, Institute of Nuclear and New Energy Technology of Tsinghua University, Beijing, China

NanSheng Xu Department of Mechanical Engineering, University of South Carolina, Columbia, SC, USA

Jin Yang Division of Engineering and Applied Science, California Institute of Technology, Pasadena, CA, USA

Satoru Yoneyama Department of Mechanical Engineering, Aoyama Gakuin University, Sagamihara, Japan

Yuelin Zhang Department of Mechanical Engineering, Aoyama Gakuin University, Sagamihara, Japan

Zhengming Zhang Collaborative Innovation Center of Advanced Nuclear Energy Technology, The Key Laboratory of Advanced Reactor Engineering and Safety of MOE, Institute of Nuclear and New Energy Technology of Tsinghua University, Beijing, China

Jiaqing Zhao Collaborative Innovation Center of Advanced Nuclear Energy Technology, The Key Laboratory of Advanced Reactor Engineering and Safety of MOE, Institute of Nuclear and New Energy Technology of Tsinghua University, Beijing, China

Chapter 1
Beyond the Airbrush: Applications of Digital Image Correlation in Vascular Biomechanics

Susan M. Lessner and John F. Eberth

Abstract Before digital image correlation (DIC) could find widespread application for strain measurement in biological soft tissues, there were a number of technical challenges that had to be addressed. The mechanical behavior of soft tissues depends significantly on hydration state; therefore, both application of a high-contrast speckle pattern and imaging must be achieved while maintaining the specimen in a fully hydrated state. Furthermore, soft tissues such as arteries typically undergo finite deformations under a physiologically relevant loading range. While hydration can be achieved by submerging the sample in a medium having appropriate osmolarity, pH, and ionic strength, imaging submerged objects introduces its own set of challenges due to refractive index changes. The issue of sample hydration also requires consideration of alternative approaches to speckle pattern creation, beyond the classic "airbrush method," since ideally the specimen must not be allowed to dry out during pattern application. For samples that are submerged, the speckle pattern must be firmly bonded to the specimen and water-resistant, in addition to deforming with the specimen, often outside the small-strain regime. In some specific instances, nature provides a hand through the presence of intrinsic fine-scale structure in the specimen that, with innovative staining or imaging techniques, can serve as a satisfactory "speckle pattern." Added to these issues is the difficulty of imaging small, often irregularly shaped specimens that can degrade rapidly over time. In collaboration with Dr. Sutton's group, my colleagues and I have developed a number of approaches to address the issues of hydrating and speckling soft tissues for measurement of local strains in blood vessels at multiple length scales, with a particular focus on the mouse aorta and carotid artery.

Keywords Digital image correlation · Soft tissues · Speckle pattern · Artery · Nuclear stain

1.1 Introduction

Digital image correlation (DIC) originally was developed at the University of South Carolina in the 1980s as a non-destructive technique to accurately measure small strains and deformations in engineering materials such as metals and ceramics. This technique relies on optically tracking displacements of a random, high-contrast "speckle pattern" on the specimen surface during loading.

1.2 What Is a "Speckle Pattern"?

The basic requirements for a "speckle pattern" suitable for DIC are that it be random, high-contrast, and scaled appropriately to the specimen and camera system in use. 'Randomness' implies that the pattern within the original subset area and its corresponding image after deformation be uniquely matched in a least squares best-fit sense. Patterns that are too regular in their intensity variation (e.g., a grid of evenly spaced black dots on a white background) or that have too small an intensity range (low contrast) can give rise to non-unique solutions during the matching process. A number of authors have developed analytical criteria for characterizing the quality of speckle patterns [1–4]. The physical size of 'speckles' can

S. M. Lessner (✉) · J. F. Eberth
Department of Cell Biology and Anatomy, University of South Carolina, Columbia, SC, USA
e-mail: susan.lessner@uscmed.sc.edu

L. Lamberti et al. (eds.), *Advancement of Optical Methods & Digital Image Correlation in Experimental Mechanics, Volume 3*, Conference Proceedings of the Society for Experimental Mechanics Series, https://doi.org/10.1007/978-3-319-97481-1_1

vary considerably, but a rule of thumb for optimum tracking of the pattern is to create speckles that cover 3–5 pixels in each direction at the image plane of the camera.

Additional considerations in optimizing speckle size based on camera parameters, focal length and desired field of view can be found in [5].

1.3 Challenges in "Speckling" Soft Tissues

In comparison to typical engineering materials, soft biological tissues present a number of challenges in creating a surface speckle pattern suitable for DIC. Perhaps most vexing of these is the issue of maintaining adequate hydration. Human soft tissues range from roughly 25% to somewhat more than 75% water by weight [6], and their mechanical behavior is highly dependent on hydration state. Some water in soft tissues is tightly bound within the extracellular matrix—for example, water molecules associated with highly negatively charged proteoglycans—and its removal leads to collapse of associated matrix architecture. Thus, soft tissue specimens cannot be allowed to dry out completely. Hydration must be maintained during both patterning and imaging, which can be challenging when high intensity lights are used to ensure good image contrast. When possible, performing both patterning and imaging in solution are preferable, but these strategies present their own sets of challenges.

Another challenge in speckling soft tissues arises from the large deformations experienced by many of these materials under loading. Elastin-rich tissues such as ligaments and elastic arteries can undergo strains of 50% or more under load and return to their initial dimensions with little obvious damage (i.e., without hysteresis in the loading-unloading curve). Any type of speckle pattern created by coating the specimen, either with paint or with some other film-forming material, must not only deform with the underlying tissue but must also resist delamination, cracking, and wrinkling.

1.4 Some Potential Solutions to Speckle Patterning of Soft Biological Tissues

Early applications of DIC to biological specimens focused on hard tissues such as bone and used classic approaches such as airbrush application of black paint to create a suitable speckle pattern. Some pioneering studies of soft tissues [7, 8] also tried this approach with some success, although samples had to be allowed to dry off to some extent to obtain a pattern. The need to maintain tissue hydration can cause smearing or spreading in patterns created using standard inks, dyes, or paints, whether applied by hand or by airbrush. In our first collaborative efforts with Dr. Sutton [9], we used toner powder (3–10 micron, Ricoh) suspended in phosphate-buffered saline (PBS) with the addition of a small amount of surfactant to permit wetting of the hydrophobic powder. This approach provided a high-contrast but somewhat sparse speckle pattern. Unfortunately, there was some tendency for the speckles to wash off as the specimen was rehydrated between image series with a saline drip.

As the advantages of DIC for strain and deformation measurements of soft tissues became evident, several groups, including our own, tested a variety of approaches to overcome the challenges of creating robust speckle patterns on these materials. In the first DIC investigation of human aorta mechanics [10], Avril and colleagues created a speckle pattern by hand-gluing a set of 300+ spherical black markers to the specimen using fast-drying cyanoacrylate glue. Glues formulated with 2-octyl cyanoacrylate rather than methyl cyanoacrylate have relatively high viscosity, limiting penetration into the tissue, and are used as surgical skin adhesives (e.g., 'Dermabond', Ethicon). This method is limited to relatively large specimens and is too time-consuming for analysis of large numbers of specimens. Murphy and colleagues used gold nanorods suspended in a collagen gel to measure deformations associated with fibroblast-mediated remodeling of the matrix [11]. In this study, the gold nanorods were visualized under darkfield microscopy. Gold or silver nanoparticles can also be imaged by cross-polarized incident light microscopy, with the extent of depolarization dependent on particle size and shape [12, 13].

In our own stereo-DIC studies of mouse carotid artery biomechanics, we ultimately settled on using fluorescent nuclear staining as a method to create a random, dense speckle pattern (Fig. 1.1) [14, 15]. To improve the contrast, we filled the vessel lumen with diluted black tissue marking dye. While we chose to use ethidium bromide as a nuclear stain, there are a wide variety of fluorescent DNA-binding dyes that would serve equally well; the choice can be based on available excitation light sources and emission filters. Recent improvements in dye chemistry have also improved quantum yields and reduced photobleaching relative to many 'classic' fluorophores. In addition, we developed methods to covalently attach fluorescent microspheres to the adventitial or endothelial surface of mouse blood vessels using carbodiimide chemistry, although ultimately we abandoned this approach in favor of simpler, less time-consuming nuclear staining.

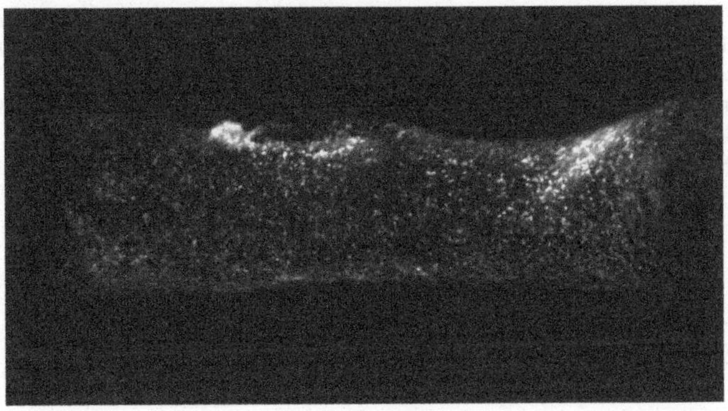

Fig. 1.1 Speckle pattern created on adventitial surface of mouse carotid artery by fluorescent nuclear staining with ethidium bromide. Vessel diameter is about 300 μm

In some fortuitous instances, nature provides an intrinsic pattern in or on the material that proves suitable for DIC image tracking. Pigmentation of the developing *Xenopus laevis* (African clawed frog) embryo is one such example. Some biological materials that lack an obvious pattern under white light imaging may reveal intrinsic surface textures when viewed at specific wavelengths or under suitable imaging modalities. For example, adventitial collagen fibers on the outer surface of large blood vessels are nearly transparent under white light, but they produce a strong second harmonic signal when visualized with coherent light at near-IR wavelengths [16].

1.5 Conclusion

Continuing advances in nanoparticle synthesis, bioconjugate chemistry, optics, and image correlation software have alleviated many of the difficulties originally faced by investigators interested in applying DIC to soft biological tissues, including arteries.

References

1. Lecompte, D., Smits, A., Bossuyt, S., Sol, H., Vantomme, J., Van Hemelrijck, D., Habraken, A.: Quality assessment of speckle patterns for digital image correlation. Opt. Lasers Eng. **44**, 1132–1145 (2006)
2. Pan, B., Lu, Z., Xie, H.: Mean intensity gradient: an effective global parameter for quality assessment of the speckle patterns used in digital image correlation. Opt. Lasers Eng. **48**, 469–477 (2010)
3. Reu, P.L., Miller, T.J., Sutton, M., Wang, Y.: Uncertainty Quantification for Digital Image Correlation. Sandia National Laboratories (SNL-NM), Albuquerque, NM (2009)
4. Wang, Y.Q., Sutton, M.A., Bruck, H.A., Schreier, H.W.: Quantitative error assessment in pattern matching: effects of intensity pattern noise, interpolation, strain and image contrast on motion measurements. Strain. **45**, 160–178 (2009)
5. Sutton, M.A., Orteu, J.J., Schreier, H.: Image Correlation for Shape, Motion and Deformation Measurements: Basic Concepts, Theory and Applications. Springer, New York, NY (2009)
6. Forbes, R., Cooper, A., Mitchell, H.: The composition of the adult human body as determined by chemical analysis. J. Biol. Chem. **203**, 359–366 (1953)
7. Zhang, D., Arola, D.D.: Applications of digital image correlation to biological tissues. J. Biomed. Opt. **9**, 691–699 (2004)
8. Zhang, D.S., Eggleton, C.D., Arola, D.D.: Evaluating the mechanical behavior of arterial tissue using digital image correlation. Exp. Mech. **42**, 409–416 (2002)
9. Sutton, M.A., Ke, X., Lessner, S.M., Goldbach, M., Yost, M., Zhao, F., Schreier, H.W.: Strain field measurements on mouse carotid arteries using microscopic three-dimensional digital image correlation. J. Biomed. Mater. Res. A. **84**, 178–190 (2008)
10. Avril, S., Badel, P., Duprey, A.: Anisotropic and hyperelastic identification of in vitro human arteries from full-field optical measurements. J. Biomech. **43**, 2978–2985 (2010)
11. Stone, J.W., Sisco, P.N., Goldsmith, E.C., Baxter, S.C., Murphy, C.J.: Using gold nanorods to probe cell-induced collagen deformation. Nano Lett. **7**, 116–119 (2007)
12. Aaron, J., de La Rosa, E., Travis, K., Harrison, N., Burt, J., José-Yacamán, M., Sokolov, K.: Polarization microscopy with stellated gold nanoparticles for robust, in-situ monitoring of biomolecules. Opt. Express. **16**, 2153–2167 (2008)
13. Sokolov, K., Follen, M., Aaron, J., Pavlova, I., Malpica, A., Lotan, R., Richards-Kortum, R.: Real-time vital optical imaging of precancer using anti-epidermal growth factor receptor antibodies conjugated to gold nanoparticles. Cancer Res. **63**, 1999 (2003)

14. Ning, J., Braxton, V.G., Wang, Y., Sutton, M.A., Wang, Y., Lessner, S.M.: Speckle patterning of soft tissues for strain field measurement using digital image correlation: preliminary quality assessment of patterns. Microsc. Microanal. **17**, 81–90 (2011)
15. Ning, J., Xu, S., Wang, Y., Lessner, S.M., Sutton, M.A., Anderson, K., Bischoff, J.E.: Deformation measurements and material property estimation of mouse carotid artery using a microstructure-based constitutive model. J. Biomech. Eng. **132**, 121010 (2010)
16. Watson, S.R., Lessner, S.M.: (Second) harmonic disharmony: nonlinear microscopy shines new light on the pathology of atherosclerosis. Microsc. Microanal. **22**, 589–598 (2016)

Chapter 2
A New Method of Fringe Pattern Analysis

C. A. Sciammarella and L. Lamberti

Abstract Fringe patterns can be optically generated by the different developed optical methods or computer generated. This study tackles down one of the more difficult aspects of classical fringe pattern analysis, pattern unwrapping. Discontinuous patterns such as the microscopic displacement/strain fields of a particulate composite and shear bands are studied.

Keywords Unwrapping of singular fringe patterns · Determination of displacements and strains · Particulate composites · Shear bands

2.1 Introduction

In previous papers [1–4], the authors analyzed different aspects of processing displacement information given as scalar functions of gray levels called fringe patterns. These fringe patterns can be optically generated by the different developed optical methods or computer generated for example as it is the case in the magnetic resonance method (MRI). The approach adopted in these papers is the utilization of the Image Analysis Science basic framework supported by the general mathematical theory of transforms. The theory of transforms deals with vector spaces and the selection of a set of basis functions that lead to simplifications of the problem under study. In these papers, different basic concepts in fringe pattern analysis have been discussed within the wider frame work of the General Theory of Signal Analysis. One of the purposes of this paper is to introduce one fundamental aspect of data retrieval that was anticipated in [1] and is elaborated in a more comprehensive way in this paper. This paper tackles down one of the more difficult aspects of classical fringe pattern analysis, pattern unwrapping. This is a subject that has been thoroughly analyzed in the literature of fringe patterns and there is a classical book that deals with the different problems that arise in the interpretation of fringe patterns and different possible solutions for these problems, including algorithms and software that can be applied in different cases [5].

The main difficulty of the process of extracting information, for example from fringes that are related to displacement fields, the object of this publication, is the presence of singularities in the analyzed displacement field. If one has a continuous displacement field with continuous derivatives up to the third order (usual convention in Continuum Mechanics), the isothetic lines, lines of projected displacements (moiré fringes) are either closed lines or they end at the boundaries of the field, never intersecting each other. This is not the case in many of the applications of practical interest in Experimental Mechanics. In some fields of application, the continuity condition is valid only piecewise. For example, to analyze displacement fields in composites it is required the use of tiling of patches of continuous areas that possibly are separated by the presence of microscopic cracks and networks of dislocations. To calculate the displacement/strain values it is important to define the singularities present in a field. In [6], it is shown that in complex fringe patterns it is possible to outline Burger-type circuits with the Burger's vector equal to the grating pitch. It should be noticed that the Burger's vector in this case is not a vector in the physical space but in the space of the projected displacements. Consequently, the analysis of patterns to obtain displacements requires a study of all the different fringe dislocations networks present in the field and the possible separation

C. A. Sciammarella (✉)
Department of Mechanical, Materials and Aerospace Engineering, Illinois Institute of Technology, Chicago, IL, USA

Department of Mechanical Engineering, Northern Illinois University, DeKalb, IL, USA
e-mail: sciammarella@iit.edu

L. Lamberti
Dipartimento Meccanica, Matematica e Management, Politecnico di Bari, Bari, Italy

L. Lamberti et al. (eds.), *Advancement of Optical Methods & Digital Image Correlation in Experimental Mechanics, Volume 3*, Conference Proceedings of the Society for Experimental Mechanics Series, https://doi.org/10.1007/978-3-319-97481-1_2

of continuous patches. This is not an easy task to fulfill when the fringe patterns have multiple singular points and complex fringe dislocation structures. Complex singular areas cannot be easily handled by self-adapting software and require the intervention of the operator performing the analysis of the singular patterns.

2.2 Concept of Phase Supporting a New Unwrapping Method

To understand the new unwrapping method, it is necessary to review some basic concepts on the analysis of fringe patterns. We are analyzing methods that use tagged carriers to obtain displacement information by comparing the initial configuration of a carrier with its final configuration. This comparison includes large displacements that imply the comparison of very different geometries generated in the process of deformation of a body. This becomes feasible because a system of reference is introduced in the observed body [3]. Figure 2.1 shows large deformations of a ring with a system of tagged carriers. Since the tagged carriers can be given a parameter number, it is always possible to compare the initial basic element with its deformed position.

A matter of notation, we will be dealing real or complex-valued functions f(**x**) defined on R_2. Ordinary case letters will represent scalar quantities, bold letters will represent vectorial quantities and we will write f(**x**) or f(x,y), the bold lower case indicating a vector quantity or we will list the low cases x and y, whichever is more convenient in context.

There is another important observation, if we write the equation of the displacement in 2D with respect to a Cartesian frame of [1],

$$\mathbf{d}_T(\mathbf{x}) = \mathbf{d}_{rb}(\mathbf{x}) + \mathbf{d}_r(\mathbf{x}) \tag{2.1}$$

In the above equation, $\mathbf{d}_T(\mathbf{x})$ is the total displacement of a point of R_2, $\mathbf{d}_{rb}(\mathbf{x})$ is the result of a rigid body motion (translation and rotation), $\mathbf{d}_r(\mathbf{x})$ is the relative motion of points of R_2. As shown in [1], it is possible by analyzing a given pattern to separate the absolute motion that requires a reference point of known position with respect to a reference frame and the relative displacements. In the analysis of deformations, one is interested in the relative motions that are composed of deformations plus rigid body rotations. Since the carrier method is insensitive to rigid body translations, the assignment of a

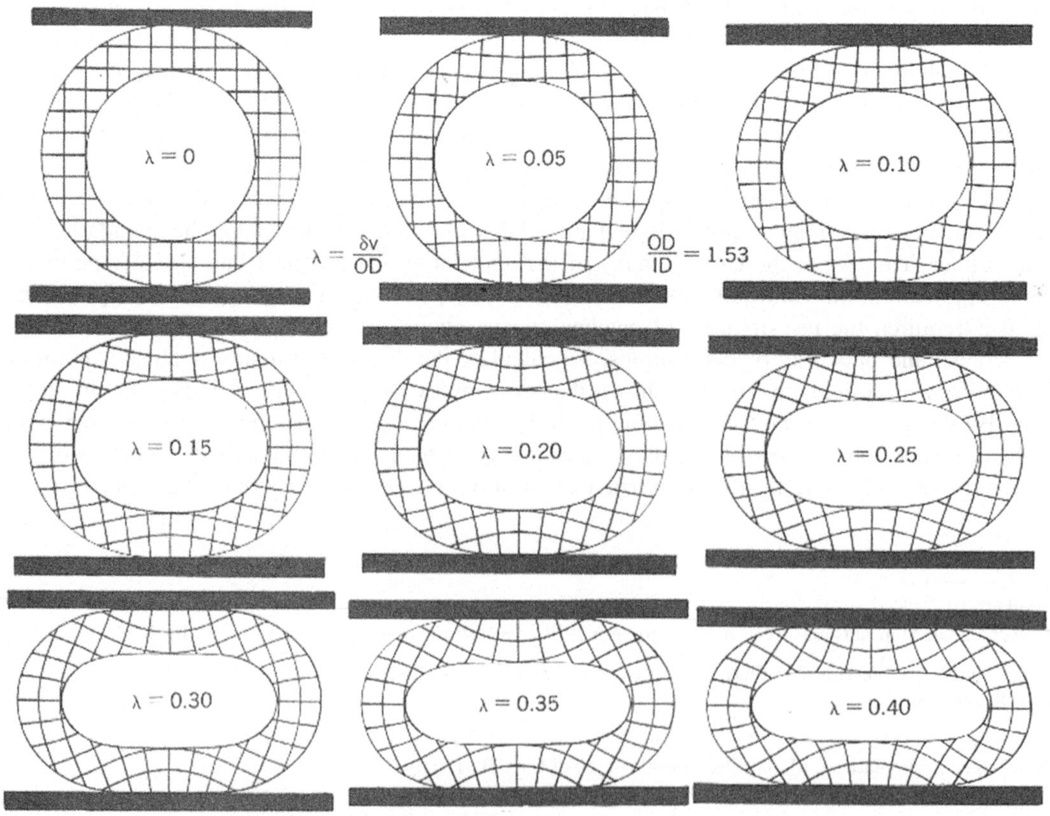

Fig. 2.1 Large deformations of a tagged body as a function of a reference parameter (Moiré Analysis of Strain, A. J. Durelli, V. J. Parks)

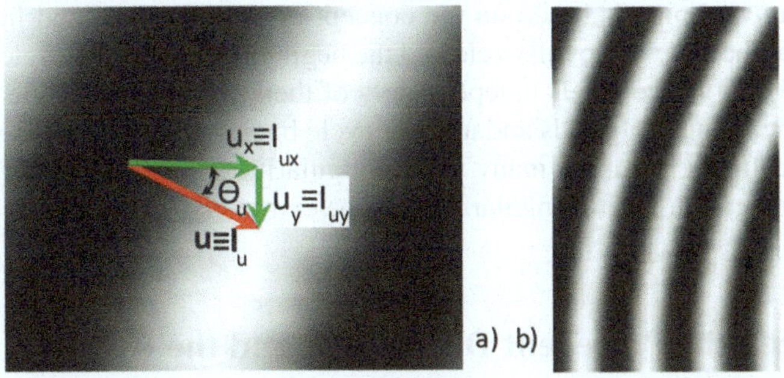

Fig. 2.2 (**a**) Enlarged view of region of pattern shown in **b**, (**b**) u pattern carrier in the initial condition parallel to the y-axis

zero displacement is done with a criterion that depends on the analyzed problem. This means that is not necessary to perform additional measurements to establish the zero order, as is the case in photoelastic fringes.

The displacement information is encoded in a carrier as levels of gray that represent the intensity at a given pixel I(**x**), a scalar field. A method of decoding the displacement information is to convert the scalar field of gray levels into a vectorial field of intensities **I**(**x**) that provide projected displacement vectors. This conversion is achieved using transforms, either the Fourier transform or the Hilbert transform [1–4]. As shown in Fig. 2.1, in the case of 2D a system of orthogonal carriers is required. The reason for this requirement is the mathematical need to define a tensorial field that is built in the mechanics of the continuum. These carriers have axes originally parallel to the coordinate axes. The carriers parallel to the x-axis give the vertical displacements or v-displacements, and the carriers parallel to the y-axis, give the horizontal displacements, or u-displacements plus the first order derivatives as it is going to be documented in this paper. After deformation, the carriers change their trajectories as shown in Fig. 2.2. The symbol ≡ indicating the co-axiality of the involved vectors.

The carrier originally parallel to the y-axis that provides the projected displacements in x-directions (u displacement) has rotated and at a point of the field the carrier makes an angle θ_u with the x axis. Locally, is defined a vector **u**(**x**) with components $u_x(\mathbf{x})$ and $u_y(\mathbf{x})$.

We need now to review the concept of phase, a pixel function, that in the literature is called local phase.

The phase concept is associated with the notion of vector. The definition of local phase or phase at a point for a 1D signal, introduces in the corresponding complex plane one additional dimension to the dimension of the physical plane, in the complex plane one has two dimensions. This notion is a fundamental concept in the Gabor's analytic signal theory [7], basic starting point of many developments in Signal Analysis and in Optics.

There is another concept of phase, the classical one in optics. This concept is the global phase that represents the optical path followed by a recorded wave front from a selected reference point where the phase is assumed to be zero, a selected departure point up to the final value at the point of arrival. The classical definition of phase in optics is an expression that measures distances through an angular variable,

$$\mathbf{d}\phi\,(\mathbf{x}) = \frac{2\pi\,\delta\,(\mathbf{x})}{\mathbf{p}} \tag{2.2}$$

where $\delta(x)$ is the optical path; p is pitch of a sinusoidal function, unit of measure utilized to evaluate a path length and convert distances into angles.

The optical path length of the light arriving at an image is given by,

$$\mathrm{L}(x) = \int_0^{S_0} \delta(x)dx \tag{2.3}$$

How these two concepts are connected to each other? Moving from the continuum model to the discrete representation, the global phase is the sum of the local phases [1],

$$\phi_g\,(\mathrm{x}) = \sum_{j=0}^{j=N} \Delta\phi_L\,(\mathrm{x}) \tag{2.4}$$

where $\phi_g(x)$ is the global phase at a pixel N, and $\phi_L(x)$ is the local phase at a given pixel.

The classical procedure of unwrapping is based on the concept of global phase, and introduces a problem: the need to secure continuity of the fringe at the end of one full cycle and the beginning of the next cycle. As it has been shown in [1], the addition of the local phases handles all the pixels independently of their position in a cycle. Thus, local phase addition does not need to recognize pixels where one cycle ends and another cycle begins. This feature is very important in the application of tagged carriers in complex patterns including many fringe singularities. The analysis of patterns observed in particulate composites is one important application where singularities can be extremely complex.

2.3 Relationship Between the Precedent Derivations and the Classical Approach for Fringe Pattern Analysis

To facilitate the understanding of the paper contents, some of the basic properties of isothetic lines are reviewed. These properties were initially introduced in [8] and are further expanded and summarized in [6].

Figure 2.3 represents an isothetic line of the family **u(x)**, the tangent to the isothetic line whose slope tgα_u gives the derivative $\partial u/\partial x$ of the displacement u, projection of the gradient vector $\nabla u(x)$on the x-axis, and the cross-derivative $\partial u/\partial y$ is the projection of the gradient on the y-axis. It can be seen that the angles θ_u (see Fig. 2.3) and α_u are related,

$$\alpha_u = \theta_u + \frac{\pi}{2} \tag{2.5}$$

Hence, we get the following relationship,

$$\frac{\partial u}{\partial x} = \text{tg}\ \alpha_u = -\cot\ g\ \theta_u \tag{2.6}$$

From Fig. 2.3, the following relationships are valid:

$$\text{tg}\ \theta_u = \frac{\frac{\partial u}{\partial y}}{\frac{\partial u}{\partial x}} \tag{2.7}$$

Also,

$$\text{tg}\ \alpha_u = -\frac{\frac{\partial u}{\partial x}}{\frac{\partial u}{\partial y}} \tag{2.8}$$

The negative sign considers the usual convention for the sign of slopes: for $\alpha < 90^o$, the slopes are positive, for $\alpha > 90^o$, the slopes are negative.

Similar relationships can be derived for the field of the v-displacements.

Fig. 2.3 Isothetic line of the u family, gradient vector and the partial derivatives with respect to the coordinate axes

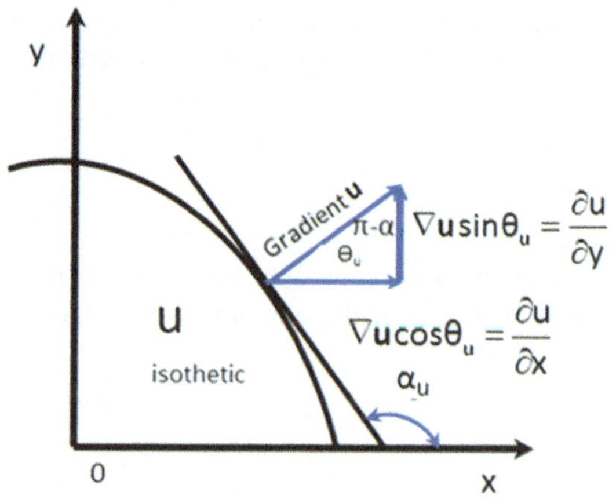

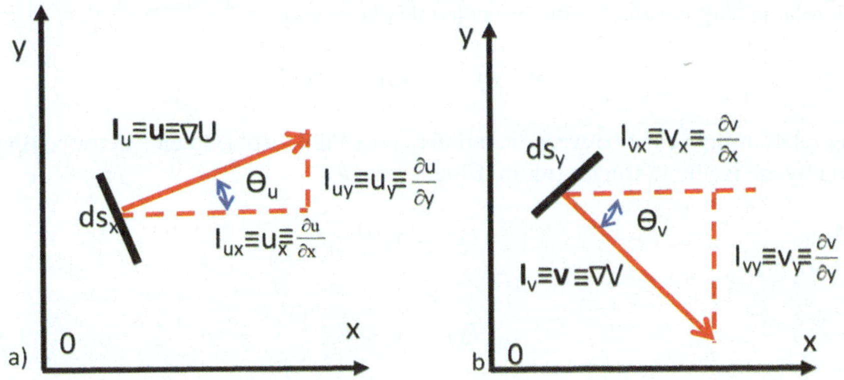

Fig. 2.4 (**a**) dS_x tangent to a *u*-isothetic line and vectors associated with this isothetic line; (**b**) dS_y tangent to a *v*-isothetic line and vectors associated with this isothetic line

Fig. 2.5 Resultant displacement vector as a sum of the two projected displacement fields

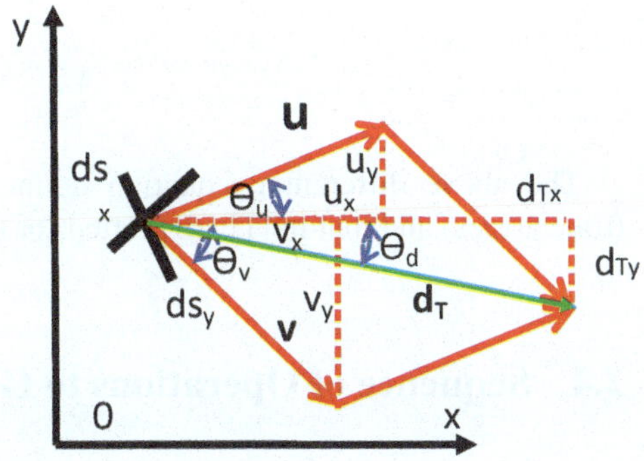

It is possible to connect the above derived equations with the more recent developments. In [4], it is proven the validity of the relationships graphically represented in Fig. 2.4 between the vectors that define the light intensity fields and the displacement fields. The vectors $\mathbf{I_u}(\mathbf{x})$, the displacement vector $\mathbf{u}(\mathbf{x})$, the vector gradient of $u(\mathbf{x})$, and the normal vector to the isothetic line $\mathbf{n}(\mathbf{x})$ are all co-axial. Similar properties apply to the field $\mathbf{v}(\mathbf{x})$, Fig. 2.4b.

The connection between the displacement field vector at a pixel and the scalar signal of intensity $I(\mathbf{x})$ that represents light intensity in terms of gray levels is done through the equation,

$$u_x(x) = \frac{\phi_u(x)}{2\pi} p \tag{2.9}$$

The quantity $\frac{\phi_u(x)}{2\pi}$ is the fringe order n. Since we are speaking of local fringe orders, n is a real number that can take values between 0 and 1, and p is the carrier signal pitch. The first derivatives of the displacement fields can be obtained from Eq. (2.6) if θ_u is a known quantity and similar equations exist for the displacement *v*.

Figure 2.5 shows the displacement vector at a pixel as the sum of the two component vectors $u(\mathbf{x})$ and $v(\mathbf{x})$ and also as the vectorial addition of the components of these two vectors.

The following vectorial equation applies,

$$\mathbf{d}_T(\mathbf{x}) = \mathbf{u}(\mathbf{x}) + \mathbf{v}(\mathbf{x}) \tag{2.10}$$

From Eq. (2.10), we have the projection equation

$$\mathbf{d}_T(\mathbf{x}) = [u_x + v_x]\mathbf{i} + [u_y + v_y]\mathbf{j} \tag{2.11}$$

and

$$tg\theta_d = \frac{u_y + v_y}{u_x + v_x} \tag{2.12}$$

Since the following relationship is valid for the projected displacements,

$$\mathbf{d}_{\mathrm{T}}(\mathbf{x}) = \mathrm{u}(\mathbf{x})\,\mathbf{i} + \mathrm{v}(\mathbf{x})\,\mathbf{j} \tag{2.13}$$

Then, the following relationships exist between the relative projected displacement vector components and the projected displacements provided by the isothetic fringes (moiré fringes):

$$\mathrm{u}(\mathbf{x}) = \mathrm{u_x} + \mathrm{v_x} \tag{2.14}$$

$$\mathrm{v}(\mathbf{y}) = \mathrm{u_y} + \mathrm{v_y} \tag{2.15}$$

There are several alternative ways to compute θ_u and θ_v, these two angles in a 2D field are essential parameters to characterize the field because they define the gradients of the corresponding displacements.

The relative displacement vector field is a vectorial field governed by the differential equation:

$$\frac{dy}{dx} = \frac{v(\mathbf{x})}{u(\mathbf{x})} \tag{2.16}$$

The above differential equation defines field trajectories that are called in [1] "trochias" from the Greek τροχιά (trajectories) and that in [1] are plotted for the case of a disk under diametrical compression.

2.4 Sequence of Operations to Get All the Displacement Information at a Given Pixel

One has two images corresponding to two orthogonal carriers or one image that has the two orthogonal carriers. There are two possible alternatives: (a) the carrier tagged to the observed surface is detected by the sensor and one applies digital moiré; (b) one has two images that contain the patterns u(**x**) and v(**x**) but does not display the original carriers tagged to the observed surface.

Let us consider the case of two separate images. We start with two square matrices of levels of gray (scalar functions), we have then,

$$\mathbf{I}_i(N,N) = \begin{vmatrix} I_i(1,1) & I_i(1,2) & \ldots & I_i(1,N-1) & I_i(1,N) \\ I_i(21) & I_i(2,2) & \ldots & I_i(2,N-1) & I_i(2,N) \\ \ldots & \ldots & \ldots & \ldots & \ldots \\ I_i(N,1) & I_i(N,2) & \ldots & I_i(N,N-1) & I_i(N,N) \end{vmatrix} \tag{2.17}$$

where i = x, y, with the coordinated expressed in pixels numerated 1,2, 3 . . . N, first index rows, second index, columns, the $I_i(N,N)$ are levels of gray expressed as integer powers of 2. We have then two matrices $\mathbf{I}_u(N,N)$ and $\mathbf{I}_v(N,N)$.

To these two matrices are applied the procedures outlined in [3], Sect. 2.8. The first step is to pre-process the two patterns, add high frequency carriers, and extend the fringes to the full field of view.

In the second step, the normals to the fringes are determined. Determining the gradient of the fringe pattern as shown in Fig. 2.6 one obtain two square matrices corresponding to the local angles $\theta_i(x)$, where i = u,v, with the coordinates expressed in pixels numerated 1,2,3 . . . N, first index rows, second index, columns.

$$\theta_i(N,N) = \begin{vmatrix} \theta_i(1,1) & \theta_i(1,2) & \ldots & \theta_i(1,N-1) & \theta_i(1,N) \\ \theta_i(21) & \theta_i(2,2) & \ldots & \theta_i(2,N-1) & \theta_i(2,N) \\ \ldots & \ldots & \ldots & \ldots & \ldots \\ \theta_i(N,1) & \theta_i(N,2) & \ldots & \theta_i(N,N-1) & \theta_i(N,N) \end{vmatrix} \tag{2.18}$$

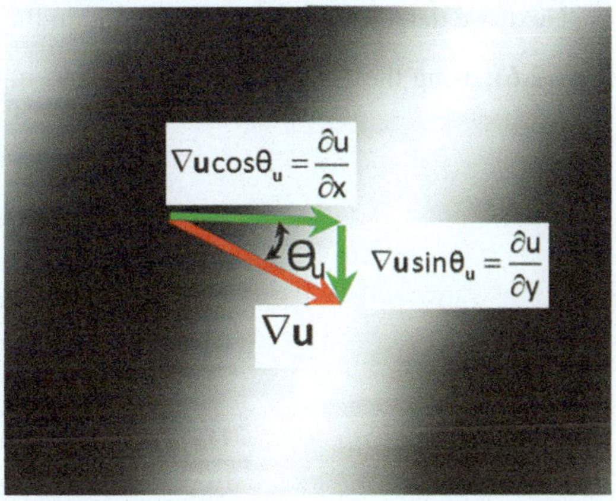

Fig. 2.6 Angle θ_u that the isothetic line u(x) makes with the x-axis defined through the local gradient

Utilizing the matrices of Eq. (2.18), one can proceed to apply, for example, the Hilbert transform. From the Hilbert transform one obtains the matrices of the local phases.

$$\tan\,\phi_i = \frac{I_{iq}}{I_i} \tag{2.19}$$

The local phases are represented by $\phi_i(\mathbf{x})$, where i = u,v.

$$\Phi_{iL}(N,N) = \begin{vmatrix} \Phi_{iL}(1,1) & \Phi_{iL}(1,2) & \ldots & \Phi_{iL}(1,N-1) & \Phi_{iL}(1,N) \\ \Phi_{iL}(21) & \Phi_{iL}(2,2) & \ldots & \Phi_{iL}(2,N-1) & \Phi_{iL}(2,N) \\ \ldots & \ldots & \ldots & \ldots & \ldots \\ \Phi_{iL}(N,1) & \Phi_{iL}(N,2) & \ldots & \Phi_{iL}(N,N-1) & \Phi_{iL}(N,N) \end{vmatrix} \tag{2.20}$$

The next step is to unwrap the phases by applying Eq. (2.4), starting from a selected origin of phases. This operation generates a matrix $\Phi_{iT}(N,N)$ that in place of containing local phases, contains total phases. From the total phases on can compute the total displacements a given pixel through Eq. (2.9), using the notation $\begin{matrix} d_i = u \\ d_i = v \end{matrix}$ for i = u, v,

$$D\,(N,N) = \begin{vmatrix} d_i(1,1) & d_i(1,2) & \ldots & d_i(1,N-1) & d_i(1,N) \\ d_i(21) & d_i(2,2) & \ldots & d_i(2,N-1) & d_i(2,N) \\ \ldots & \ldots & \ldots & \ldots & \ldots \\ d_i(N,1) & d_i(N,2) & \ldots & d_i(N,N-1) & d_i(N,N) \end{vmatrix} \tag{2.21}$$

From Eq. (2.14), the modulus of the vector $\mathbf{d}_T(\mathbf{x})$ can be computed from the matrices (2.21)

$$\|\mathbf{d}_T\| = \sqrt{d_u^2 + d_v^2} \tag{2.22}$$

One generates a matrix $\mathbf{D}_T(N,N)$

$$\mathbf{D}_T\,(N,N) = \begin{vmatrix} d_T(1,1) & d_T(1,2) & \ldots & d_T(1,N-1) & d_T(1,N) \\ d_T(21) & d_T(2,2) & \ldots & d_T(2,N-1) & d_T(2,N) \\ \ldots & \ldots & \ldots & \ldots & \ldots \\ d_T(N,1) & d_T(N,2) & \ldots & d_T(N,N-1) & d_T(N,N) \end{vmatrix} \tag{2.23}$$

Matrix (2.23) together with matrix (2.18) is utilized to plot the field of trochias or trajectories of the displacement vectors.

Next, we derive the matrices that contain the first derivatives of the displacements u and v. From Eq. (2.18) and applying Eq. (2.6), using the notation $d_i = \begin{matrix} u \\ v \end{matrix}$ for $i = u, v$ we get,

$$\frac{\partial D_i}{\partial x}(N,N) = \begin{vmatrix} \frac{\partial d_i}{\partial x}(1,1) & \frac{\partial d_i}{\partial x}(1,2) & \ldots & \frac{\partial d_i}{\partial x}(1,N-1) & \frac{\partial d_i}{\partial x}(1,N) \\ \frac{\partial d_i}{\partial x}(21) & \frac{\partial d_i}{\partial x}(2,2) & \ldots & \frac{\partial d_i}{\partial x}(2,N-1) & \frac{\partial d_i}{\partial x}(2,N) \\ \ldots & \ldots & \ldots & \ldots & \ldots \\ \frac{\partial d_i}{\partial x}(N,1) & \frac{\partial d_i}{\partial x}(N,2) & \ldots & \frac{\partial d_i}{\partial x}(N,N-1) & \frac{\partial d_i}{\partial x}(N,N) \end{vmatrix} \tag{2.24}$$

$$\frac{\partial \mathbf{D}_i}{\partial y}(N,N) = \begin{vmatrix} \frac{\partial d_i}{\partial y}(1,1) & \frac{\partial d_i}{\partial y}(1,2) & \ldots & \frac{\partial d_i}{\partial y}(1,N-1) & \frac{\partial d_i}{\partial y}(1,N) \\ \frac{\partial d_i}{\partial y}(21) & \frac{\partial d_i}{\partial y}(2,2) & \ldots & \frac{\partial d_i}{\partial y}(2,N-1) & \frac{\partial d_i}{\partial y}(2,N) \\ \ldots & \ldots & \ldots & \ldots & \ldots \\ \frac{\partial d_i}{\partial y}(N,1) & \frac{\partial \varphi_i}{\partial y}(N,2) & \ldots & \frac{\partial d_i}{\partial y}(N,N-1) & \frac{\partial d_i}{\partial y}(N,N) \end{vmatrix} \tag{2.25}$$

2.5 Flow Diagram of the Computation Process

The flow diagram of the computation process utilized to obtain the displacement components, displacement vectors and first derivatives of the displacements is shown in Fig. 2.7. The original inputs are scalar fields representing levels of gray. The application of the transform operations converts the scalar field of intensities into vector fields of intensities that yield finally displacement vectors and gradients of the displacement field.

2.6 Application Example I: Microscopic Displacement Field of a Particulate Composite

Microscopic patterns obtained from a metallic particulate composite [9, 10] are chosen to illustrate the procedure to piecewise analyze complex patterns. In view of the heterogeneity of the observed material, the images of the patterns contain a variety of singular patterns, extremely difficult to analyze utilizing traditional techniques of pattern unwrap. The examples show that with the new method one can handle highly complex patterns with standard routines. A brief description of the patters generation is provided in the following paragraphs.

A cross-grating p = 833 nm (1200.05 1/mm) was engraved in a tensile specimen using photo-resists. The measurements were carried out in a small region of a tensile specimen made from a plate of Al 356 reinforced with silicon carbide (SiC) particles. The optical set up was secured to an Instron machine to minimize noises during testing. Table 2.1 provides all the relevant parameters of the optical system and the CCD camera. An unloaded image of the grating in the region under observation was grabbed and stored in the computer memory. The specimen was loaded and images of the deformed grating were taken at increasing loads.

In Table 2.1, the first column is the size of the observed region, the second column is the pitch of the printed grating, r is the optical resolution in the plane of the image, Δz is a measure of the axial resolution of the image, m is the magnification of the microscope and NA is the numerical aperture of the microscope.

The patterns shown in Fig. 2.8 are not optically recorded patterns but computer-generated patterns utilizing pixel interpolation techniques. The patterns were printed utilizing a computer algorithm for fringe multiplication that reduced the pitch of the printed grating from p = 833 nm to p = 55.6 nm, that is a reduction by a factor of 15. Another observation that needs clarification is the concept of resolution utilized in the interpretation of the images different from the value r given in Table 2.1. Following a method often utilized in electron microscopy, an alternative definition of resolution is adopted. Resolution is defined as the minimum distance that can be measured in the image plane [11]. Figure 2.8 shows the *u* and *v* patterns of the displacements. It is possible to see the complexity of the two patterns, fringe singularities in the two patterns are different. The corresponding Burger vectors of the singular fringes depend on the analyzed patterns. To overcome the difficulties in the pattern analysis presented in [10], the derivatives of the displacement field were determined through utilizing energy considerations.

The specimen was pulled in tension, Fig. 2.9. The images shown in Fig. 2.8 correspond to a uniform tension specimen in the elastic regime, with and axial strain of $\varepsilon_y = 0.001675$ in the quasi-elastic regime. Locally, the computed strains resulted about two orders of magnitude greater than the macroscopic strains. The large plastic strains caused the premature fracture

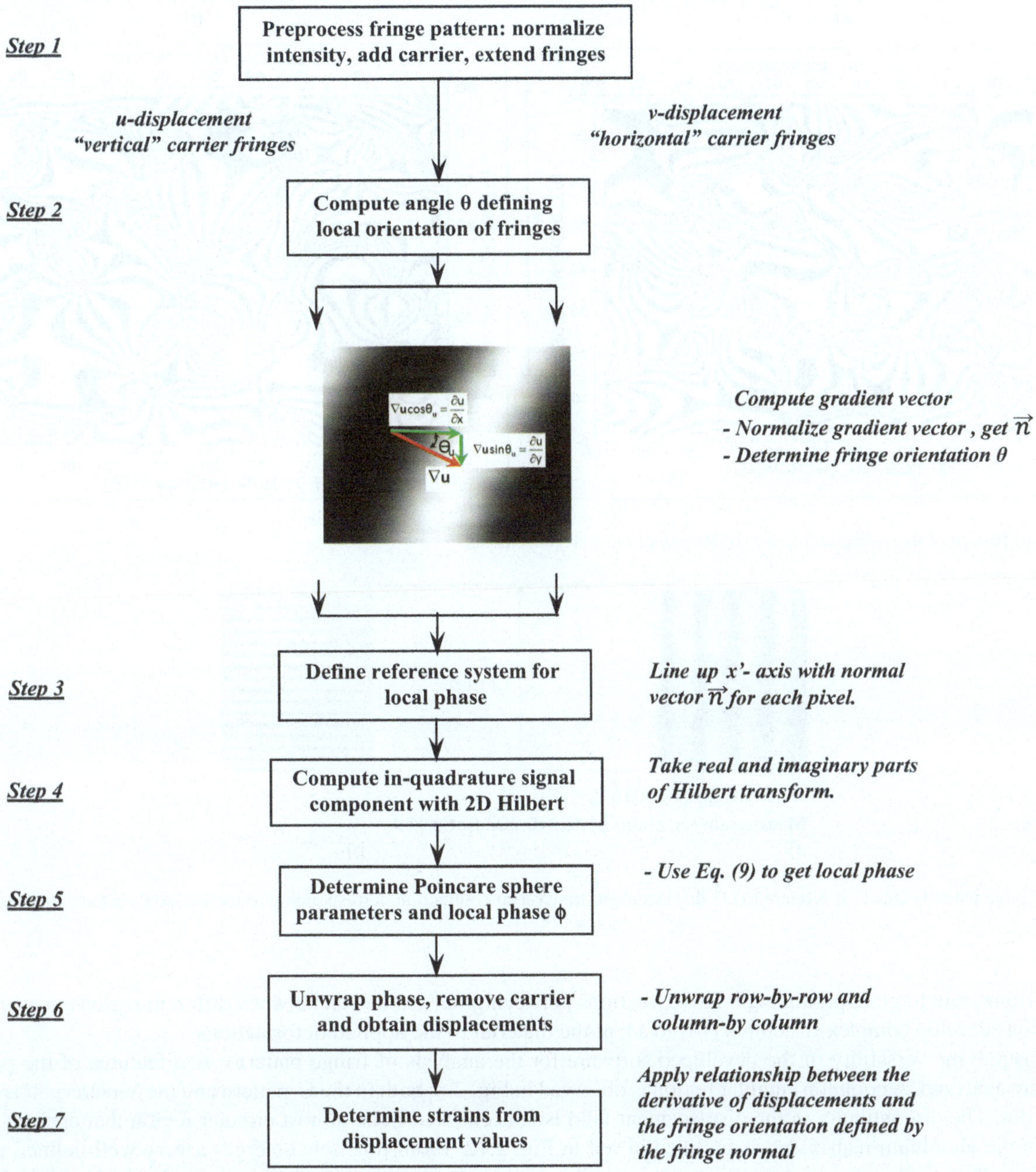

Fig. 2.7 Flow diagram of the computation of the computation process

Table 2.1 Parameters of the optical system and camera

Obs. reg (μm)	p (μm)	r (μm)	Δz (μm)	mag.	NA
102 × 85	0.833	0.398	1.17	86.27	0.678

λ (white light), pixels = 512 × 480 * Δx = 0.199 μm, Δy = 0.165 μm
Values are obtained through calibration of the camera

of the specimen when compared to a finite element numerical model based on the J_2 flow rule [12]. The strain concentrations were produced by the irregular shapes of the utilized silicon carbide reinforcing particles (Fig. 2.10).

To help understand the images shown in Fig. 2.8, is useful to look at Fig. 2.11, which is a micrograph [13] of Al 356 alloy. It shows aluminum dendritic grains and inter-dendritic regions of Al intermetallic phases, it contains also Si particles. The observed fringes correspond to a single grain. The effect of the plastic deformations produces sub-grains that experience relative rotations with respect to each other. As mentioned in [13, 14], crystal orientation of plastically deformed aluminum

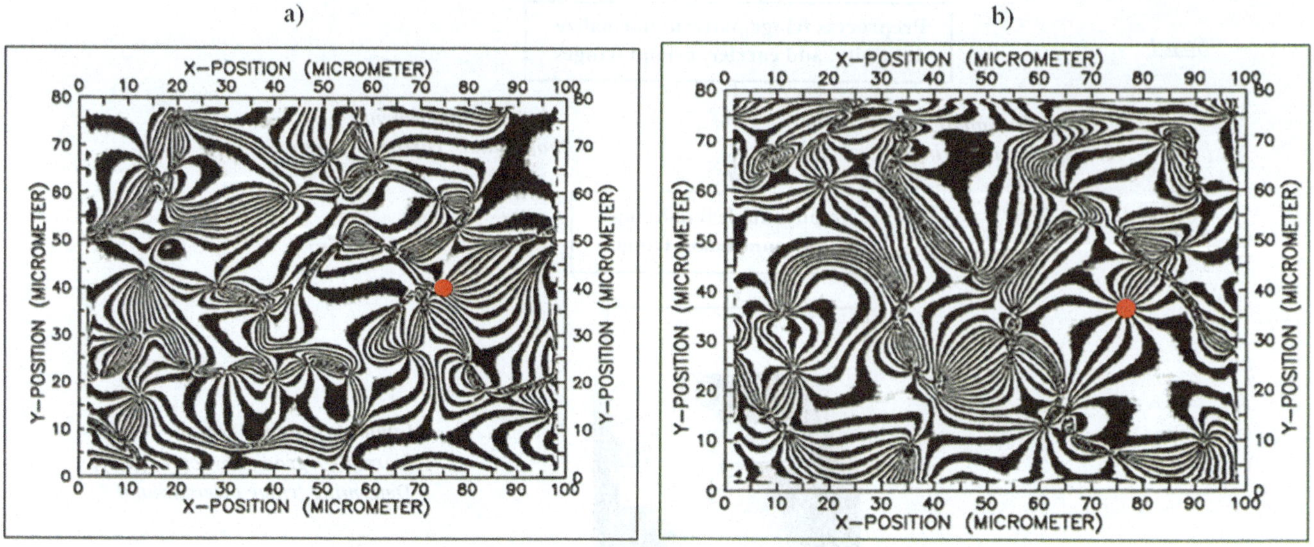

Fig. 2.8 (**a**) Pattern of the u-displacements. (**b**) Pattern of the v-displacements

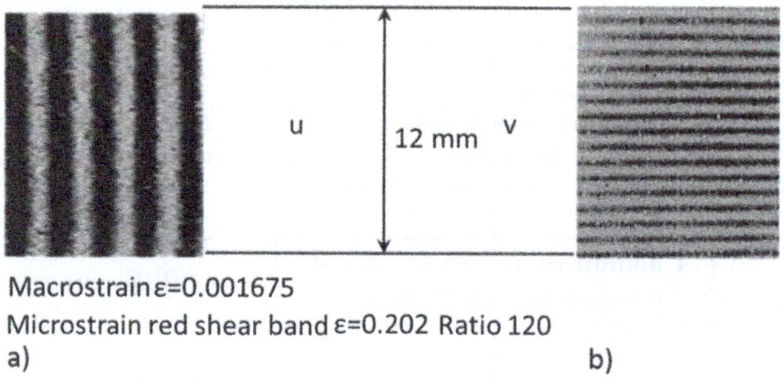

Fig. 2.9 Fringe patterns tensile specimen: (**a**) U-displacement, transversal contraction corresponding to the Poisson's effect; (**b**) V-pattern axial elongation

changes from grain to grain producing relative rotations between grains, and even between different regions in a same grain. Dislocations develop complex structures [14] to adapt the material to the applied deformations.

To illustrate the versatility of the developed software for the analysis of fringe patterns, two features of the patterns of Fig. 2.8 are analyzed. A common singular region is observed in Fig. 2.8: both in the U-pattern and the V-pattern, it is indicated by a red dot. The discontinuity in the displacement field is concentrated in an almost circular region that appears as a hole present in the aluminum matrix and is outlined in red in Fig. 2.12. From this hole emerges a very well-defined pattern of fringes providing displacement information corresponding to the two projected displacement fields, the U and V fields. The inserts in the Fig. 2.9 shows the macroscopic fringes of the tensile specimen.

Applying to the fringe patterns at the edges of the hole shown in red in Fig. 2.12 the method described in the preceding sections one obtains local derivatives of the displacement field. These derivatives are introduced in the Almansi Eulerian strain tensor components,

$$\varepsilon_x^E = 1 - \sqrt{1 - 2\frac{\partial u}{\partial x} + \left(\frac{\partial u}{\partial x}\right)^2 + \left(\frac{\partial v}{\partial x}\right)^2} \tag{2.26}$$

$$\varepsilon_y^E = 1 - \sqrt{1 - 2\frac{\partial v}{\partial y} + \left(\frac{\partial v}{\partial y}\right)^2 + \left(\frac{\partial u}{\partial y}\right)^2} \tag{2.27}$$

$$\left(\varepsilon_{xy}^E\right) = \arcsin \frac{\frac{\partial u}{\partial y} + \frac{\partial v}{\partial x} - \frac{\partial u}{\partial x}\frac{\partial u}{\partial y} - \frac{\partial v}{\partial x}\frac{\partial v}{\partial y}}{\left(1 - \varepsilon_x^E\right)\left(1 - \varepsilon_y^E\right)} \tag{2.28}$$

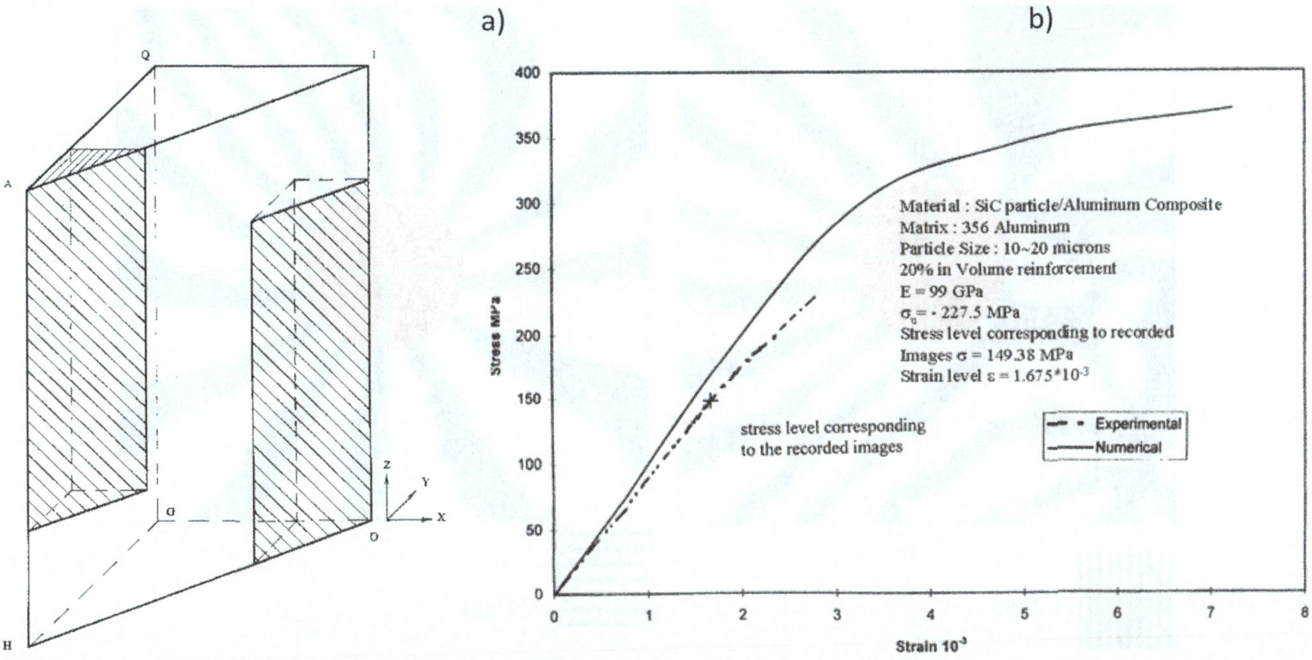

Fig. 2.10 (**a**) Basic cell used in the finite element study of a particulate composite [12]. Because of the symmetry involved only 1/8 of the cell is represented. (**b**) Stress-strain curve of the finite element together with the experimentally measured values of the tensile specimen used in the experiment. Notice the earlier fracture of the specimen when compared to the one predicted by the finite element results (average stress critetion)

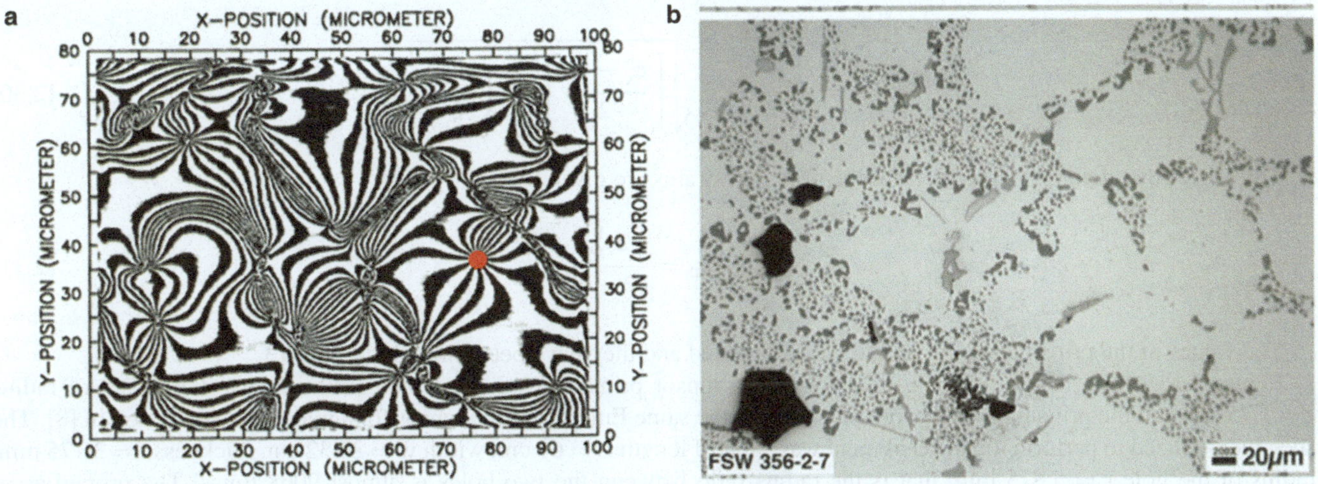

Fig. 2.11 (**a**) V-pattern of the observed region. (**b**) Micrograph of a region of Al 356 surface (the size of the micrograph 511 × 380 μm)

In [3], it is shown that to set forth the analytic function approximation it is necessary to add to the image carrier fringes to generate high density fringes. The effect of the carrier fringes depends on the pitch of the original carrier, in this case a 55.6 nm, this is a pitch that in the scale of the picture provides a high strain sensitivity. The fringes resulting from the body rotation experienced by the region under observation provide the required additional carrier fringes. The effect of these additional fringes is automatically removed by the computation of the non-linear strain tensor. The computed strains form the 2D strain tensor,

$$\begin{bmatrix} e_x^E & e_{xy}^E \\ e_{xy}^E & e_{x_y}^E \end{bmatrix} \tag{2.29}$$

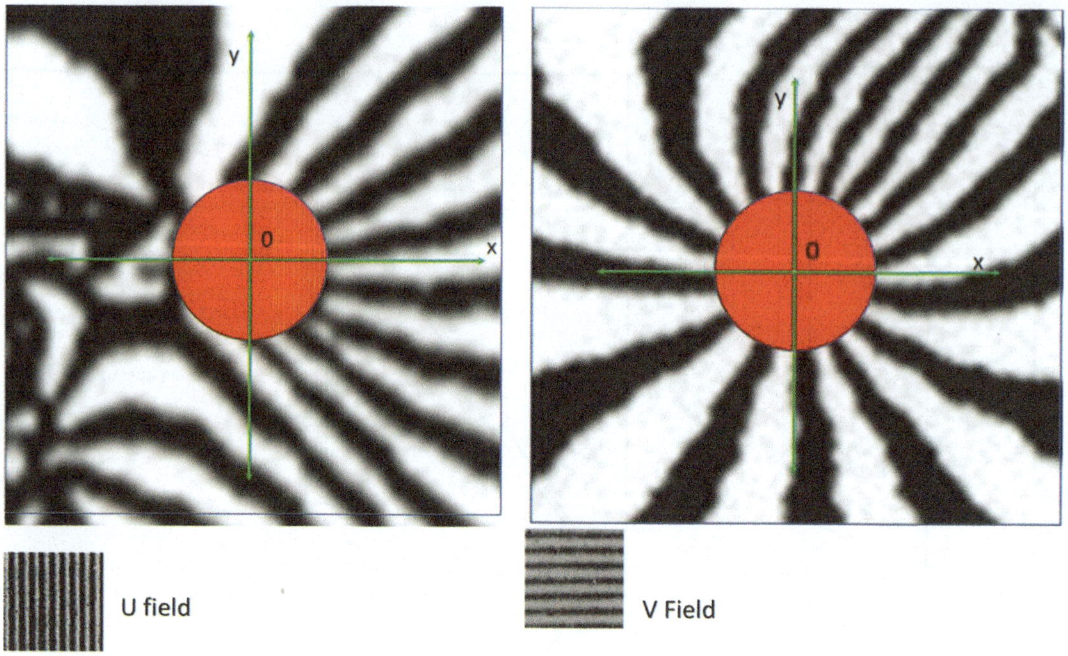

Fig. 2.12 U-field and V-field of a dislocation indicated in Fig. 2.8 with a red dot. Diameter of the hole 3.5 μm

The principal strains of the Eulerian tensor can be computed as

$$\boldsymbol{\varepsilon}_{p1,2}^{E} = \frac{e_x^E + e_y^E}{2} \pm \sqrt{\left[\frac{e_x^E - e_y^E}{2}\right]^2 + \left(\frac{\gamma_{xy}^E}{2}\right)^2} \tag{2.30}$$

The angles that define the directions of the principal strains are given by,

$$\mathrm{tg}\ 2\theta_p = \frac{2\varepsilon_{xy}^E}{e_x^E - e_y^E} \tag{2.31}$$

The values of the principal strains, $e_{p_1}^E$, $e_{p_2}^E$ are computed and they have been plotted in Fig. 2.13.

Figure 2.13 shows the comparison of the Eulerian tensor principal strains of the circular locus of Fig. 2.8 of radius $r = 1.75$ μm and the principal strains corresponding to the same Eulerian strain tensor for a macro hole analyzed in [8]. The specimen utilized to perform the measurement was a bar of length $L = 61$ cm, width $w = 20.32$ cm, thickness $t = 3.175$ mm, radius of the hole $r = 5.875$ mm, that is the radius ratio between the two holes is almost 9068 times. The properties of the material of the bar measured in standard tensile tests are the following, yield stress 151.68 MPa, modulus of elasticity $E = 68.95$ GPa, Poisson's ratio $\mu = 0.33$, maximum tensile strength greater than 158 MPa, close to the Al 356 aluminum alloy.

To obtain the stress distribution in the bar with the hole, the Prandlt-Reuss equations were numerically solved by an iteration algorithm utilizing the strain distributions retrieved from moiré patterns where carrier fringes were added [8]. The values of the strains of the microscopic hole (Fig. 2.13) have a trend like those observed in the macro-hole. Particularly, it is observed that the transition from tensile to compressive stresses are very closely located in both cases. A stress strain curve for the Al 356 alloy was taken from the literature was adopted to make comparisons between the two holes. In [13], there is a detailed analysis of the properties of the Al 356 alloy including the tensile test plotted in Fig. 2.14.

In Fig. 2.15a, there are plotted the principal stresses around the hole taken from [8], together with the principal stresses in the microscopic hole utilizing the principal strain values shown in Fig. 2.13 and the stress-strain curve of Fig. 2.14. Figure 2.15b has been added to explain the sign change of the family of principal stresses around the hole from maximum principal stresses to algebraic minimum principal stresses. The free body diagram of Fig. 2.15b corresponds to the elastic solution of a bar with a hole. It shows that in order to satisfy equilibrium along the axis of symmetry of the hole, bending stresses are present. Along the boundary at approximately an angle of 60° there is a singular point of zero stress and zero strain. As it

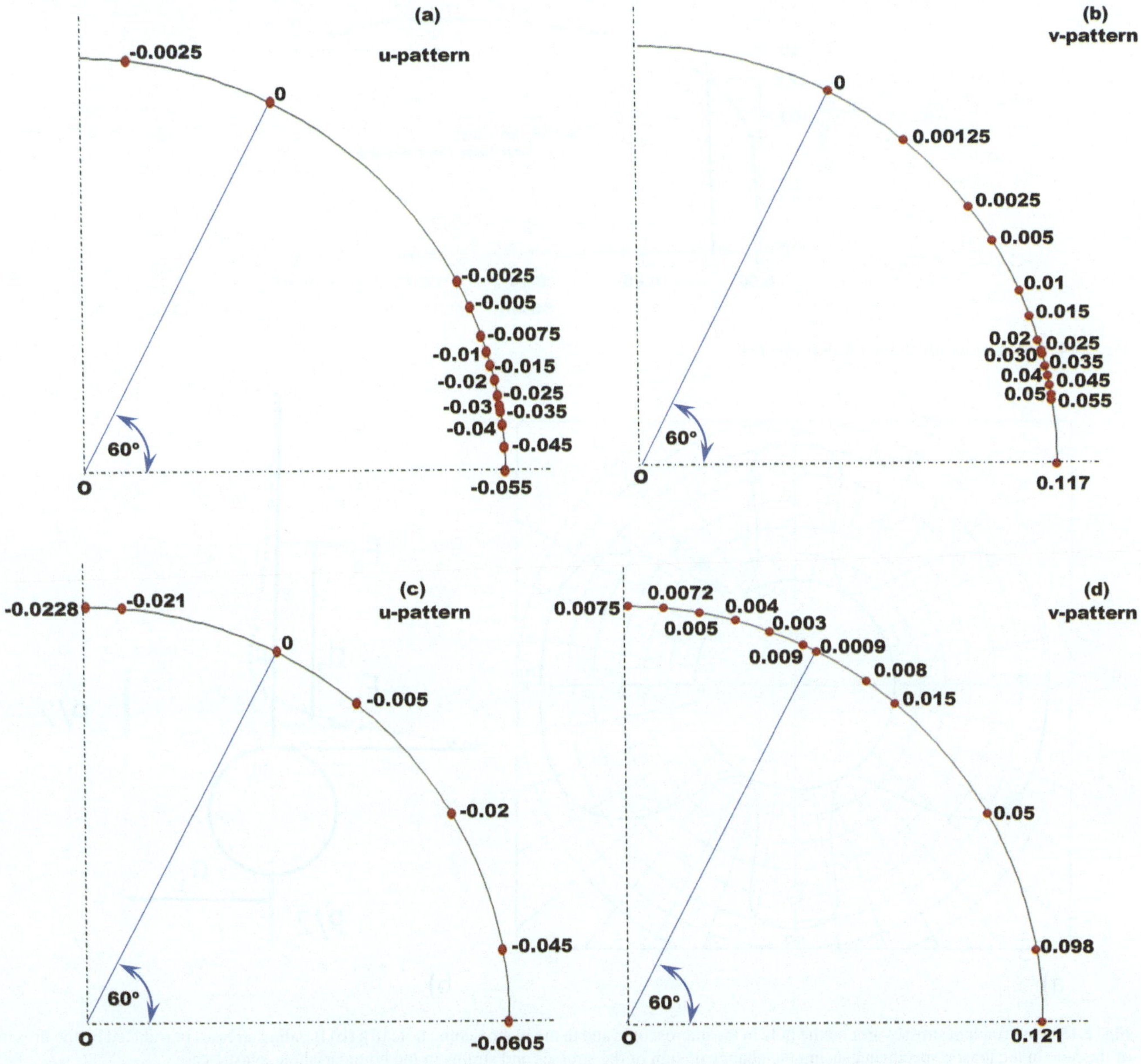

Fig. 2.13 Strain distribution along the hole boundary: (**a**) Minimum principal strain for the large hole; (**b**) Maximum principal strain for the large hole; (**c**) Minimum principal strain for the 3.5 μm diameter hole; (**d**) Maximum principal strain for the 3.5 μm diameter hole

can be seen in Fig. 2.13a, the transition of the principal strains from the maximum algebraic value to the minimum algebraic value occurs around the point of zero stress (Fig. 2.15a).

After this preliminary introduction, it is possible to discuss the relationship between the deformation field around the macro-hole and the micro-hole. If one looks at Fig. 2.13a, the maximum principal strains in the region of small deformations are close in value and signs but the larger hole shows a more concentrate rate of change that the microscopic hole. Fig. 2.13b shows a similar behavior for the minimum principal strains.

Let us consider the plots of Fig. 2.13: for the microscopic hole, the circular region corresponds to an Eulerian description of the medium, that is a configuration in the deformed state under observation. The fact that there is a transition from compression to tension at the edge of the circular hole indicates that the region inside the circular locus is an actual opening. For the macro-hole, the deformation change caused by plasticity (it will transform the circular hole into and elliptic one) has been neglected (in actuality, the ellipticity of the hole is small), a Lagrangian representation of the geometry has been adopted for the comparison, while the plotted deformations and stresses are Eulerian.

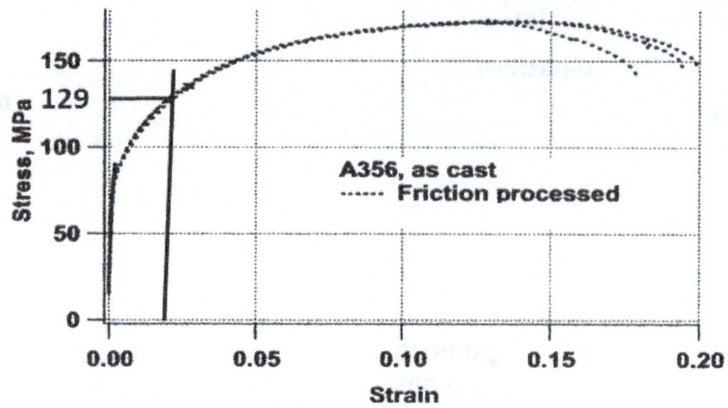

Fig. 2.14 Stress strain of Al 356 taken from [13]

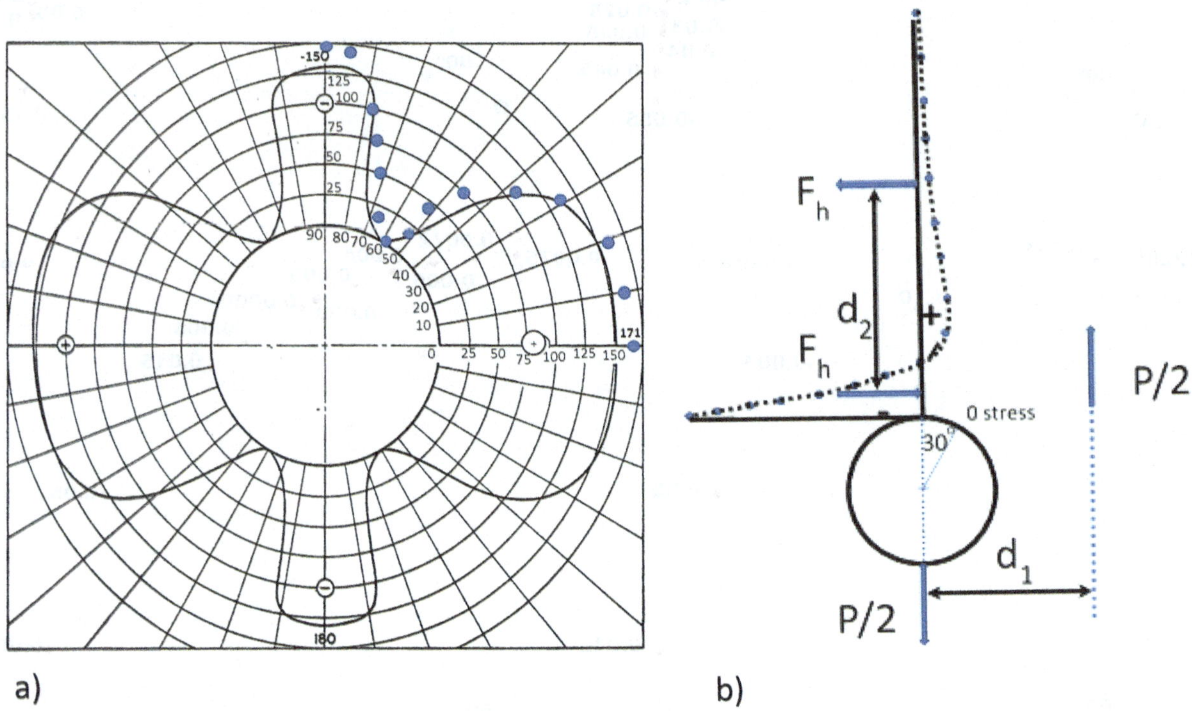

Fig. 2.15 (**a**) Principal stresses around the hole in the microscopic and in the macroscopic hole [8]; (**b**) Bending stresses introduced by the presence of the hole in the tensile specimen causing the change of sign of the stresses and strains in the boundary hole, elastic case

The presence of a hole on the surface of the specimen indicates the formation of a spherical dimple in the volume of the material. In [15], it is possible to verify that, for tensile specimens, in fractured specimens' micrographs there are visible many quasi-spherical and spherical dimples.

2.7 Application Example II

Another region of interest is observed in the v-pattern, Fig. 2.16. Unlike the first example, the presence of discontinuities does not coincide in the two patterns. The shear band has the length of 6.28 μm. The derivatives at mid-point of the shear band have been computed. The following values have been obtained,

$$\begin{bmatrix} \frac{\partial u}{\partial x} = 0.06112 & \frac{\partial u}{\partial y} = 0.06112 \\ \frac{\partial v}{\partial x} = 0.1441 & \frac{\partial v}{\partial y} = 0.1441 \end{bmatrix} \tag{2.32}$$

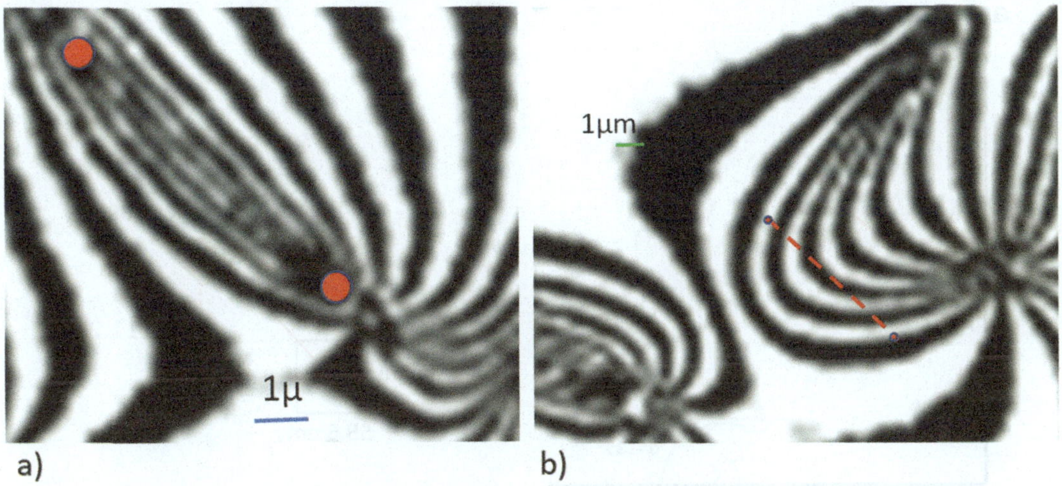

Fig. 2.16 Enlargement of a region of the pattern of Fig. 2.8: (**a**) The *v*-pattern shows a shear band, the red dots are sources of the shear band; (**b**) The *u*-pattern shows the position of the dots of the *v*-pattern and the position of the shear band

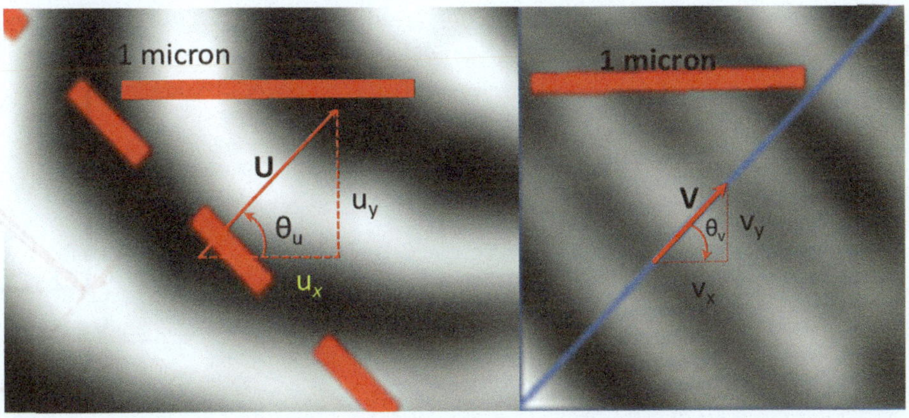

Fig. 2.17 Enlargement of the local patterns of Fig. 2.16 at the mid-point between of the two singularities shown as red dots

The sign of the direct derivatives or derivatives in the principal direction cannot be directly determined from the isothetic fringes. The sign can be determined through the carriers change of pitch or other additional information. In this case, we know that the *v*-displacements correspond to the extension of the specimen, a positive quantity. The *u*-displacements relate to the contraction of the specimen due to the Poisson's effect, a negative quantity. We need to determine the sign of the cross-derivative. This sign comes from Eq. (2.8), since $\frac{\partial u}{\partial x}$ is negative, $\frac{\partial u}{\partial y}$ is positive. We know that in the region of analysis the grad(*v*) is a positive quantity, then both derivatives are positive.

Computing the Almansi Eulerian strain tensor one obtains,

$$\mathbf{E} = \begin{bmatrix} e_x^E = 0.0501 & e_{xy}^E = 0.2236 \\ e_{xy}^E = 0.2236 & e_y^E = 0.14193 \end{bmatrix} \tag{2.33}$$

Replacing values in Eq. (2.30), one obtains the principal strains $e_{p1}^{E} = 0.169$ and $e_{p2}^{E} = 0.02485$. The directions of the principal strains are $\theta_{p1} = +39.2°$, $\theta_{p2} = -50.8°$.

We need to find the local displacement corresponding to the analyzed region, this is done applying Eq. (2.9) for the u displacements, and the corresponding equation for the v displacements. Figure 2.17 shows in a larger scale the two patterns of interest.

It is necessary to get information that is not provided by the patterns but comes from knowledge of the actual displacement field.

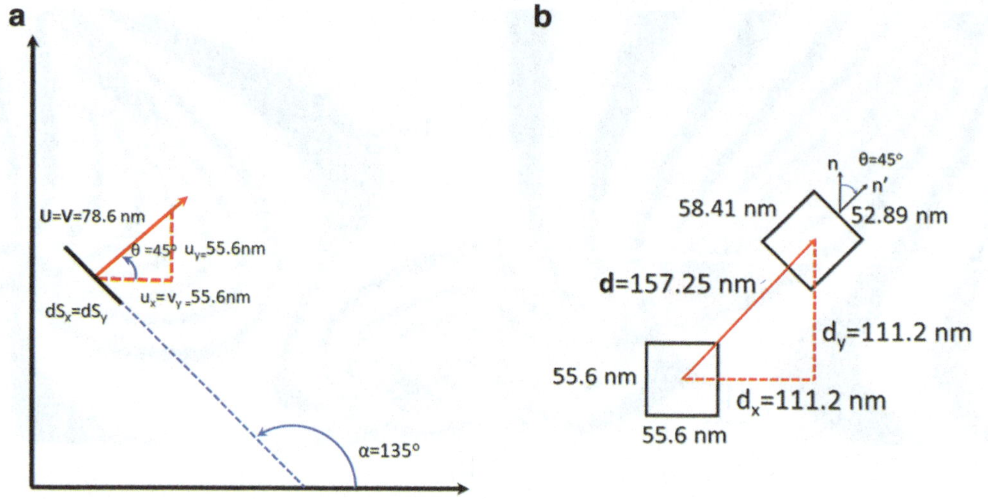

Fig. 2.18 (**a**) Vectors **U**(**x**) and **V**(**x**); (**b**) Pixel in its original position and dimensions and pixel after deformation, final dimensions and displacement vector **d**(**x**) of the pixel

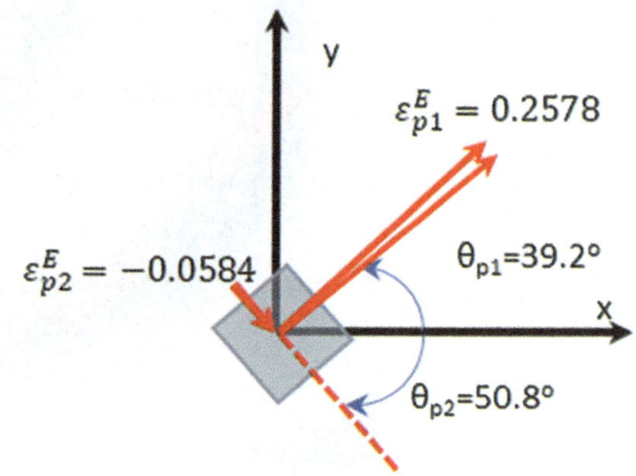

Fig. 2.19 Principal strains of the Almansi-Eulerian strain tensor, showing the effect of the rigid body rotation $\Omega = 2.16^{\circ}$ of the principal strains

The displacement vectors arrows, Fig. 2.17, give the direction of the displacements, an elongation corresponding to the V-pattern and a contraction that corresponds to the U-pattern.

The angles θ_u and θ_v indicate the sense of the rotation of the normals of the fringes. It can be seen that the normals to the fringes, the perpendiculars to the corresponding isothetic lines, coincide in direction and sense in both patterns. In view of the relationships between the corresponding vectors illustrated in Figs. 2.4 and 2.5 and utilizing the steps illustrated in Fig. 2.7 one gets the information displayed in Fig. 2.18.

There is an additional piece of information that is required to complete the analysis of the region of the pattern under study. The rigid body rotation of the principal strains that is computed applying the equation,

$$\mathrm{tg}\,\Omega = \frac{\frac{1}{2}\left[\frac{\partial v}{\partial x} - \frac{\partial u}{\partial y}\right]}{\sqrt{1+\frac{\partial u}{\partial x}}\sqrt{1+\frac{\partial v}{\partial y}}} \tag{2.34}$$

The principal strains are plotted in Fig. 2.19.

To complete the study of the analyzed region of 100×80 μm the derivatives and principal strains and orientations of the Almansi-Eulerian strain tensor have been computed for a raster of points in the observed region of 100×80 μm. The averages of the projections in the x–y directions have been computed and the obtained values are, respectively, $\boldsymbol{\varepsilon}^E_x =$ 0.0005928 and $\boldsymbol{\varepsilon}^E_y = 0.00156$, thus giving the Poisson's ratio,

$$\nu = \frac{0.0005928}{0.00156} = 0.38 \tag{2.35}$$

As previously mentioned, the actual strain experimentally measured for the load corresponding to the recorded images is $\varepsilon_{\mathrm{y}}^{\mathrm{E}} = 0.001675$. The difference between measured and computed $\boldsymbol{\varepsilon}_y^E$ is $\Delta\varepsilon_{\mathrm{y}}^{\mathrm{E}} = 11.5\ \mu\varepsilon$ or a percent difference of +7.4%. In view of the complexity of the strains distribution on the observed region this is a good agreement. Also, the computed transverse strains agree with a possible value of Poisson's ratio taking into consideration the presence in the region of observation of areas that are in the plastic regime. In the analyzed region very, high strains and strain gradients are present.

These effects are observed at an earlier stage of the loading of the specimen, Fig. 2.10, indicating that the interaction of the aluminum matrix with the reinforcing particles created earlier damage that resulted in a brittle behavior of the composite. Irregular shapes of the reinforcing particles provide an explanation for this behavior of the composite resulting from regions of high local deformations due to the interaction of matrix and reinforcing particles.

Figures 2.8 and 2.9 and the information presented in preceding sections provide a clear depiction of effect of the scale factor in the analysis of displacement fields and illustrates the power of the continuum hypothesis in the mechanics of solids.

2.8 Discussion and Conclusions

The paper content is an extension of preceding papers [1–4], a synthesis of a new developed model to extract displacement information from fringe patterns in 2-D fields. It merges early developments in the theory of isothetic lines, or moiré fringes, with more recent advances based on the property of FT and Hilbert transforms of yielding vectorial fields from scalar signals containing displacement information under the form of levels of gray. This property is at the core of the notion of monogenic phasor that contains at a given pixel in 2-D or at a given voxel in 3D all the information necessary to get displacement vectors and first derivatives of field. A light intensity vector $\mathbf{I}(\mathbf{x})$ associated with each one of the two orthogonal carriers is obtained. This vector is co-axial with a projected displacement vector, also co-axial with the gradient vector of the scalar field and with the normal $\mathbf{n}(\mathbf{x})$ to the isothetic line. By introducing a displacement vector associated with a pixel, the new processing method deals with all the image pixels in the same way. No distinction is made of pixels as starting point of a carrier signal or ending pixels as it is done in traditional fringe unwrapping, a traditional source of errors and serious difficulties when fringe singularities are present. Although remedies for these problems exist, the actual application is very difficult to make them self-correcting.

To illustrate the utilization of the proposed method, the analysis of a feature of a very complex fringe pattern is performed. The utilization of classical approaches required a very labor-intensive process as shown in [10]. Utilizing the described approach and the developed software, the data processing is a straightforward routine. Although no independent evaluation of the obtained results is available, a comparison of the analyzed feature, a microscopic hole subjected to tension in an aluminum matrix, with a macro-hole of a similar aluminum yields a stress distribution around the hole perimeter that is comparable in both cases despite the different scales. The stress distribution is determined by geometry and the requirement of satisfying equilibrium conditions. The deformation of the material is a complex process because of the formation of singularities, consequence of the structure of dislocations that the material adopts to accommodate to the externally imposed deformations. In [17], it has been observed that at the level of atomic distances strains up to $\mathrm{e}_{\mathrm{p1}}^{\mathrm{E}} = 0.130$ have been measured in the area of a dislocation forming a pseudo-cavity of roughly 2 atomic distances, 0.6 nm dimension. The initiation the rupture of atomic bonding is a local event at a scale that, compared to the diameter of the microscopic hole, is $1.5\mathrm{x}10^{-5}$ smaller.

A new method of fringe pattern analysis opens the possibility of greatly simplifying the analysis of the complex patterns that are generated by the research of a large variety of very important technical problems.

References

1. Sciammarella, C.A., Lamberti, L.: Basic models supporting experimental mechanics of deformations, geometrical representations, connections among different techniques. Meccanica. **50**, 367–387 (2015)
2. Sciammarella, C.A., Lamberti, L.: Mathematical models utilized in the retrieval of displacement information encoded in fringe patterns. Opt. Lasers Eng. **77**, 100–111 (2016)
3. Sciammarella, C.A., Lamberti, L.: Generalization of the Poincare sphere to process 2D displacement signals. Opt. Lasers Eng. **93**, 114–127 (2017)
4. Sciammarella, C.A., Lamberti, L.: Determination of displacements and their derivatives from 3D fringe patterns via extended monogenic phasor method. Opt. Lasers Eng. **104**, 117–125 (2018)
5. Ghiglia, D.C., Pritt, M.D.: Two Dimensional Phase Unwrapping. Wiley, New York, NY (1998)
6. Sciammarella, C.A., Sciammarella, F.: Experimental Solid Mechanics. Wiley, Chichester (2012)

7. Gabor, D.: Theory of communications. Part 1: the analysis of information. J. Inst. Electr. Eng. II Radio Commun. Eng. **93**(26), 429–441 (1946)
8. Sciammarella, C.A.: Theoretical and Experimental Study on Moiré Fringes. PhD Dissertation, Illinois Institute of Technology, Chicago, IL (1960)
9. Sciammarella, C.A., Sciammarella, F.M.: Study of damage in particulate composites. In: Gdoutos, E.E. (ed.) Recent Advances in Experimental Mechanics, pp. 121–132. Kluwer Academic, Dordrecht (2002)
10. Sciammarella, C.A., Sciammarella, F.M., Kim, T.: Strain measurements in the nanometer range in a particulate composite using computer-aided moiré. Exp. Mech. **43**(3) (2003)
11. Williams, D.R., Barry Carter, C.: Transmission Electron Microscopy, A Textbook for Materials Science, 2nd edn. Springer, New York (2009)
12. Sciammarella, C.A., Nair, S.: Micromechanics study of particulate composites. In: Proceedings of the SEM Spring Conference on Experimental Mechanics, pp. 188–189 (1998)
13. Santella, M.L., Engstrom, T., Storjohann, D., Pan, T.Y.: Effects of friction stir processing on mechanical properties of the cast aluminum alloys Al319 and Al356. Script Mater. **53**, 201–206 (2005)
14. Roy, K.: Behavior of commercial purity aluminum recrystallization alloys. In: Monteiro, W.A. (ed.) Light Metal Alloys Applications (2014)
15. Kristoffersen, M., Børvik, T., Westermann, I., Langseth, M., Hopperstad, O.S.: Impact against X65 steel pipes—an experimental investigation. Int. J. Solids Struct. **50**(20-21), 3430–3445 (2013)
16. Dal Corso, F., Bigoni, D.: The interaction between shear bands and rigid lamellar inclusions in ductile metal matrix. Proc. R. Soc. Lond. A. **465**, 143–163 (2009)
17. Sciammarella, C.A., Sciammarella, F.M., Lamberti, L.: Merging experimental evidence and molecular dynamics theory to develop efficient models of solids. In: Yoshida, S., Lamberti, L., Sciammarella, C.A. (eds.) Advancement of Optical Methods in Experimental Mechanics. Springer, Berlin (2017)

Chapter 3
A Review: Optical Methods That Evaluate Displacement

Cesar A. Sciammarella

Abstract In the literature of methods utilized to recover displacement information encoded in images as gray levels, scalar fields, converting them into vectorial fields there are available alternative methods. This paper will focus in three methods, the most widely used method the image correlation method (DIC), the HARP method that has evolved from the field of bio-engineering and a more recent method that we can call optical signal analysis method (OSA) that has its foundation on the application on the theory of Optical Signal Analysis. Basic aspects of these three methods are presented in this Part I of the paper. To simplify a very complex subject, 2D versions of these method are discussed.

In Part I, the basic premises and developments of three different approaches to the experimental analysis of full field kinematics of the continuum are presented. This second part is devoted to an evaluation of certain performance characteristics of these methods by looking to the actual results obtained from their application to specific cases. A disk under diametrical compression is a specimen that presents some typical features that provide grounds for comparison of spatial resolution and accuracy because it has a known theoretical solution and can be modelled by a well-defined FE approach. Furthermore, numerous studies of this specimen exist due to its application to the Brazilian test to determine the fracture load of concrete. In the case of the HARP method there are solutions in the literature of heart studies that are available for comparison purposes. A fair comparison of different methods is a formidable task. The high complexity of each of the different methods, the number of variables in play in each of the steps of a given approach, algorithms details, selection of scales makes it very difficult to provide definitive answers. The observation extracted from the results of this Part II of the paper can be considered as a first step in a required more comprehensive study. No matter the pointed difficulties, the obtained result provides a glimpse to basic aspects of spatial resolution, noise content coming from random errors and biased errors.

Keywords Displacement information · Optical signal analysis · Digital image correlation · Harmonic phase method

3.1 Part I Theory

3.1.1 Introduction

The subject matter of the paper encompasses a vast number of developments that have evolved in time and cover more than a century of contribution of many researchers in many different disciplines. The aim of the review is not to bring a historic account of different contributions by individuals but to introduce an analysis of the fundamental ideas that have finally lead to tools that are currently in use. The basic concepts that underpin present methods will be presented as a body of knowledge that have converged to actual applications.

We consider real- or complex-valued functions f(x) defined on **R**n, where n is an orderly list of integers. Ordinary case letters will represent scalar quantities, bold letters will represent vectorial quantities. It will be written f(**x**) or f (x,y,z) where the bold lower case indicates a vector quantity, or we will list low-case variables, whichever is more convenient in the context of the discussion.

C. A. Sciammarella (✉)
Department of Mechanical, Materials and Aerospace Engineering, Illinois Institute of Technology, Chicago, IL, USA

Department of Mechanical Engineering, Northern Illinois University, DeKalb, IL, USA
e-mail: sciammarella@iit.edu

L. Lamberti et al. (eds.), *Advancement of Optical Methods & Digital Image Correlation in Experimental Mechanics, Volume 3*,
Conference Proceedings of the Society for Experimental Mechanics Series, https://doi.org/10.1007/978-3-319-97481-1_3

The study of displacement fields utilizing image technology whether the image is the result of an optical set up or any other process that encodes displacement information into an image is a subset of Image Analysis, a branch of a well-developed discipline Signal Analysis. In Image Analysis terms, the recovery of information from levels of gray, a scalar field into displacements and displacements derivatives, vectorial fields, is a process that included many different developments. One should realize an important aspect that characterizes this application of Image Analysis: the process of displacement information retrieval is done in the context of a mathematical model of material media, the Continuum Mechanics model.

This model assumes that at, least piece-wise, one has a continuous displacement fields with continuous derivatives up to the third order (usual convention in Continuum Mechanics). Displacement fields are vectorial fields. Gradient fields are characterized by the existence of a scalar field, called potential field and a vector field results from the application of the operation gradient to the scalar field. Calling

$$\mathbf{V_{ec}}\,(\mathbf{x}) = \text{grad}\ \psi\ (\mathbf{x}) \tag{3.1}$$

where $\mathbf{V}_{ec}(\mathbf{x})$ is a vector field and $\Psi(\mathbf{x})$ is a potential function.

The basic equations derived in the literature of the experimental determination of displacement fields are based on Eq. (3.1) and displacement fields have properties that are defined by the characteristics of gradient fields. Vector fields are far more complex that those satisfying continuity conditions, this complexity is manifested by the presence of singularities in the field. This is the case in many of the applications of practical interest in Experimental Mechanics. In many applications, the continuity condition is valid only piecewise. For example, to analyze displacement fields in materials near fracture, it is required the use of tiling of patches of continuous areas that are separated by the presence of microscopic cracks and networks of dislocations. These features of the material manifest as singularities of the observed fringe patterns. In complex fringe patterns, it is possible to extend the dislocation concept of Burger vectors utilized in the science of materials, but this time applied to fringe patterns. Consequently, the analysis of patterns to obtain displacements requires a study of all the different fringe dislocations networks present in the field and the possible separation of continuous patches. This is not an easy task to fulfill when the fringe patterns have multiple singular points and complex fringe dislocation structures.

3.1.2 The Concept of Phase

We begin the analysis of displacements with the simplest case, the 2D case. This is an approximation because our physical space is 3D but, as it is done in Continuum Mechanics, the 2D approach helps understanding the more complex case of 3D signals. It is a practical approach that gives excellent results in very many cases of technical and scientific interest and can be also applied for 3D analysis by projecting displacement information into planes.

In optical sciences, it is usual to represent the optical signal as formed by two components, an amplitude that defines the intensity of the signal and a phase that represents the optical path of the signal. Mathematically a phasor is a complex number that in the case of a 1D signal is expressed by,

$$\overset{\Rightarrow}{\mathbf{I}_{\mathbf{sp}}}\ (\mathrm{x}) = \mathbf{I_{sp}}\,(\mathrm{x})\,\mathbf{e}^{2\pi \mathbf{j}\phi(\mathrm{x})} \tag{3.2}$$

In Fig. 3.1 that represents graphically Eq. (3.2), I_p is called the in-phase component of the intensity, I_q is the in-quadrature component, I_{sp} is the amplitude of the intensity signal and φ is the phase. In the complex plane, the following relationship apply,

$$\overset{\Rightarrow}{\mathrm{I}_{\mathrm{sp}}}\,(\mathrm{x}) = \mathrm{I}_p\,(\mathrm{x})\,\overset{\Rightarrow}{\mathrm{i}} + \mathrm{I}_q\,(\mathrm{x})\,\overset{\Rightarrow}{\mathrm{j}} \tag{3.3}$$

$$\phi\,(\mathrm{x}) = \mathbf{arctg}\frac{\mathrm{I}_{\mathbf{q}}\,(\mathrm{x})}{\mathrm{I}_{\mathbf{p}}\,(\mathrm{x})} \tag{3.4}$$

$$\left\|\overset{\Rightarrow}{\mathbf{I}}_{\mathbf{sp}}\,(\mathrm{x})\right\| = \sqrt{\mathrm{I}_{\mathbf{p}}^2\,(\mathrm{x}) + I_{\mathbf{q}}^2\,(\mathrm{x})} \tag{3.5}$$

Fig. 3.1 Complex plane representation of a phasor

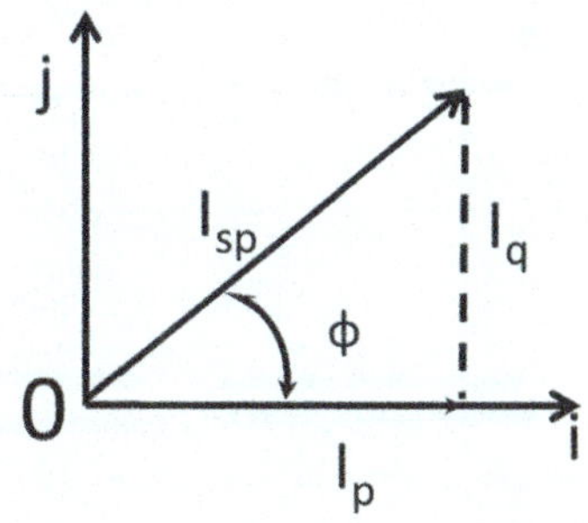

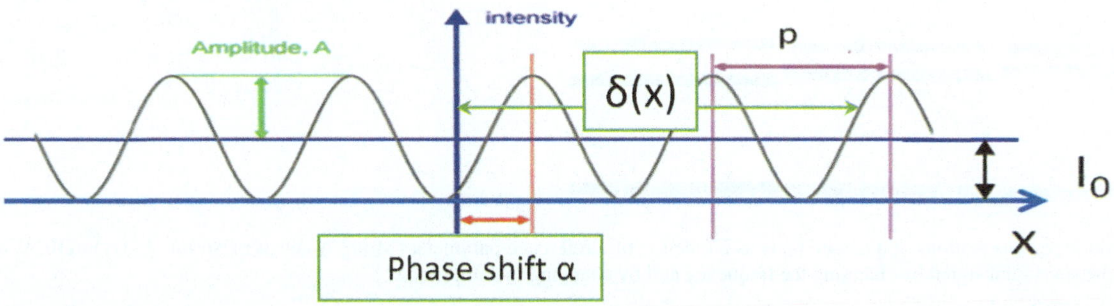

Fig. 3.2 Sinusoidal signal of pitch p

The adopted notation in Eq. (3.2) and following equations introduces the symbol ⇒ that indicates that these relationships apply to the complex space of Fig. 3.1 and not to the physical space.

Let us apply the Eqs. (3.3) to (3.5) to the sinusoidal signal of Fig. 3.2 and define the phase for this signal. The definition of phase in optics is an expression that measures distances through an angular variable,

$$\phi\,(\mathrm{x}) = \frac{2\pi\delta\,(\mathrm{x})}{\mathrm{p}} \tag{3.6}$$

If we know the total angle that corresponds to the distance $\delta(x)$ from the origin of coordinates, we can compute $\delta(x)$ or vice versa.

We can introduce now the concept of local phase, a very important concept in image analysis. This concept is related to the fact that in image analysis the space is quantified, in 2D, by a square element of area of side Δx, a pixel, or in the 3D space by a voxel, a cube of side Δx. In two dimensions the local phase is defined,

$$\Delta\phi_{\mathrm{L}}\,(\mathbf{x}) = \frac{2\pi\Delta\,(\mathbf{x})}{\mathrm{p}} \tag{3.7}$$

To this point, we have analyzed a sinusoidal signal that has constant pitch. The actual signals corresponding to fringe patterns are not sinusoidal signals but phase (frequency modulated) signals of changing pitches. To develop a consistent model of these signals, it is necessary to introduce the concept of local phase that implies as an approximation at the level of a pixel, the signal can be assumed to have local constant pitch and that a global phase corresponding to a given distance is given by,

$$\phi_{\mathrm{g}}\,(\mathrm{x}) = \sum_{\mathrm{j=0}}^{\mathrm{j=N}} \Delta\phi_{\mathrm{L}}\,(\mathrm{x}) \tag{3.8}$$

Where $\phi_g(x)$ is the global phase at a pixel N, and $\phi_L(x)$ is the local phase at a given pixel.

The classical procedure of unwrapping is based on the concept of global phase. It introduces a problem, the need to secure continuity of the fringe at the end of one full cycle and the beginning of the next cycle. Additionof the local phases handles all

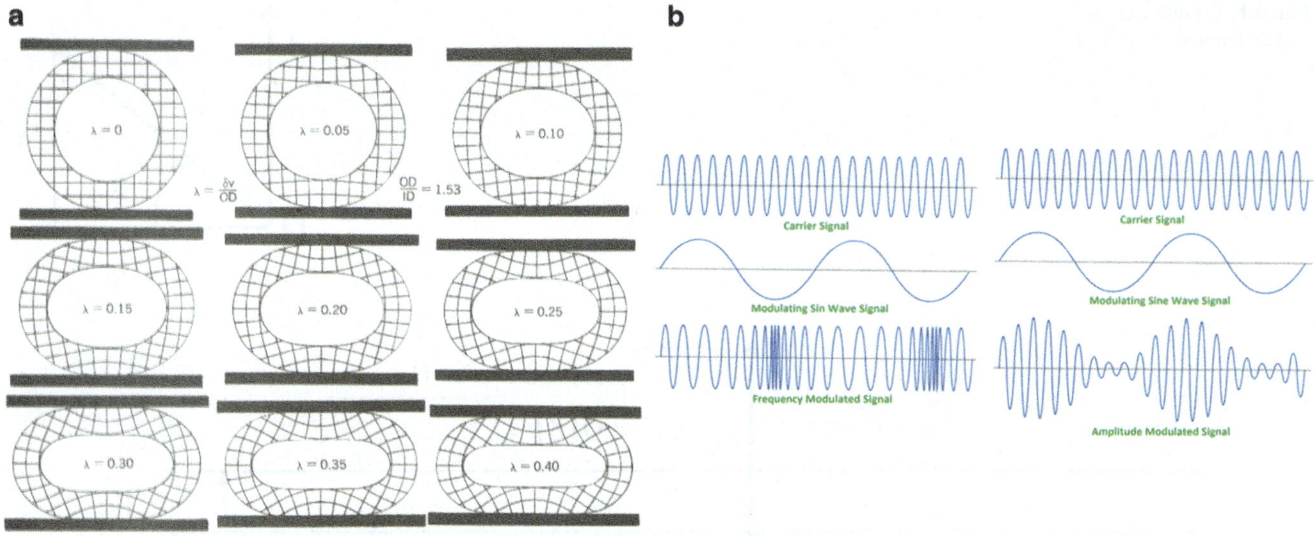

Fig. 3.3 (**a**) Large deformations of a tagged body as a function of a reference parameter (Moiré Analysis of Strain, A. J. Durelli, V. J. Parks). (**b**) Change of the sinusoidal signal by changing the frequency and by changing the amplitude

the pixels independently of their position in a cycle. Thus, local phase addition does not need to recognize pixels where one cycle ends and another cycle begins. This feature is very important in the application of tagged carriers in complex patterns where there are present many fringe singularities.

3.1.3 Method of the Tagged Carriers

To measure displacements the method of tagged carriers utilizes generally a carrier, a known sinusoidal signal. This is not mandatory choice, but is a common starting point that fits well in the general theory of Signal Analysis. This signal is fixed to a surface in a reference state and the evolution of this signal, as the analyzed body is deformed, is recorded and the difference between the initial signal and the final or deformed signal is computed. This comparison includes large displacements that imply very different geometries generated in the process of deformation of a body. This comparison becomes feasible because a system of reference is introduced in the observed body. Figure 3.3 shows large deformations of a ring with a system of tagged carriers. Since the tagged carriers can be given a parameter number, it is always possible to compare the initial basic element with its deformed position. Figure 3.3b illustrates two separate effects that can introduce information into a carrier, either frequency modulation or amplitude modulation, of course one can have a combination of both effects. In the case of deformations, one can see by looking at Fig. 3.3a that the ideal sinusoidal signal is frequency modulated. In actual signals, amplitude modulation cannot be avoided.

The initial signal is a sinusoidal carrier but after modulation it is transformed into a modulated frequency signal that will include multiple harmonics. Figure 3.4a shows in one dimension the harmonics of a frequency modulated signal. Figure 3.4b shows that by filtering it is possible to separate the background term from the harmonics of the signal. Each harmonic has its own phase and its own amplitude, and the signal is the sum of many phasors.

In the literature of isothetic lines (moiré fringes), the fringes are represented by Eq. (3.9),

$$I(x) = I_0 + I_1 \cos \phi(x) \tag{3.9}$$

To have a consistent model with the actual recorded signals, the concept of local phase must be adopted and Eq. (3.9) must be transformed in the one-dimensional case into,

$$I(x) = I_0(x) + I_1(x) \cos \phi(x) \tag{3.10}$$

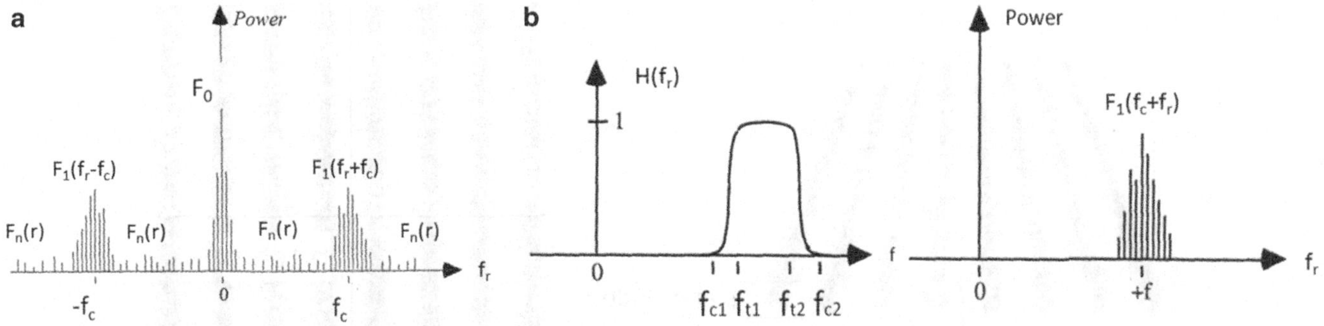

Fig. 3.4 (**a**) Spectrum of a frequency modulated function and (**b**) filter to remove the first harmonic of the different frequencies contained in this harmonic

Equation (3.10) indicates that at a local pixel, the frequency modulated function can be represented as a sinusoidal signal. This expression is only an approximation, and to be valid, certain conditions must be satisfied. These conditions will be analyzed later in the paper.

3.1.4 2D Signals

In preceding paragraphs, we have illustrated the carrier as a one-dimensional signal. However, the signals of practical interest are 2D and 3D signals. Figure 3.5 illustrates a 2D signal and it shows that the signal has the same basic properties that the 1D signal but has a very important additional parameter that is the orientation of the signal. The orientation is given by the direction of the normal **n(x)**, the angle θ that the normal makes with the reference x-axis in a Cartesian coordinate system. This parameter has a fundamental role in the actual recovery of the displacement information encoded in the pattern. Consequently, the accurate recovery of the angle θ is one of the critical steps on fringe data processing besides the recovery of the other parameters that are necessary to obtain the components of the displacement field.

The angle θ has a very important property, it is defined by the grad Ψ, the scalar potential function that corresponds to the sinusoidal levels of gray and has two components derivatives,

$$\text{grad}\ \psi\ (\mathbf{x}) = \frac{\partial \psi\ (\mathbf{x})}{\partial \mathrm{x}}\mathbf{i} + \frac{\partial \psi\ (\mathbf{x})}{\partial \mathrm{y}}\mathbf{j} \tag{3.11}$$

Continuum Mechanics tell us that the deformation field is a tensor field and in 2D is defined by four derivatives, hence in Cartesian coordinates we need as illustrated in Fig. 3.3, two orthogonal system of carrier fringes, each system providing two derivatives.

In Cartesian coordinates, there are two projected displacements families: u(x,y), the displacements projected on the x-axis, and v(x,y), the displacements projected on the y-axis. Each of these two functions can be represented by Monge's expressions, $z_u = u(x,y)$ and $z_v = v(x,y)$ as two surfaces in space. Then, the relative displacement vector is given by:

$$\mathbf{u}_r\ (\mathbf{x}_i) = \mathrm{u}\ (\mathrm{x},\ \mathrm{y})\ \mathbf{i} + \mathrm{v}\ (\mathrm{x},\ \mathrm{y})\ \mathbf{j} \tag{3.12}$$

The subscript r indicates relative displacements.

The modulus of this vector is:

$$|\mathbf{u}_r| = \sqrt{u^2 + v^2} \tag{3.13}$$

The direction of the vector is:

$$\theta_{\mathbf{xd}} = \mathrm{arctg}\frac{\mathrm{v}}{\mathrm{u}} \tag{3.14}$$

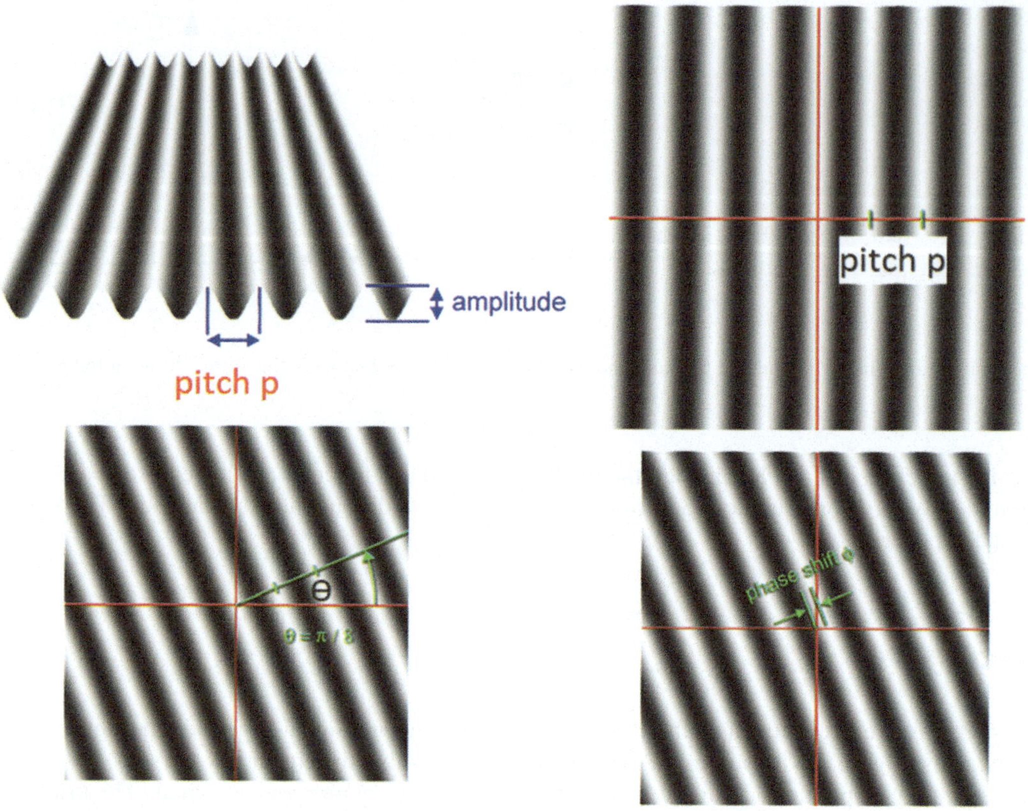

Fig. 3.5 Two-dimensional signal

The subscript d in Eq. (3.14) has been added to distinguish this angle from the angle θ shown in Fig. 3.5 corresponding to the projected displacement vector.

3.1.5 Differences Between the Theoretical Model and the Recorded Signals

Equation (3.10) is the model introduced to represent the optical signal that contains the displacement information and this model corresponds to information encoded in a pixel. The actual signal is not only phase modulated but it is also amplitude modulated by the optical phenomena involved in the recording of the signal. The amplitude modulation has important consequences in the process of recovery of the information contained in the fringe pattern. The correction required to take care of this problem will be discussed later in this paper.

Figure 3.6a shows the u-pattern of a disk under diametrical compression, Fig. 3.6b is the same pattern with carrier fringes added. Figure 3.7a shows the FFT of the u-pattern shown in Fig. 3.6a. It is possible to see that the FFT has a number of superimposed harmonics that cannot be separated by filtering harmonics in the frequency space. Figure 3.7b shows the FFT of the disk without the fringe pattern that it is also visible in Fig. 3.7a. Figure 3.7c shows the FFT of the pattern of Fig. 3.6b and an enlarged version of the FFT that corresponds to the fringe pattern that contains the u-displacement information.

The answer to the question of the feasibility of separating the different harmonic components contained in a signal is given by the Bedrosian-Nuttal's theorems [1–3]. These theorems indicate that the recovery of a given harmonic from the FFT if feasible if the harmonics do not overlap in the FFT. Figure 3.7 provides a graphical visualization of the referred theorems. The FT of the disk fringes is separated of other harmonics by digitally introducing a high frequency carrier.

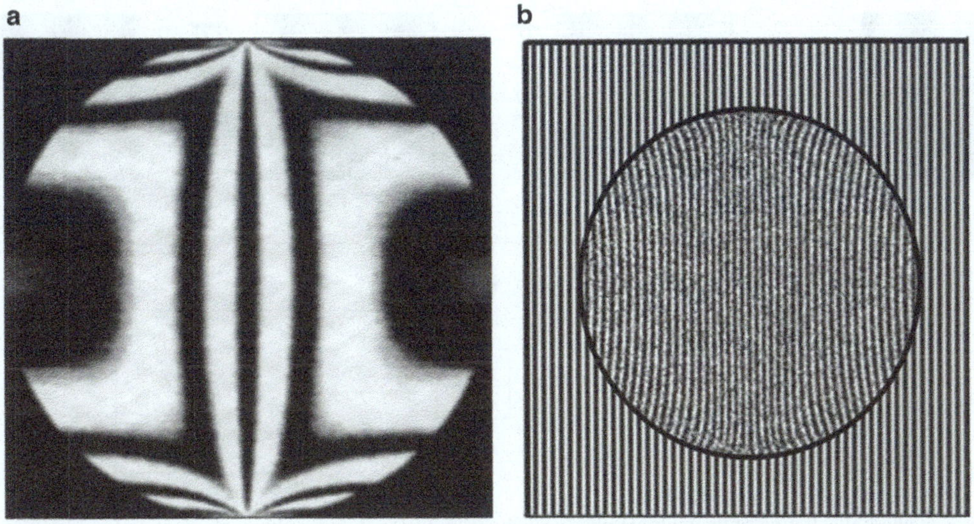

Fig. 3.6 Disk under diametrical compression u-displacement (a) without a carrier, (b) with added carrier

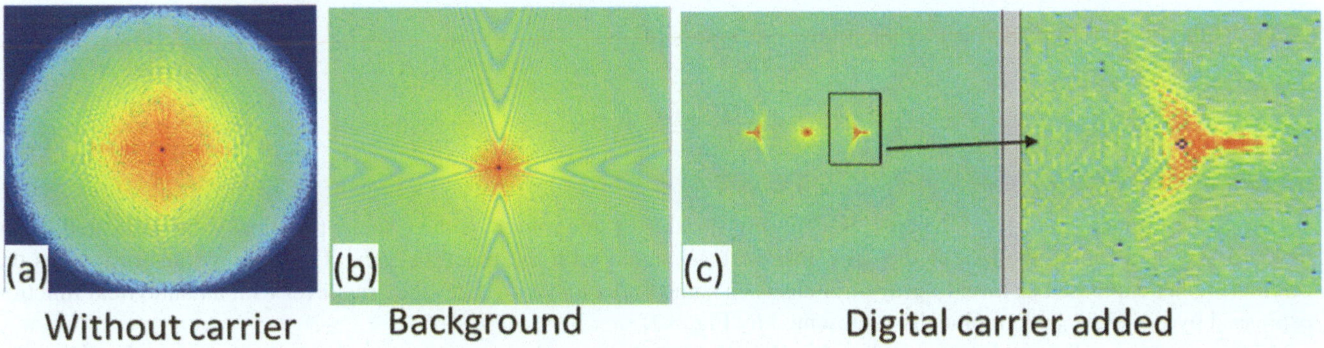

Fig. 3.7 (a) FFT of the u-pattern shown in Fig. 3.6a; (b) FFT of the image of the disk without fringe pattern; (c) FFT of the pattern of Fig. 3.6b

3.1.6 Solutions to the Representation of the Signal by a Sinusoidal Phasor

3.1.6.1 The Multiple Phase Approaches

In the preceding section, it is introduced one condition required for the recovery of the displacement information contained in fringe patterns through the model represented by Eq. (3.10). Historically, in the initial developments of fringe pattern analysis, two different approaches were introduced. One was based on the Hilbert transform and was called in-quadrature phase method, the other the multiple-phase method. In the current literature, the two methods are considered as non-related methods and the multiple phase method has been the predominantly and successfully utilized in actual applications [4].

One can experimentally record multiple patterns with phase differences introduced in the process of recording images and assemble a system of equations, for example applying Eq. (3.10), with the unknowns I_0 (**x**), I_1(**x**) and Ψ(**x**) and solve it for Ψ(**x**). Solutions have been developed for dynamic cases where the acquisition of simultaneous images becomes a more complex problem. In this way, the wrapped phase of the image is obtained as illustrated in Fig. 3.8, four different images are utilized to get an image that allows the recovery of the displacement information encoded in the phase of the optical signal.

3.1.6.2 The In-Quadrature Images Approach

The in-quadrature method was introduced utilizing the Hilbert transform as a tool. It converts a cosine signal into a sine signal without changing the signal amplitude [5].

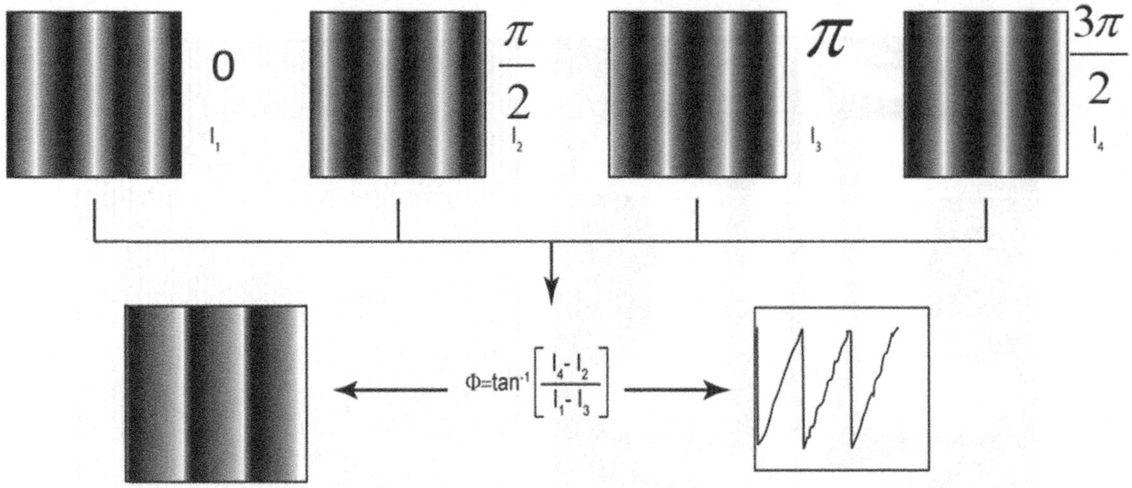

Fig. 3.8 Four phases method: illustration of the multiple-phases methods to deal with the problem of adopting Eq. (3.10) as a solution to extract displacement information from fringe patterns

The Hilbert transform is defined by the following expression,

$$I_q(x) = \frac{1}{\pi} PV \int_{-\infty}^{\infty} \frac{I_p(\eta)}{x - \eta} d\eta = \frac{1}{\pi}\left(I_p(x) * \frac{1}{x}\right) \quad (3.15)$$

The Hilbert transform of $I_p(\eta)$, where η is a dummy variable of integration, is the Cauchy principal value indicated in Eq. (3.15) by PV and can be thought as the convolution of the signal with the function $1/\pi x$. The Hilbert transform takes the original signal, expressed as level of gray or intensity in some scale, and associates the gray level with an analytical function expressed by Eq. (3.3), and graphically represented by Fig. 3.1.

At this point we will consider the problem posed by the fact that actual patterns are not only frequency modulated but also amplitude modulated. Amplitude and simultaneously phase modulated signals do not have a uniquely defined analytic signal representation. Amplitude and phase signals FT's can overlap and, consequently, phase and amplitude recovery information is not possible unless the effect of the mixing of different harmonics is minimized. One condition that is necessary to satisfy, has already been discussed and this condition is contained in the Bedrosian theorems.

The amplitude modulation FT components must be such that they do not overlap with the FT components of the frequency modulation. This means that changes of amplitude as a function of the coordinates of the patterns must be very small when compared with the changes of the signal caused by deformation.

An intuitive picture of the concept of local phase defined in Eq. (3.10) for a 1D signal is provided by Fig. 3.9. One has the 2D complex plane associated with the 1D signal. The definition of instantaneous frequency provides the argument to introduce the concept of analytical signal,

$$f_x(x) = \frac{1}{2\pi}\frac{d}{dx} \arg \vec{\mathbf{I}}_{sp}(x) \quad (3.16)$$

In Eq. (3.16), $\vec{\mathbf{I}}_{sp}(x)$ is given by Eq. (3.3) and the argument of $\vec{\mathbf{I}}_{sp}(x)$ by Eq. (3.4).

In Fig. 3.9a, the full line represents the trajectory of the motion of a point moving along with changing velocity in the complex plane, red vector tangent to the trajectory. Under certain restrictions, the trajectory of the local signal can be assumed to be a circle. Thus, a simple harmonic function is utilized to define instantaneous phase. Periodic motions are represented by a point moving with constant speed along a circular path. The corresponding mathematical expression is,

$$S(x) = A_0 \cos\phi(x) \quad (3.17)$$

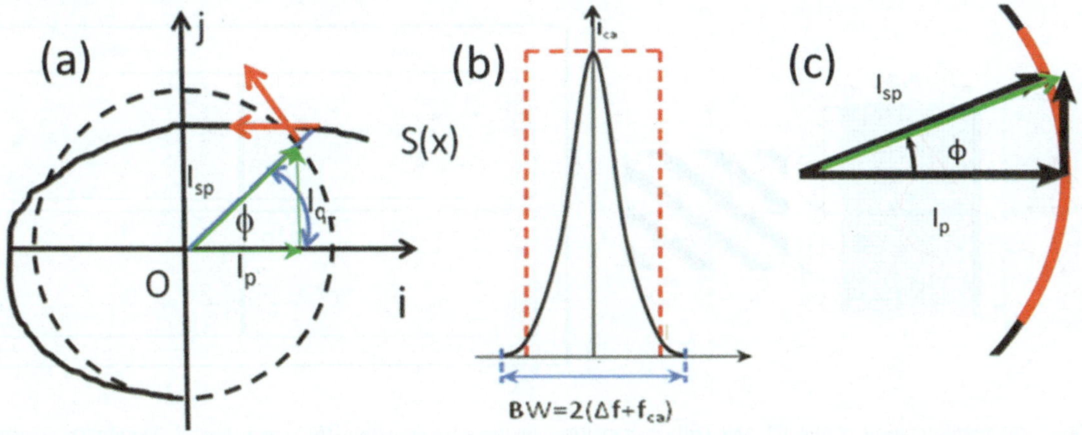

Fig. 3.9 (**a**) Simple harmonic motion as a model to define instantaneous frequency; (**b**) resulting narrow band signal; (**c**) approximation of the final local signal

In Eq. (3.17), A_o is the radius of the trajectory circle; the angle swept by the rotating vector is $\phi(\mathbf{x})=\omega x$, where ω represents a constant angular frequency corresponding to $f_x=\omega/2\pi$. Adding to the frequency modulated signal that contains the displacement information $\phi(\mathbf{x})$ a high frequency carrier, the equation of the signal becomes,

$$\Psi(x) = [2\pi f_{ac} x + \phi(x)] \tag{3.18}$$

If $f_{ca} >> f_x$ where f_x is the frequency of the signal, this addition produces a narrow band signal, shown in Fig. 3.9b, where the band width is $B = 2(f_{ac} + f_x)$.

The condition that the frequency of the carrier is much higher than that of signal allow us to make the approximation,

$$\mathbf{I_p} e^{i\Psi(x)} = \mathbf{I_p}\left[\overset{\Rightarrow}{\mathbf{i}} + [\Psi(x)]\overset{\Rightarrow}{\mathbf{j}}\right] \tag{3.19}$$

Then,

$$\mathbf{I_p}(x)\,\Psi(x) = \mathbf{I_q}(x) \tag{3.20}$$

and finally, as an approximation

$$\mathbf{I_{sp}}(x) = \sqrt{\mathbf{I_p^2}(x) + \mathbf{I_q^2}(x)} \tag{3.21}$$

The preceding developments have a simple intuitive graphical representation. The modulated signal can be assumed to be generated by a rotation vector of amplitude I_p and angular position $\Psi(x)$, green vector of Fig. 3.9c. The local changes of phase are small quantities such the tangent function has to a certain number of significant figures the same values that the arc function. This derivation takes us back to Fig. 3.1 and using Eq. (3.10), a sinusoidal function, as the carrier of the displacement function and to the analytical function theory representing displacement fields.

3.1.7 Role of the Transforms

As mentioned before, the recorded signal is a scalar signal and to get the displacement field, a vectorial field, it is necessary to apply a tool that converts the scalar signal into a vectorial field.

There are two equivalent alternatives to perform this operation [4, 5]. One is the Fourier Transform through the numerical procedures provided by the Fast Fourier Transform (FFT) and the other is the Hilbert transform. These two alternatives are very similar although different numerical algorithms are utilized. In the FFT, negative frequencies are introduced in the frequency spectrum, although they do not have a direct physical meaning. In the Hilbert transform, negative frequencies are removed. Figure 3.10 graphically represents the fundamental aspects of the transforms.

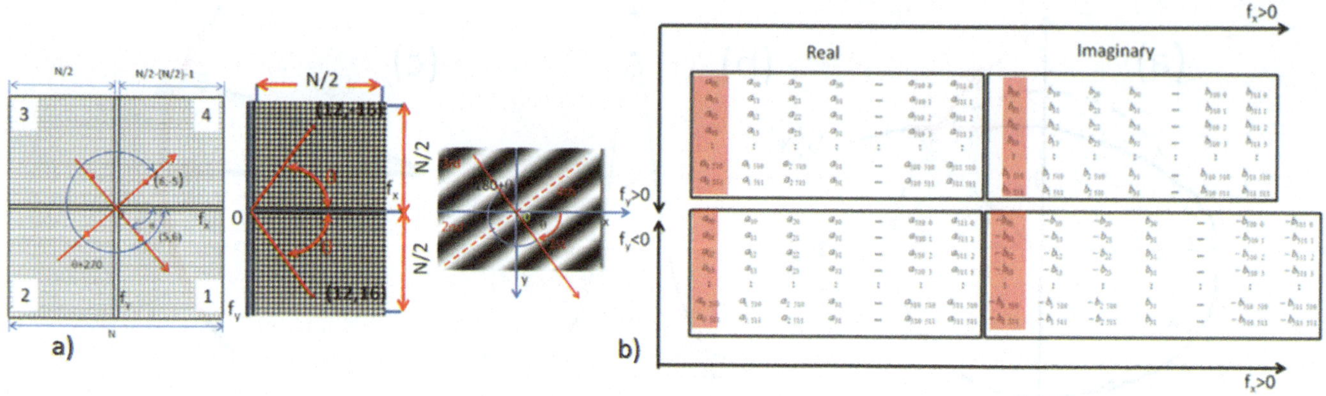

Fig. 3.10 (**a**) Graphical representation of the FT and Hilbert transform in the corresponding frequency spaces, symmetry conditions and a sinusoidal fringe pattern; (**b**) Table of coefficients assuming that the image has 1024 × 1024 pixels; the columns outlined in light red will correspond to the coefficients of the zero order that are present in the FT and are not present in the Hilbert transform

The FT uses a representation in the frequency space where to the 4 quadrants of the physical space correspond 4 quadrants in the physical space. The introduction of negative frequencies brings redundant information that are expressed in symmetry conditions. The FT is processed in the frequency space and generates coefficients that are assembled in different manners and Fig. 3.10b shows one of the many possible arrangements in the memory of a computer. Figure 3.10a shows also the equivalent frequency space for the Hilbert transform. Although the Hilbert transform as shown by Eq. (3.15) is computed in the physical space, there is also the alternative of performing the necessary computations in the equivalent frequency space. The important consequence is that both transforms produce the analytical function corresponding to a given harmonic of the spectrum,

$$\overset{\Rightarrow}{\mathbf{I}_{sp}}(x) = \mathbf{I_p}(x)\overset{\Rightarrow}{\mathbf{i}} + \mathbf{I_q}(x)\overset{\Rightarrow}{\mathbf{j}} = a_{ij}\overset{\Rightarrow}{\mathbf{i}} + b_{ij}\overset{\Rightarrow}{\mathbf{j}} \tag{3.22}$$

The transform components are complex numbers ($a_{ij} + jb_{ij}$) that can be arranged in a square matrix where "i" indicates the rows and "j" the columns as shown in Fig. 3.10b. The columns outlined in light red will correspond to the coefficients of the zero order that are not present in the Hilbert transform. The zero order is a singularity in the Hilbert transform since it will produce an infinity.

Summarizing, both the FT and the Hilbert transform provide the analytic functions that transform scalar fields of intensities into vectorial field of intensities. In the case of 1D signal, the representation of the signal in the complex plane is shown in Fig. 3.9, a circle as it can be concluded from preceding developments. It is essential to remember that this representation is an approximation that must satisfy all the restriction that have been previously outlined. It must be pointed out that the preceding derivations provide only the vector intensity but not its orientation in the physical space that must be obtained by separated procedures. In 2D, the graphical representation of the signal that contains the information on displacement vectors in 2D is a sphere, that is known in Photoelasticity as Poincare sphere [5].

The signal that contains the information is called monogenic signal and is a phasor represented in Fig. 3.11,

$$\overset{\Rightarrow}{\mathbf{M}_s^r} = I_{sp}\exp\left[\cos\phi\cos\theta\overset{\Rightarrow}{i} + \cos\phi\sin\theta\overset{\Rightarrow}{j} + \sin\phi\overset{\Rightarrow}{k}\right] \tag{3.23}$$

In this phasor appears the angle θ that provides the direction of the signal (isothetic line). It is necessary to point out that Eq. (3.23) represents a signal that is valid at a given point of the 2D continuum, a pixel in the quantified space of a sensor.

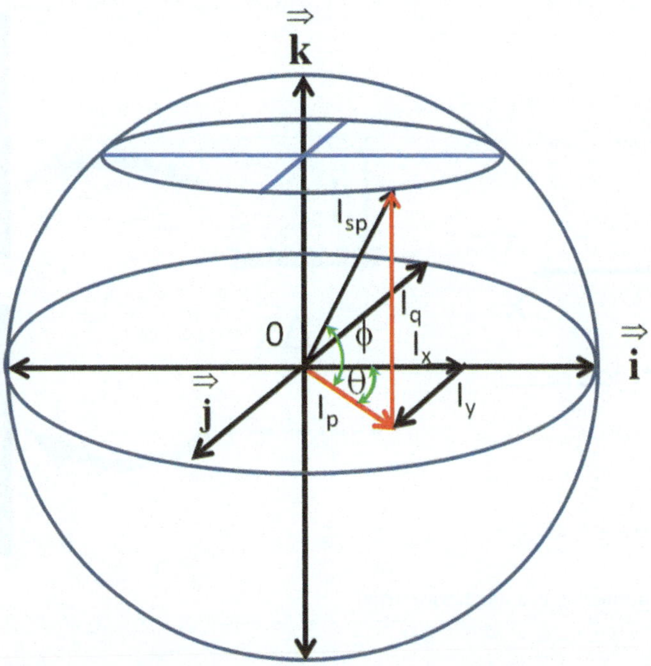

Fig. 3.11 Poincare sphere that represents the monogenic function for a 2D space

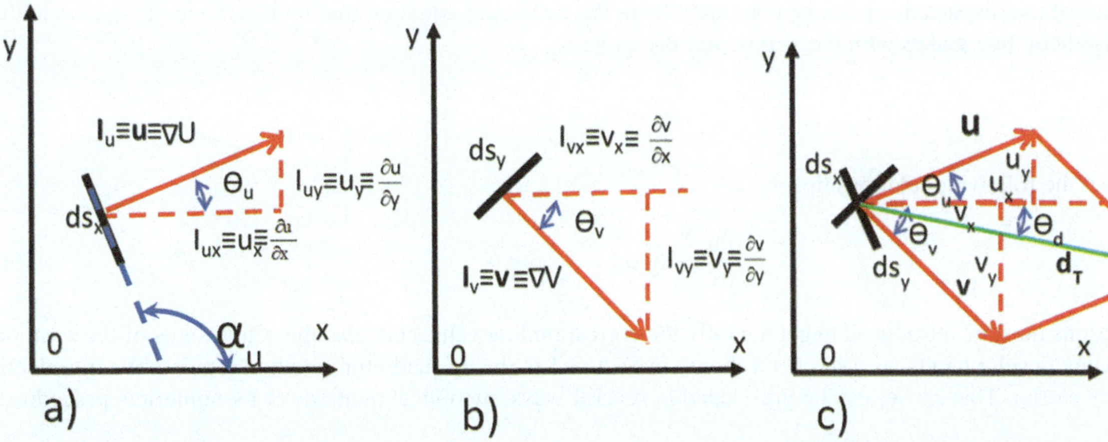

Fig. 3.12 (**a**) U-intensity vector; (**b**) V-intensity vector; (**c**) Addition of the U-intensity vector and the V- intensity vector

3.1.8 Intensity Vectors, Displacement Vectors, Gradient Vectors

Procedures to obtain from scalar recordings of gray levels, intensity vectors have been outlined. The next step is to establish a relationship between intensity vectors and displacement vectors [6]. This relationship is outlined in Fig. 3.12; in the figure. isothetic lines are represented by their tangent at a given point. Figure 3.12a shows the vector $I_u(\mathbf{x})$, the displacement vector $\mathbf{u}(\mathbf{x})$, the vector gradient of $u(\mathbf{x})$, and the normal vector to the isothetic line $\mathbf{n}(\mathbf{x})$ and all these vectors are co-axial. Figure 3.12b indicates that similar properties apply to the field $\mathbf{v}(\mathbf{x})$. Figure 3.12c shows the addition of the vectors displacement to get the total displacement vector $\mathbf{d_T}(\mathbf{x})$ and the angle $\theta_d(\mathbf{x})$ that this vector makes with the x-axis.

The connection between the displacement field vector phase as defined in Eq. (3.4) at a pixel is done through the equation,

$$u_x(\mathbf{x}) = \frac{\phi_u(\mathbf{x})}{2\pi} p \tag{3.24}$$

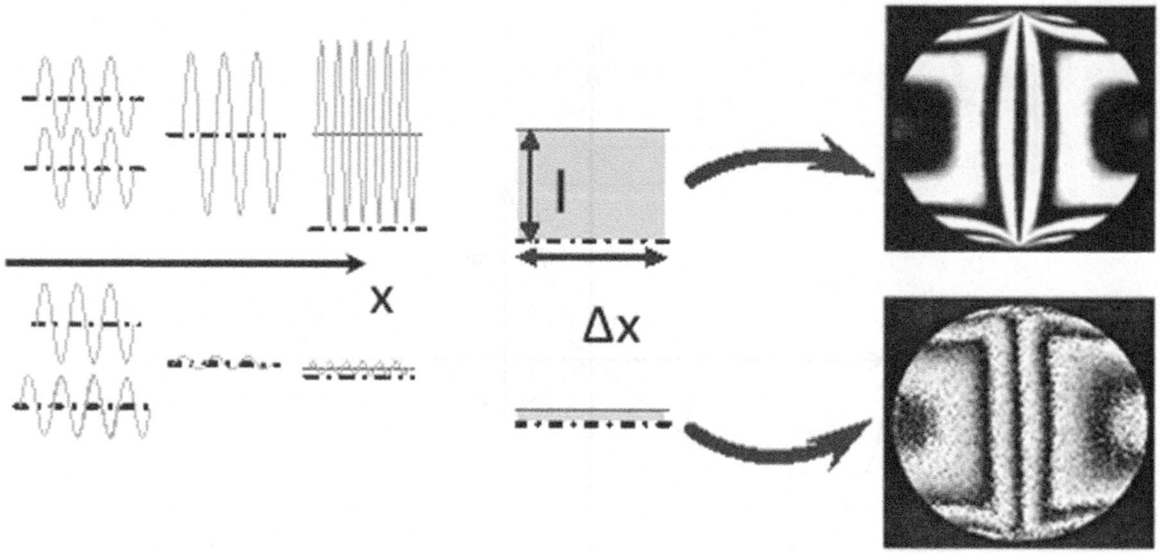

Fig. 3.13 Encoding displacement information in a rough wave front

The quantity $\frac{\phi_u(x)}{2\pi}$ is the fringe order n. Since we are speaking of local fringe orders, n is a real number that can take values between 0 and 1, and p is the carrier signal pitch.

The derivative of the displacement can be computed from the following equation that relates the angle α_u, angle that the tangent of the isothetic line makes with the x-axis, and the angle θ_u,

$$\alpha_u = \theta_u + \frac{\pi}{2} \tag{3.25}$$

Hence, we get the following relationship,

$$\frac{\partial u}{\partial x} = \text{tg}\ \alpha_u = -\text{cotg}\ \theta_u \tag{3.26}$$

Sign conventions must be introduced to get not only the corresponding values but also the actual signs of the components.

In the preceding developments, as it has been shown in Sect. 3.1.6, the in-quadrature method requires the introduction of a high frequency carrier. This carrier can be introduced in several ways, by optical methods or by numerical procedures.

3.1.9 Encoding Displacement Information Using Rough Wave Fronts

Figure 3.13 illustrates two fringe patterns that contain the same displacement information, horizontal components of a disk under diametrical compression: one obtained using moiré interferometry, coherent interference of two smooth waves; the other through speckle interferometry, the interference of two rough wave fronts. The difference between patterns is a consequence of the utilized wave fronts.

There is an important difference in this process, in the case of a deterministic signal if a rotation is not introduced between the two wave fronts, the relative position of the two patterns does not change the results. This is not the case for random wave fronts, only if one superimposed locally very similar wave fronts the moiré effect produced by the optical correlation of the two patterns cannot take place. This effect in this case is limited to distance called correlation distance, a small quantity limited by a statistic quantity, speckle size. When the displacement between the wave fronts is larger than the correlation distance, the superimposition of the two rough wave fronts generates another speckle pattern. Figure 3.14 illustrates the loss of the dynamic range of the random signal with respect to the deterministic signal due to the statistic intensity distribution.

There is an alternative form to recover displacement information when the displacement exceeds the correlation distance, this process has been called speckle photography. Initially was based on the generalization of the classical Young's fringes effect. The classical speckle photography utilizes a carrier signal that is produced by the random interference fringes produced

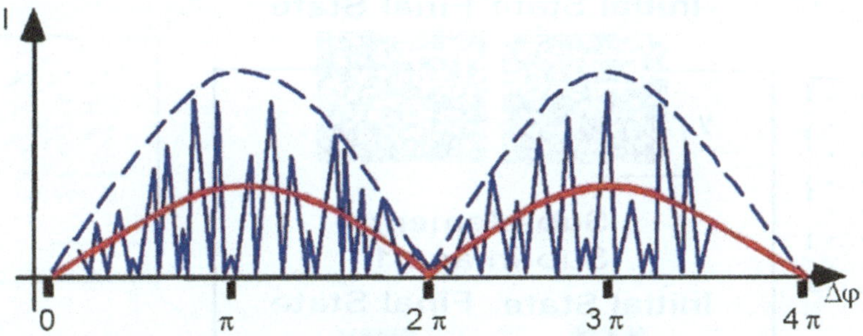

Fig. 3.14 Smooth wave front and rough wave front, red profile statistical average of the rough wave front

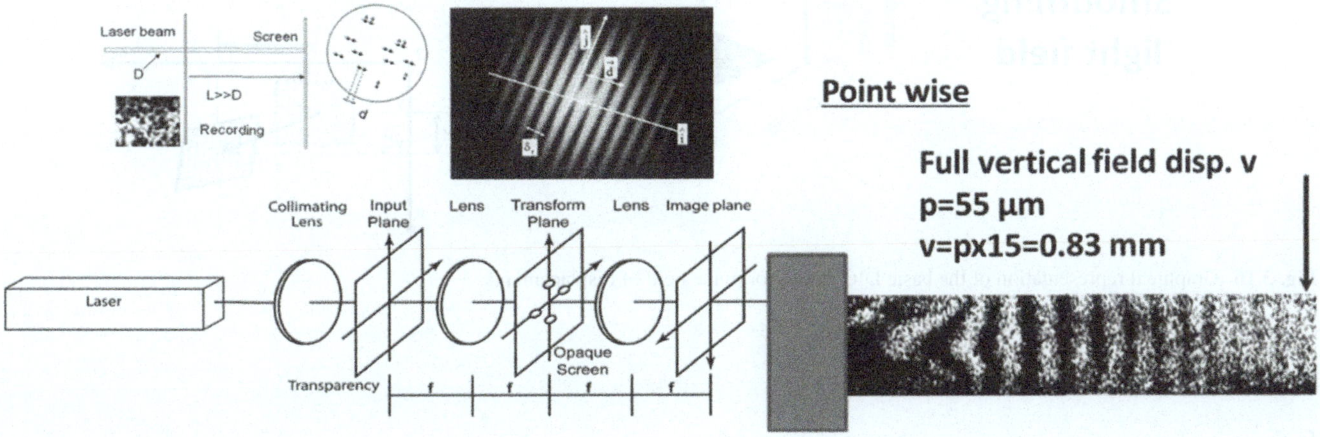

Fig. 3.15 Point-wise speckle interferometry providing both the magnitude and the orientation of the displacement vector d. Speckle photography of the vertical displacements of a cantilever beam, equivalent pitch $p = 55\ \mu m$, total number of fringes to the tip 15, vertical displacement 0.83 mm

by the surface roughness. Later, it was realized that one can as well create a random pattern that modulates the intensities reflected by the surface of a body by applying random pattern to the surface by any of the many possible mechanisms provided by the current technology. This process is referred in the literature as white light speckle. White light speckle is a methodology identical in many aspects to speckle photography, it has however one advantage. In speckle photography, it is necessary to assume that the formed speckles remain practically unchanged when the body is deformed, that is the surface carries the speckles as it is deformed; this assumption has limitations that do not affect the white light speckle. What is the difference between moiré with incoherent illumination and the so called white light speckle? Incoherent light moiré as in the case of moiré interferometry utilizes a deterministic signal; white light speckle utilizes a random signal. In most of the white light speckle applications a different fringe processing methodology has evolved from that utilized in traditional speckle photography. Why random signal and not a deterministic signal? The difference arises from practical considerations. To put a deterministic signal on an arbitrary surface is not an easy proposition. There was on the beginning of the development of methods that measure displacements many diverse attempts to do it. It was proven that is feasible but not very practical for general applications. To apply a random reflectivity signal can be as simple as spraying a surface with a nozzle that deposits random drops.

Figure 3.15 illustrates local point-wise interferometry that gives magnitude and orientation of the displacement vector at a given point. Figure 3.15 illustrates also speckle photography providing the vertical displacements of a cantilever beam with a concentrated load applied at the tip. It produces isothetic lines that project the displacement vector in the vertical direction. The pattern has been generated by optical means as illustrated in Fig. 3.15. However, the same process can be done by numerical software computation.

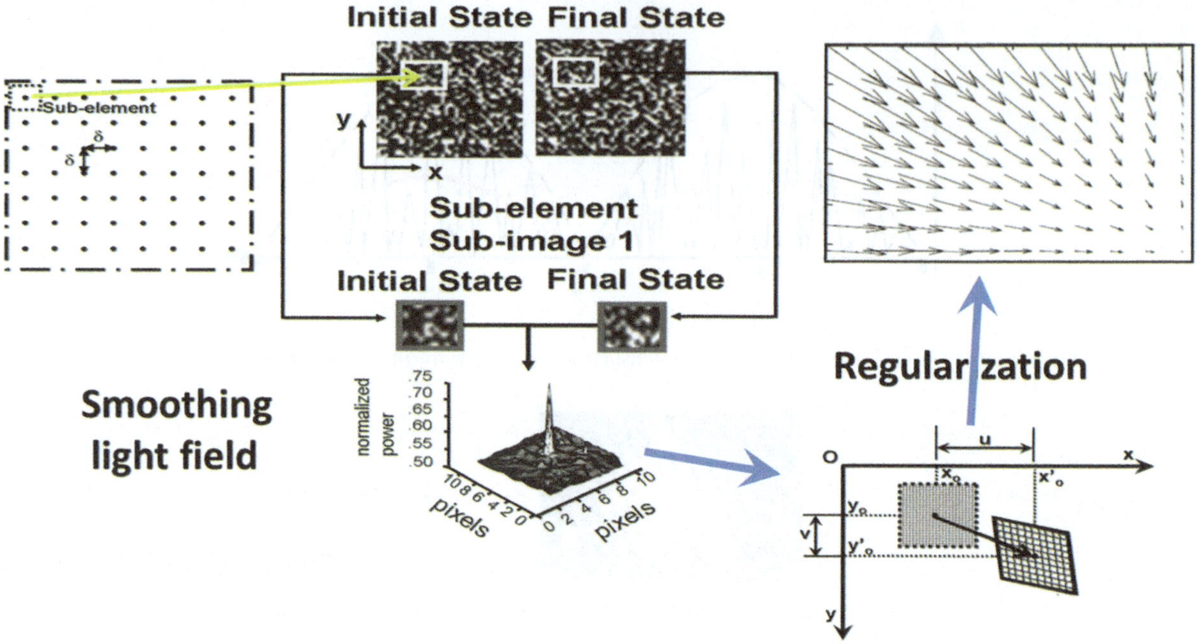

Fig. 3.16 Graphical representation of the basic DIC model for plane field of displacements

3.1.10 Digital Image Correlation

In the preceding sections, a review of the properties of fringe patterns providing displacement information from the point of view of optical signals. Many of the properties reviewed have general validity no matter what approach for fringe pattern processing. Now we will concentrate in the currently most utilized method of displacement fringe pattern processing, digital image correlation [7, 8].

Aspects of getting a random carrier on a surface also were presented. The conclusion is that, from the practical point of view, the advantage of speckle photography with artificial speckles compared to incoherent moiré patterns is the simplicity of tagging a random carrier compared to engraving a deterministic signal, lines or dots. However, what characterizes discrete image correlation DIC is not the use of artificial speckles, a feature in common with the earlier unfolding of white light photography. The basic assumption of DIC is the connection of the displacement vector of a point of the continuum with the vector light intensity of the point and with the gradient of the vector field. In preceding sections, it has been mentioned that these two properties are mathematically correct in the context of the kinematics of the continuum and provide a connection between the scalar field of intensities recorded in an image and the vector displacement field. These relationships are formally contained in the theory of the monogenic phasor [5].

Although the basic approach of DIC can be applied to any type of signal, it evolved from the analysis of displacement information encoded in a random signal. To make relative comparison of DIC methods with methods based on the general theory of image analysis, we will start with a simple case, the 2D analysis of displacements taking into consideration that in the general case DIC includes many optical aspects that are common to the two approaches. In Fig. 3.16, one has a surface that for the sake of simplicity is assumed to be a plane and is viewed in the direction normal to the surface. Furthermore, it can be assumed that a telecentric system is used to get the image of the surface thus separating the effects of image formation in the case of optical recording from the problems of image correlation. Two separate images of speckles are recorded, the initial or reference image and the deformed image. From these two images the displacement field vectors are determined and unlike the developments of preceding sections, the splitting of the optical signal into amplitude and phase is not invoked.

The basic idea behind DIC is to track the position of an element of area as the deformation takes place. In the surface one has a distribution of intensities, a random scalar signal $I_i(x,y)$. A displacement field is applied to the surface and a final distribution of intensities $I_f(x,y)$ is obtained. Furthermore, it is assumed that the changes in the image are primarily a function of the displacement field. It is further assumed that, as occurs in all experimental cases, there is noise. Noise indicates all the changes of the image intensity that are not caused by the displacement field and are assumed to be less significant than the effects of the displacement field.

The displacement field is defined by the vector function (see Fig. 3.12),

$$d_T\,(x, y) = u\,(x, y)\ i + v\,(x, y)\ j \tag{3.27}$$

From the preceding assumptions, the following scalar function is obtained.

$$I_f\,(x_i + u, y_i + v) = I_i\,(x_i, y_i) + \Delta I\,(x_i + \Delta u, y_i + \Delta v) + I_n \tag{3.28}$$

In Eq. (3.28), ΔI is the change of the scalar field caused the local deformation of the analyzed surface. The functions u and v are the continuum mechanics functions that have been defined previously. The problem to be solved is to find u(x,y) and v (x,y) knowing the scalar functions $I_i(x,y)$ and $I_f(x + u,y + v)$.

It is assumed, as it is always in the case with speckles, that the random pattern is carried by the surface without changes in the intensity. One can formulate the problem as an optimization problem, that is find the best values of these two functions that minimize or maximize a real function, the objective function of the optimization process. In mathematics, optimization indicates an algorithm that helps to choose the best components that satisfy a given mathematical model from a set of available data. In its simplest form, optimization involves the process of maximizing or minimizing a real function. This is exactly the problem that must be solved in DIC. One criterion is the minimum squares; the difference of the intensities of the two images must be minimized as a function of the experienced displacements. If the minimum squares is adopted, calling Φ (u,v) the optimization function.

$$\Phi\,(u, v) = \iint \left[I_f\,(x_i + u,\ \ y_i + v) - I_i\,(x_i, y_i)\right]^2 dxdy \tag{3.29}$$

For small u(x,y) and v(x,y), the above expression can be expanded in a Taylor series and limiting the expansion to the first order and using vectorial notation,

$$\Phi\,(D) = \iint \left[I_f(r) - I_i(r) + d_T(r) \bullet \nabla I_f(r)\right]^2 dxdy \tag{3.30}$$

In the above equation, r is the spatial coordinate, $\mathbf{d}_T(r)$ is the displacement vector and ∇ is the gradient operator. Eq. (3.30) tells us that the gradient of I_f provides the following information, the displacement information is associated with the gradient of the intensity distribution. Since I_f is a scalar function (light intensity), the gradient is a vector. Hence, the information can be retrieved following the gradient function of the light intensity. Imagine that one has a small rigid body motion that corresponds to a rotation and a translation. Furthermore, great care is taken to insure two almost identical quality images. If the correlation of the corresponding images in two dimensions is applied, the vectors will all reflect the rigid body motion, in this case,

$$\Phi\,(D)_{\,min} \rightarrow_{max} \iint \mathbf{I_i}\,(\mathbf{r}) \otimes \mathbf{I_f}\,(\mathbf{r} + D)\ dr \tag{3.31}$$

where the symbol ⊗ Indicates the correlation operation.

Equation (3.31) applies to this case and the minimum of the objective function corresponds to the maximum of the correlation function. In all other cases the problem is far more complex, and the cross correlation is not directly the solution of the problem. The minimization of the objective functions is then a central problem of the image digital correlation technique. In the technical literature, there is a large variety of approaches to this problem. One can utilize criteria other than the minimum squares. Furthermore, even if one selects the minimum square format for the objective function there are many ways that the problem can be approached. There are two possibilities to perform the required operations in DIC, one can utilize the coordinate space or the Fourier space. How the general scheme of DIC can be implemented? This aspect of the problem will be addressed by utilizing the approach that is most commonly adopted in the DIC literature.

The region of interest (ROI) shown in Fig. 3.16 is subdivided in sub-samples of size of $N_p \times N_p$ pixels and the raster of dots indicate the position of the centroids that form a regular mesh of δ × δ pixels. The parameter δ defines the mesh formed by the centers of each sub-sample used to analyze the displacement. Given a certain sub-element, a search in the neighborhood of the element is implemented such that a normalized form of Eq. (3.29) is minimized yielding a correlation peak as illustrated in Fig. 3.16. In this way is determined the displacement of the centroid of the element in the deformed condition.

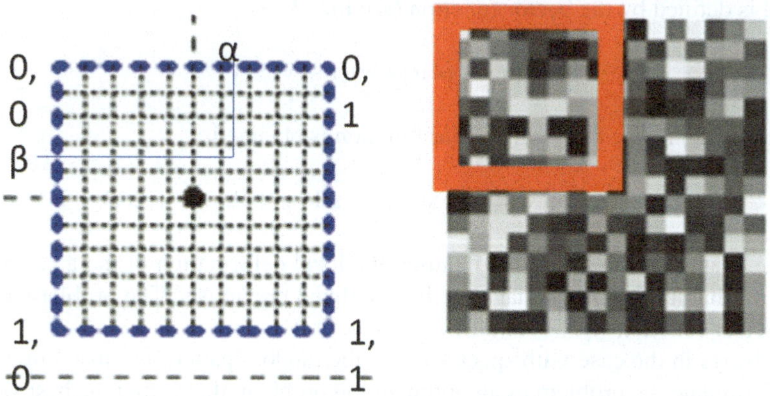

Fig. 3.17 Local coordinate system in a sub-image. Pixel locations are indicated by the parameters α and β

The actual distribution of intensities in an image is a discrete quantized distribution; the quantum value is the level of gray of the sensor system. This distribution is not smooth enough for the operation required by the optimization technique. Consequently, to perform the required operations the intensity experimental values are smoothed. A common interpolation utilized is the bicubic spline. A bicubic spline interpolation is a 2D version of the cubic interpolation utilized to interpolate points by segmentation of a curve in 2-D space. In image processing, bicubic interpolation is preferred over other alternative methods because it provides smoother results.

$$\mathrm{I}\,(\mathrm{x},\mathrm{y}) = \sum_{\mathrm{i}=1}^{3} \sum_{\mathrm{j}=1}^{3} \mathrm{a_{ij}}\ \mathrm{x^i y^j} \tag{3.32}$$

Figure 3.17 shows the pixel content of a sub-element and the bicubic interpolation yields Eq. (3.32), an expression for the levels of gray contained in one sub-element. Conditions of continuity between sub-elements are imposed.

The most frequently used approach is known as local DIC; it utilizes a reference-subset centered at each calculation point as shown in Fig. 3.17, then gets the deformed-subset using a local shape function. To obtain the deformed shape in a general case, the perturbation method is utilized. The perturbation method yields the sought solution of a problem as a power series in terms of a perturbation parameter that quantifies the deviation from a problem that has an exact solution. The leading term of the power series is an exact solution for a simplified form of the problem. The necessary additional terms quantify the deviation of the actual problem from the solution of the simplified form of the problem. The symbol $\Psi(\mathbf{d_T})$ represents the solution of the problem. The solution is the sum of the leading term, $\mathbf{d_{T0}}(\mathbf{C})$ plus additional terms: $\mathbf{d_T}(\ell)$ indicates a linear approach to the displacement vector, $\mathbf{d_T}(\mathrm{q})$ indicates a quadratic solution and one can utilize successive higher order terms.

$$\Phi\,(\mathbf{d_T}) = \mathbf{d_{T0}}\,(\mathbf{C}) + \mathbf{d_{d_{T1}}}\,(\ell) + \mathbf{d_{T2}}\,(\mathbf{q}) + \ldots \tag{3.33}$$

In Eq. (3.33), $\mathbf{d_{T0}(C)}$, indicates a constant, $\mathbf{d_{T1}}(\ell)$ indicates a linear term, $\mathbf{d_{T2}(q)}$ indicates a quadratic term. The higher order terms of the power series become smaller as the order of the terms increase.

Taking the unknown quantities corresponding to the linear shape function, the unknowns can be represented as a vector quantity with six components,

$$\mathbf{d_T} = \left[\mathbf{u}, \mathbf{v}, \mathbf{u_x}, \mathbf{u_y}, \mathbf{v_x}, \mathbf{v_y}\right]^{\mathbf{T}} \tag{3.34}$$

Where the symbol, i with i = x,y, represents the partial derivatives; this result agrees with Fig. 3.12. This is an interesting conclusion for the DIC method. In DIC and taking into consideration the spatial resolution of the field, the adopted number of pixels in a sub-element is directly connected with the selected shape function. This is the conclusion of most papers investigating the DIC accuracy. The local DIC approach by adopting a given shape function enforces local continuity and this continuity is independent of the neighbor sub-elements. Hence, there is not continuity of the vector field in the global sense. The local DIC was the approach of the initial literature of DIC and it is also the approach adopted by many current commercial programs. Later appeared a global approach that analyzes the ROI by connecting the sub-samples with nodes of a Finite Element mesh; in this fashion continuity is insured in the global sense. The preceding argument applies to full field continuous ROI's, many problems of great interest in the mechanics of solids are only piece-wise continuous and the in the general literature of DIC there is a variety of techniques developed as special solutions of complex cases. In Fig. 3.16, it has been added an arrow with the denomination "regularization" word that encompasses the methodologies to insure smooth vector field as the vector field that is illustrated in Fig. 3.16.

All the elements required for the solution of the problem of obtaining the components of the displacement vector have been outlined. We return to Eq. (3.29), it is a non-linear integral equation that must be solved. There are many ways in numerical analysis to solve this equation. In numerical analysis, the Newton–Raphson method, can be utilized to obtain successively better approximations to the roots (or zeroes) of the integral eq. (3.29). Of the many possible ways of solving Eq. (3.29), experience coming from applications has shown that Newton–Raphson seems to have the best performance compared to alternative solutions. The solution is obtained with an initial guess close to the true root and successive approximations are obtained by a process of iteration. Of course, there are many different algorithms and selection of parameters that can be utilized to reach the desired objective. Furthermore, this process includes an efficient search in the neighborhood of the given sub-element to get the highest correlation.

From the previously given summary of DIC, it becomes clear that from an intuitive point of view DIC gives an easy to understand picture of the deformation of the continuum field. However, to implement this basic scheme as a practical tool is by no means an easy task. The same can be said concerning judging the correct selection of all the different elements involved in a given package of software. When we are talking of DIC we are referring not to a single approach but to many different versions that have been developed theoretically or a consequence of practical applications in a period of more than 20 years since publications on DIC began to appear. In these 20 years, there is a basic bifurcation, local DIC and global DIC. Hence, to make fair comparisons of given packages is a very demanding and complex task as well to evaluate DIC in a fair way with other alternative methods.

There is a fundamental aspect in the process of decision making, selection of sub-element number in a given ROI that determines the number of pixels in a sub-element and the selected local shape function. It is known that as the number of pixels in a sub-element increases the correlation values improve. However, this improvement does not mean more accurate results if there is not a match with the shape function. The whole process depends on the scale factor of the ROI that determines the gradients present in the field. If the size of the sub-element is such that the gradient present in the field is small for a given sub-set as we have pointed out in the development of the monogenic approach, one gets close to the local phase concept and to the linear shape function with the six unknowns shown in Eq. (3.34) and illustrated in Fig. 3.16. Higher degree polynomials need to be introduced for increasing number of pixels in a sub-set but since this type of subset involves high order derivatives and derivatives are increasingly noisy, the obtained results are less precise. This fact has been verified in many papers investigating this topic. Similar observations can be applied to global approach as compared the local approach.

Besides this fundamental aspect, there are a number of details and parameters to be selected in the different steps that are involved. In the literature of DIC there are not many papers that provide an actual fair evaluation of the different choices that one can make in applications involving complex fields of displacements with large gradients. For proper evaluation it is required a simultaneous verification with a known theoretical solution or with a reliable Finite Element numerical solution. DIC methods have boomed in the last three decades due to the simplicity and workableness of commercial packages. The users of these packages must be aware of the previously discussed aspects of DIC to obtain correct results.

3.1.11 HARP Method to Get Displacements and Displacement Derivatives

The HARP (Harmonic Phase) method was developed to measure deformations of tissues in MRI images [9–13]. The method is based in the introduction of carrier planes, operation that in the MRI literature is referred as tagging. The planes can be introduced in the volume of the analyzed tissues and in each coordinate plane one can obtain a system of orthogonal carriers thus providing complete information to get the strain tensorial field. The argument utilized by the developers of the method apparently was arrived at independently from the literature on optical methods in Experimental Mechanics. It is basically like the argument invoked in the DIC but applied not to a sub-element resulting from tessellation of the ROI, [12]. It is the equivalent to the concept of local phase introduced in preceding sections, at the time of this reference, the authors did not have this concept and what they aimed to do is to distinguish the unwrapped phase values that they considered an unknown to the phase at a point of coordinates x and y, that they wanted to determine. This choice implies that the intensity of a pixel is given in 2D by Eq. (3.2). Similarly, to the argument represented by Eq. (3.29), it is assumed that in the process of deformation the local intensity is preserved, that is tagged to the analyzed surface.

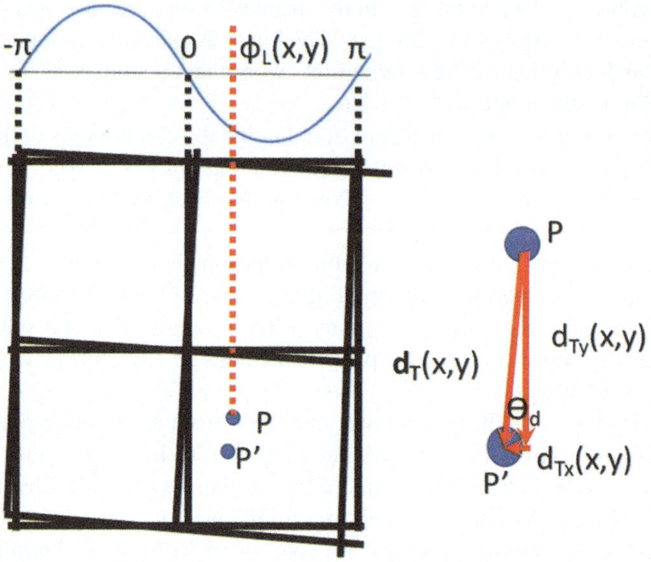

Fig. 3.18 Graphical representation of the concept of local phase and its role in the localization of a given point P as it moves to another position P′ and assuming preservation of intensity

Since the representation of the intensity is given by Eq. (3.2) and the displacement information is contained in the local phase of the signal, it is assumed that the local phase is preserved as the medium deforms, Fig. 3.18. Then,

$$\phi_{\mathbf{L}}(\mathrm{P}) = \phi_{\mathbf{L}}\left(\mathrm{P}'\right) \tag{3.35}$$

The preservation of the local phase implies,

$$\phi_{\mathbf{L}}(\mathrm{P}) - \phi_{\mathbf{L}}\left(\mathrm{P}'\right) = 0 \tag{3.36}$$

This implies, in the notation of Fig. 3.18, the equality of the phases at point P and P′ because the carrier fringe is engraved in the tissue. In the HARP method, Eq. (3.36) is solved utilizing the Newton-Raphson method. The Newton-Raphson iteration using the notation of Fig. 3.18 becomes,

$$\begin{bmatrix} \mathbf{x}^{\mathbf{n}+1} \\ \mathbf{y}^{\mathbf{n}+1} \end{bmatrix} = \begin{bmatrix} \mathbf{x}^{\mathbf{n}} - \dfrac{\phi_{\mathbf{Lu}}\left(\mathbf{x}_{\mathbf{P}'}\right) - \phi_{\mathbf{Lu}}(\mathbf{x}_{\mathbf{P}})}{\nabla\phi_{\mathbf{Lu}}\left(\mathbf{x}_{\mathbf{P}'}\right)} \\ \mathbf{y}^{\mathbf{n}} - \dfrac{\phi_{\mathbf{Lv}}\left(\mathbf{y}_{\mathbf{P}'}\right) - \phi_{\mathbf{Lv}}(\mathbf{y}_{\mathbf{P}})}{\nabla\phi_{\mathbf{Lv}}\left(\mathbf{y}_{\mathbf{P}'}\right)} \end{bmatrix} \tag{3.37}$$

Notice that the HARP method like the DIC method tracks the total displacement vector $\mathbf{d}_{\mathrm{T}}(\mathbf{x})$ that is co-axial with the phase vector $\boldsymbol{\phi}_{\mathrm{T}}(\mathbf{x})$, and not the projected displacement vectors.

Making the same assumption that it is done in DIC and utilizing Newton-Raphson iteration one can compute the position of the point P′ corresponding to P and obtains the components of the displacement vector between to exposures. In the original paper [12], since the authors did not realize the relationship stablished in the section about tagged carriers, they developed a tracking procedure to get the trajectory of the points.

3.1.12 Conclusions

The goal of the paper is to present an overview of currently available methods that recover relative displacement information from recorded levels of gray in an image. The coupling of Continuum Mechanics kinematics and optics through the theory of Optical Signals is presented. From this coupling, a method to obtain displacements and displacement derivatives is outlined.

To facilitate the presentation of the basic ideas, a 2D field case is adopted. However, the method has been generalized to 3D fields [6]. This method overcomes major difficulties present in older forms of data processing known in the literature as moiré methods or speckle methods depending on the type of wavefront utilized. It requires only the recording of the initial configuration of the region of interest (ROI) and the deformed configuration, an enormous advantage in the case of dynamic applications. It does not depend on the loci type of fringes, closed fringes versus fringes ending at the region of interest boundaries and eliminates the classical fringe unwrapping procedures with all the complexities that it introduces and replaces it by a summation of local displacements. In this way, it can analyze complex fringe singularities.

There are othermethods of recovering displacement information that also present solutions to the previously pointed out limitations of the older forms of fringe pattern processing. One is the most prevalent in usage method that is referred to as digital image correlation (DIC). The other method comes from bio-engineering field and has evolved mostly separately from the field of experimental mechanics applied to solids, known as the HARP method. Basic aspects of these two methods are discussed on some detail so that comparisons among all the methods can be made, not a simple task in view of the very large possible pathways and algorithms that are involved in each one of these methods.

For reasons of a more manageable presentation of the subject matter of the paper, the paper has been split in two parts. One, called Part I Theory, includes the material summarized in this last section. The, other called Part II Applications, presents analysis of images in order to make comparisons that are limited to certain aspects of these in view of the enormous difficulties of making fair comparisons of very complex subjects.

3.2 Part II Applications

3.2.1 Introduction

In Part I, the basic foundations of the currently utilized methods in the recovery of displacement information encoded as levels of gray in a sensor or computer generated have been reviewed. To get a comparison and an evaluation of these different approaches, examples of actual applications are presented in this second part.

3.2.2 Examples of Applications and Comparisons

3.2.2.1 Methods Based on the Standard Theory of Optical Image Analysis

Having presented the most utilized procedures to extract displacement information in 2D fields, let us go over examples and bring out some conclusions and comparisons. Let us consider the case of a disk under diametrical compression. This is a specimen that can be considered as a standard to compare different experimental solutions with the theoretical elasticity solutions and FEM solutions, Fig. 3.19. Table 3.1 provides the parameters that define the images given in Fig. 3.19. Figure 3.20 shows that, in order to obtain the displacement fields, the in-quadrature method was applied to carrier fringes of frequency order N = 128 and a process of extension of the patterns beyond specimen contour was utilized. In Fig. 3.21, experimental data match well with the FE solution over the whole region of interest. The largest difference between 2D Hilbert and windowed FFT is only 0.0818 orders (i.e. 0.1 μm near the center of the disk), which confirms the good agreement between methods.

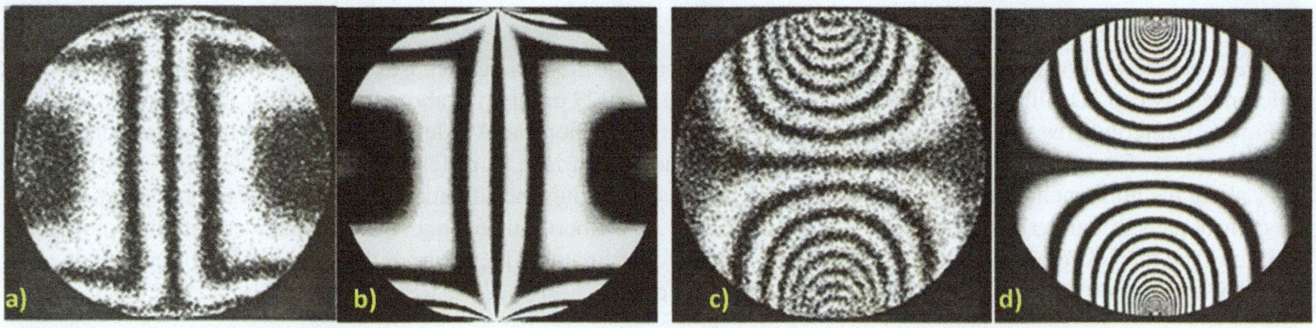

Fig. 3.19 (**a**) Recorded *u*-pattern; (**b**) Computer generated *u*-pattern; (**c**) Recorded *v*-pattern; (**d**) Computer generated *v*-pattern

Table 3.1 Details of tested specimen shown in Fig. 3.19

Material	Aluminum
Diameter	60 mm
Thickness	6.35 mm
Young's modulusPoisson's ratio	70 GPa$\nu = 0.336$
Grating Pitch	1.222 μm
Applied Load	3300 N

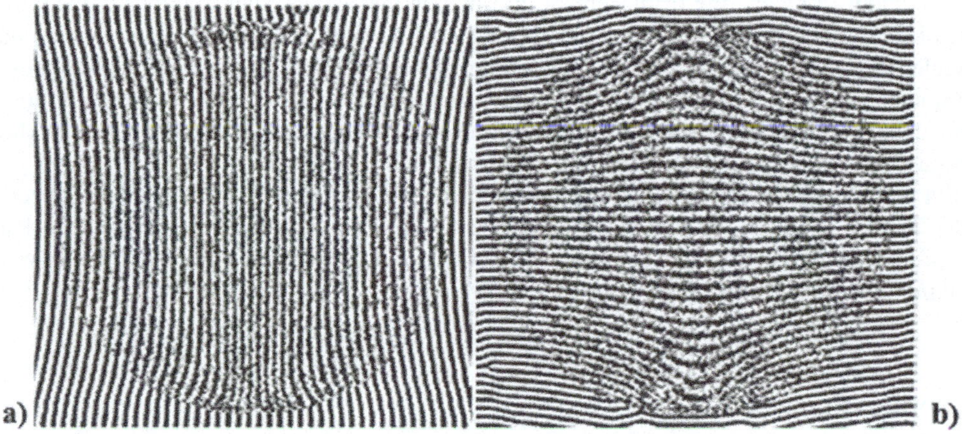

Fig. 3.20 *u*-extended (**a**) and *v*-extended (**b**) carrier fringes for the patterns of Fig. 3.19a, c

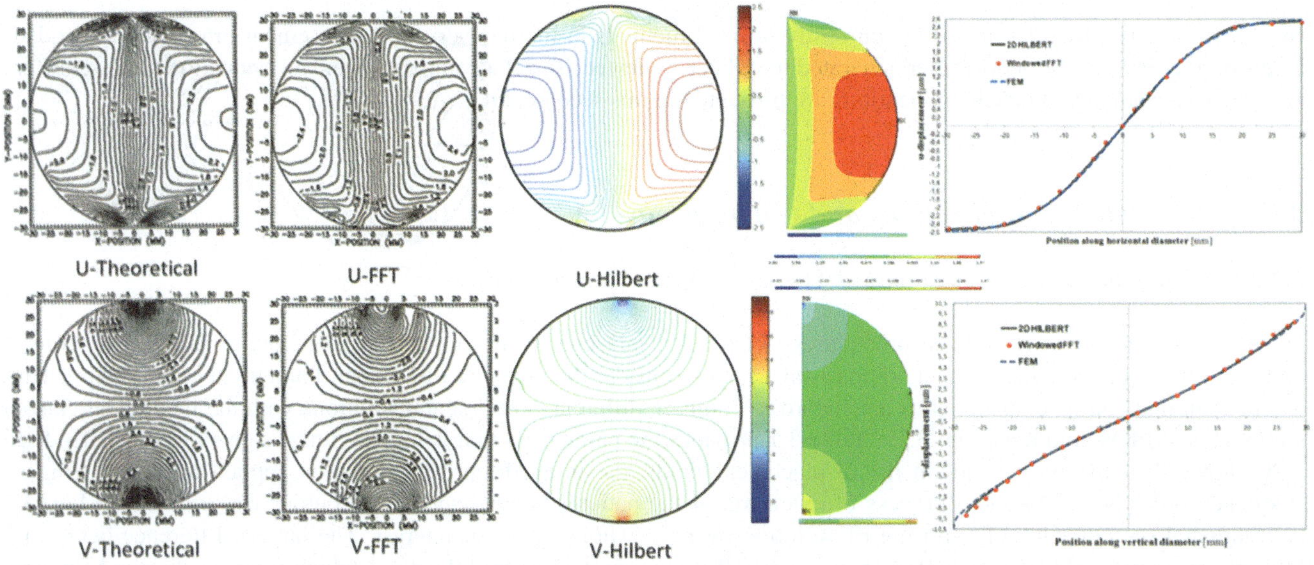

Fig. 3.21 Theoretical *u*-*v* displacements (elasticity solutions); (**b**) Experimental *u*-*v* displacements, windowed FFT; (**c, d**) Experimental *u*-*v* displacements, 2D Hilbert transform. (**e**) Finite element. (**f**) Comparison of results given by the windowed FFT, the FEM, and the 2D monogenic signal approach along disk horizontal and vertical axes. Displacement values are expressed in μm

The concentration of fringes at the loading point in the case of the *v*-pattern causes a loss of resolution. The maximum difference between 2D Hilbert and windowed FFT processing is 0.327 orders (i.e. about 0.4 μm) near the top side of the disk.

Figure 3.22 compares strains obtained by different methods: Gabor and Morlet transforms and the strains obtained by differentiation of wrapped phase in the frequency space. The comparison of different methods is done only for the values of ε_x along the horizontal diameter of the disk. Since values of contour lines do not match, data at the same locations along the control path are numerically extracted. In the case of the 2D Hilbert transform, displacements were fitted by polynomial functions and then differentiated with respect to control path coordinates. The values of the Gabor and Morlet derivatives are such that at the scale of representation do not show differences.

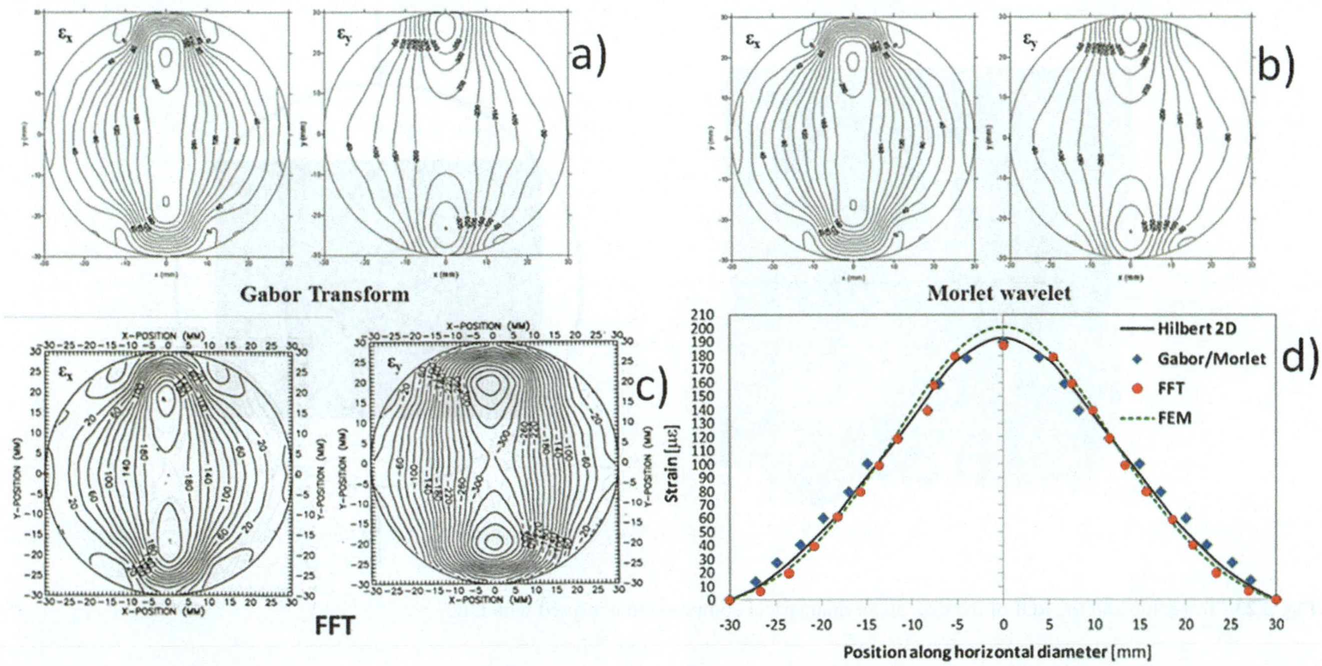

Fig. 3.22 (**a–c**) Strains ε_x and ε_y (expressed in $\mu\varepsilon$) of the disk under diametrical compression obtained from (**a**) Gabor transform, (**b**) Morlet transform and (**c**) the FFT by differentiation in the frequency space; (**d**) Comparison of strain distributions along the disk horizontal diameter

Table 3.2 Material properties of the tested specimen of Fig. 3.23

Material	Polycarbonate
Diameter	155 mm
Thickness	10 mm
Young modulus	2.46 GPa
Poisson's ratio	0.45

The agreement between the different methods is excellent (maximum strain is 190 $\mu\varepsilon$ for Gabor/Morlet, 188 $\mu\varepsilon$ for windowed FFT, 193.5 $\mu\varepsilon$ for Hilbert 2D and 201 $\mu\varepsilon$ for the finite element simulation). Due to the scale effect, the gradients at the contact region are such that for the scale selected to show the overall pattern the values for ε_y in the contact region (large gradients) give results that do not agree with Gabor/Morlet while data for ε_x agree much better with the Gabor/Morlet patterns. The wavelet methods provide more refined procedures to generate better filters that handle the scale effect. The average of all the measurements is 193.12 microstrain and the standard deviation is 2.86 microstrain.

3.2.2.2 Digital Image Correlation

In Sect. 3.2.2.1, a disk under diametrical compression was utilized to evaluate the sensitivity and accuracy of the monogenic phasor method in the 2D case. In this section, the same type of specimen is used to make an evaluation of the DIC method. The example is taken from the literature on the topic [14]. The tested specimen parameters are indicated in Table 3.2.

The test set up is shown in Fig. 3.23 and the analyzed area is indicated by the black rectangle shown in the picture. The disk was sprayed with white paint and the speckles introduced by spraying the disk with black paint. The camera pixel size is 98 μm and the size of the ROI is 1000 × 1234 pixels. The displacement field is obtained applying the DIC method based on equation a normalized form of Eq. (3.29) of Part I of this paper. The correlation process is defined by the parameter δ shown in Fig. 3.16 of Part I of this paper. In this example, the selected value was $\delta = 16$ pixels or in microns 1568 μm. Figure 3.24 shows the u-displacements corresponding to the two methods being compared and Fig. 3.25 shows the v-displacements. Figures 3.24b and 3.25b respectively correspond to Fig. 3.21 u-Hilbert and Fig. 3.21 v-Hilbert. Figures 3.24c and 3.25c correspond to a theoretical elasticity solution of the disk used to compare with the DIC results [14].

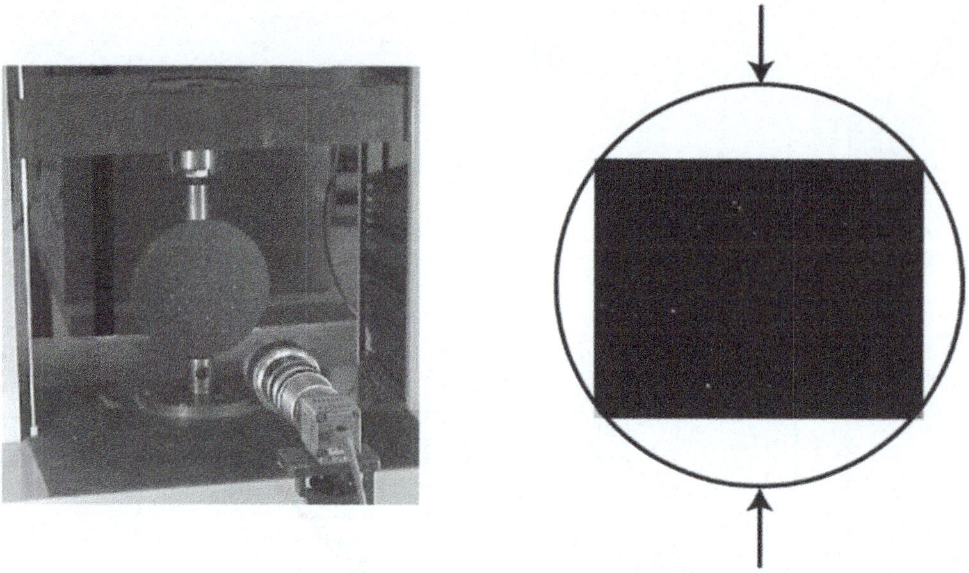

Fig. 3.23 Test setup and the ROI of the disk under diametrical compression analyzed with DIC

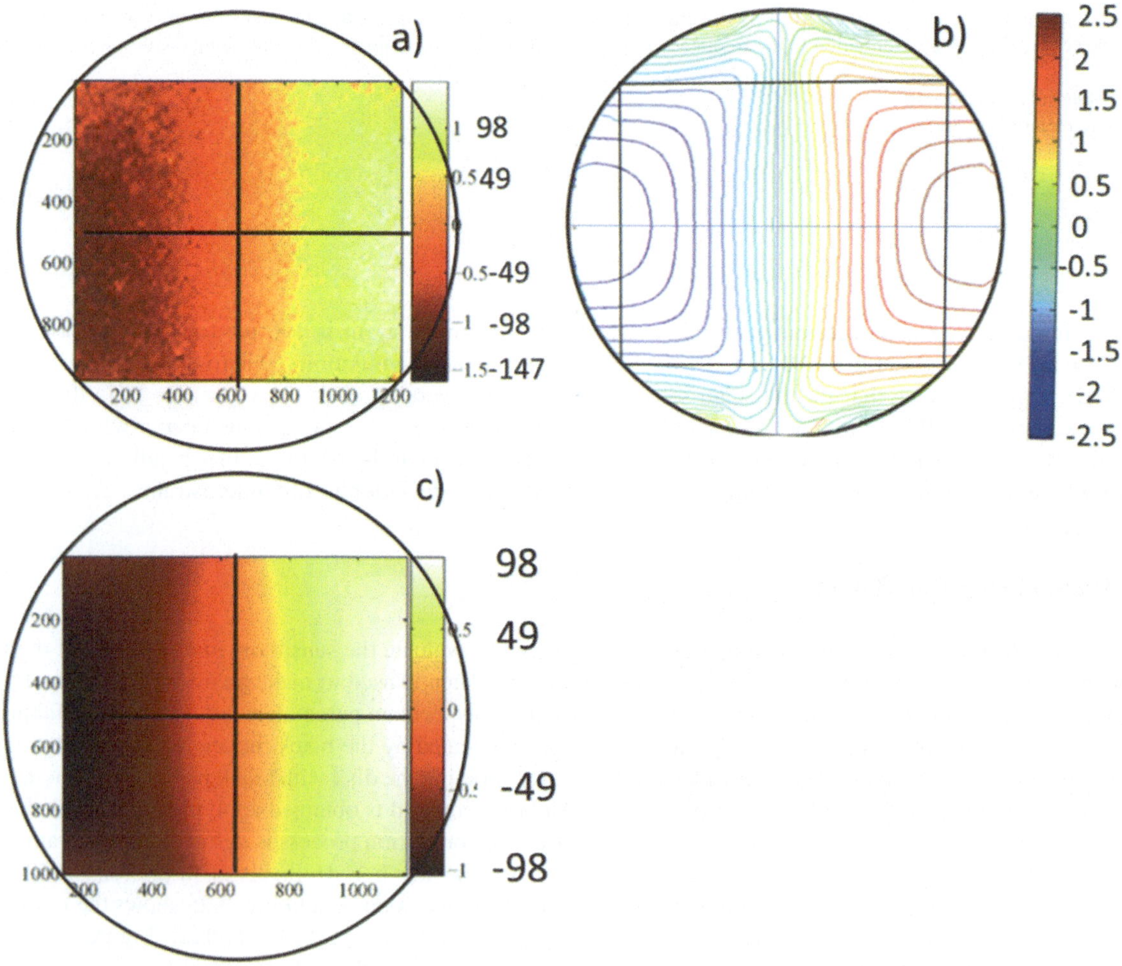

Fig. 3.24 (**a**) u-DIC displacements of the polycarbonate specimen; (**b**) u-displacements of the metallic disk (Hilbert); (**c**) u-displacements given by the elasticity solution with improved elastic parameters

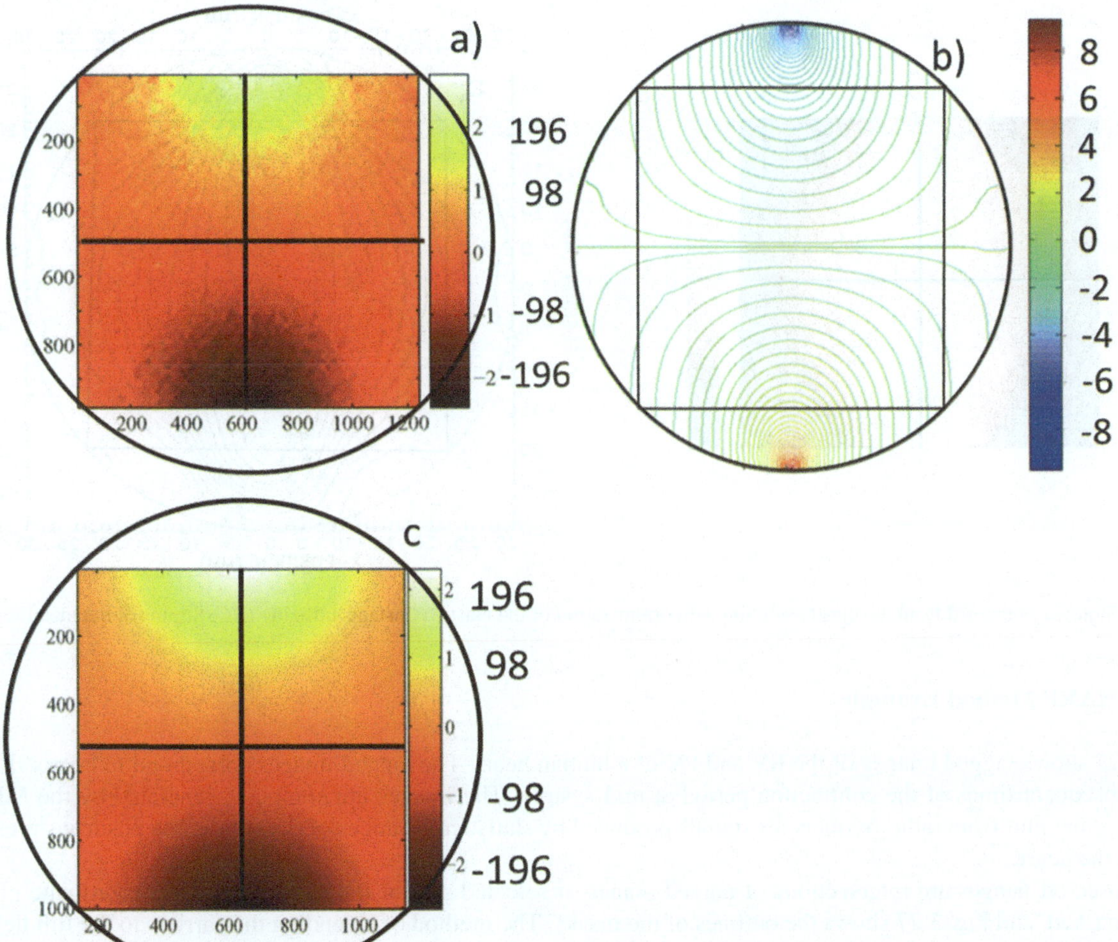

Fig. 3.25 (**a**) *v*-DIC polycarbonate specimen. (**b**) v-Displacements of the metallic disk (Hilbert); c) *v*-Displacements given by the elasticity solution with improved elastic parameters

In [14], to the solution obtained with the Kolossov-Muskhelishvili potentials is added an optimization technique of the elastic parameters based on displacement values determined by the DIC measurements [15]. In Figs. 3.23 and 3.24, are added the outlines of the disk to provide the position of the studied region respect to the boundaries of the disk.

To fairly compare different results, it is necessary to take in consideration the following arguments. According to the theory of elasticity for small deformations, since we are dealing with a simply-connected region, the obtained results expressed in dimensionless form are independent of the actual elastic constants. The above conclusion requires for the obtained solutions having identical boundary conditions. The displacements of a disk are influenced by the actual contact stresses at the edges of the disk where the compression forces are applied. This effect is very important near the region of application of the compression force and it is reduced as the distance from the region of application of the force is increased. If the flattening of the disk upon application of the compression is not excessive, measured displacement should show trends in good agreement despite the difference in order of magnitude of the displacements.

In Fig. 3.24, one can see that there is a good agreement in the trends of plots (b) and (c), the agreement of (a) with (b) and (c) is only fair. Figure 3.25 shows a better agreement between the compared plots. However, in both cases, the DIC patters are noisy. Strains are not provided in [14]. An approximate estimate of the agreement between the strains shown in the two patterns can be made by considering some points that can be identified (see Fig. 3.26). This comparison will be done in the Discussion section of this paper.

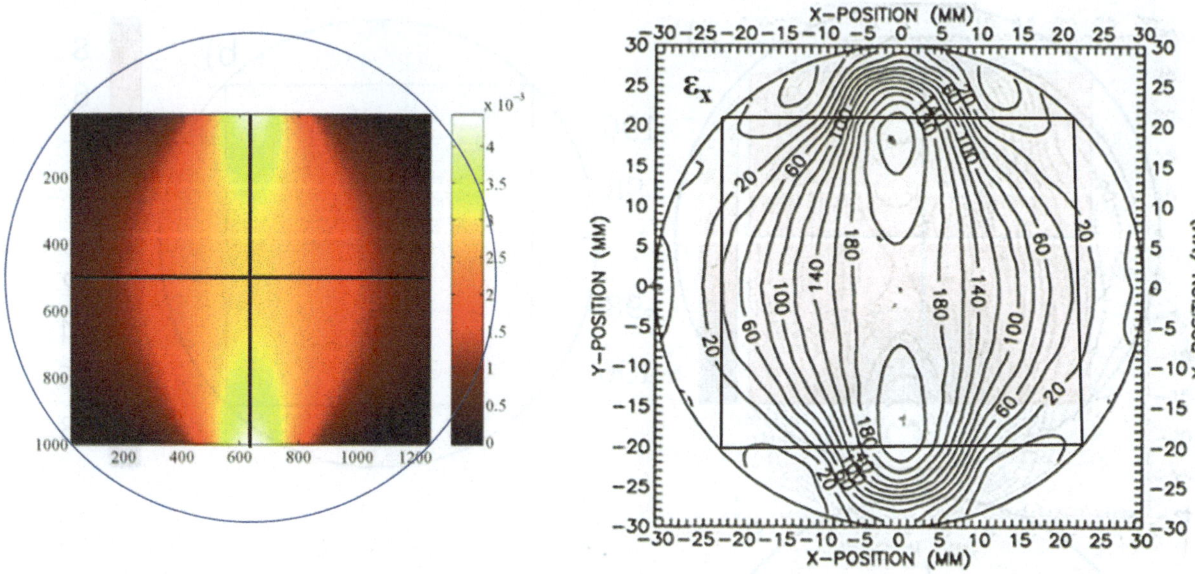

Fig. 3.26 Strains ε_x obtained from the elastic solution with optimization of the elastic constants utilizing DIC. Strains of the metallic disk

3.2.2.3 HARP Method Example

Figure 3.27 shows tagged images of the RV and LV of a human heart. The tagged images correspond to hearts' RV and LV in two subsequent times of the contraction period at mid-systole. The images are computer generated by the MRI system data processing unit from radio frequencies signals produced by statistical changes of the spin of the positrons present in the tissues of the heart.

The observed fringes are intersections of tagged planes of selected spatial frequencies with the coordinate planes. The ROI are masked, and Fig. 3.27 shows the outlines of the masks. The method of extending the carriers to the full field of view is applied. To improve spatial resolution, additional carrier lines are added to the image. As pointed out before, two different states of deformation of the heart can be compared since the displacement information is carried by the tagged carriers and the initial frequencies of the carriers are known. Figure 3.28 was obtained from the data contained in the carrier fringes shown in Fig. 3.27 for each of the corresponding planes.

Figure 3.29 shows the resultant displacement vectors and indicates some torsion of the walls of the ventricle.

Figure 3.30 shows the Eulerian component of the simplified Almansi strain tensor along the coordinate axis.

The strain analysis of a cross-section of the myocardium carried out in [13] makes it possible to compare the results of the HARP method with the carrier fringe analysis method presented in this paper. Since the quality of the image in [13] is not as good as the images in Fig. 3.27, contours of the walls are irregular.

Figure 3.31 provides a comparison between the carrier method and the HARP method. Due to the quality of the images, it was not possible to compare the total field. Only regions where the carriers could be recovered were compared. Finite differences of were utilized in [13] to compute radial strains tracking the deformations in the radial direction. In the carrier method radial strains were computed using the simplified Almansi strain tensor and strain tensor transformations to get radial components. While in the cine HARP point tracking was utilized by connecting one image with the next, in the carrier method the obtained values come from one image that contained the initial carrier information frequency and the deformed frequency information. It is possible to see that by comparing areas of the patterns in orders of magnitude and signs of the deformations there is a fair agreement. This conclusion is interesting if one considers the whole process of point tracking and finite differences method to compute radial deformations involved in the HARP method compared to the straightforward procedure applied in the carrier method.

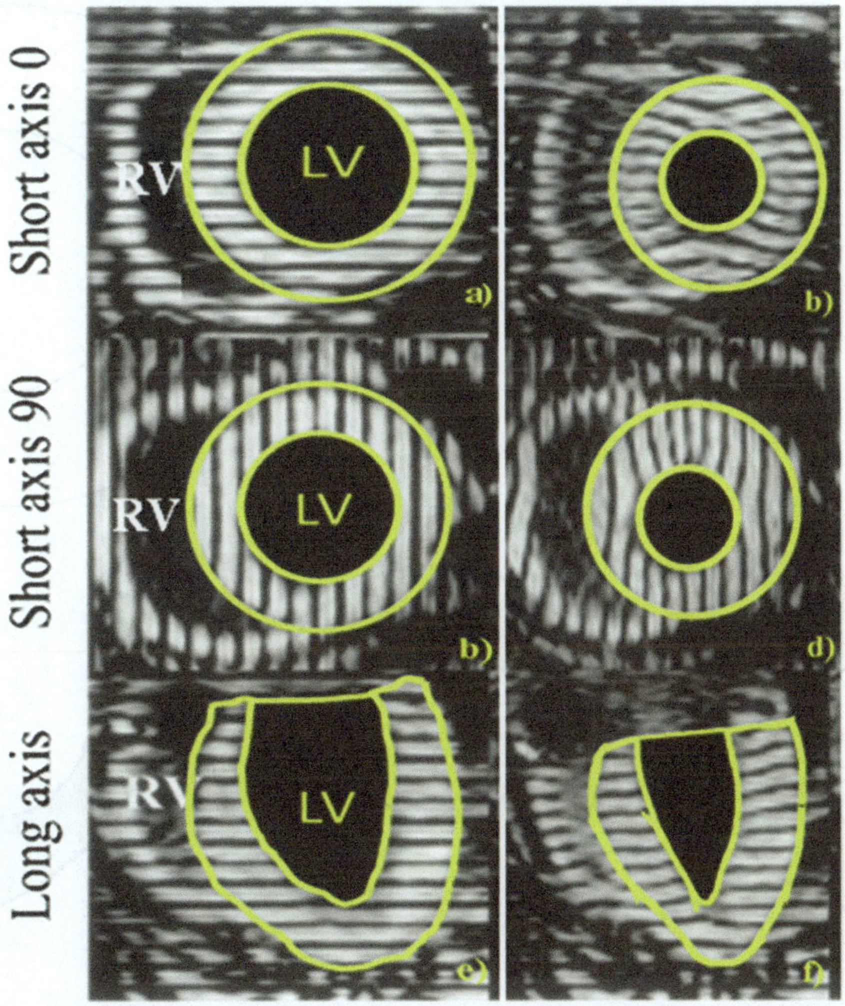

Fig. 3.27 Short-axis (0° and 90°) and long-axis MRI images of the right ventricle (RV) and left ventricle (LV). Reference configurations correspond to figures (**a**), (**c**) and (**e**)

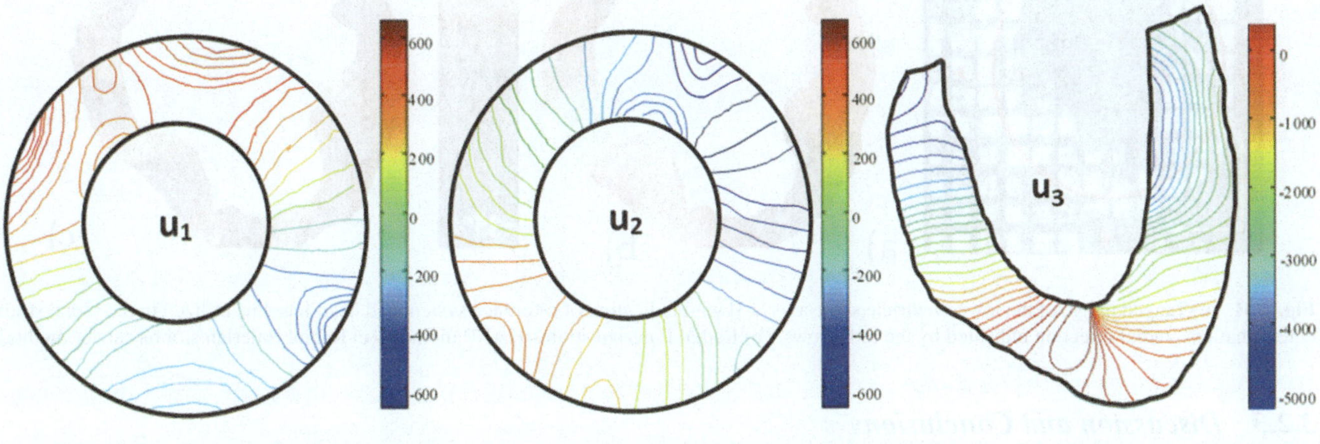

Fig. 3.28 Contour plots of displacement components (expressed in μm) determined for the left ventricle

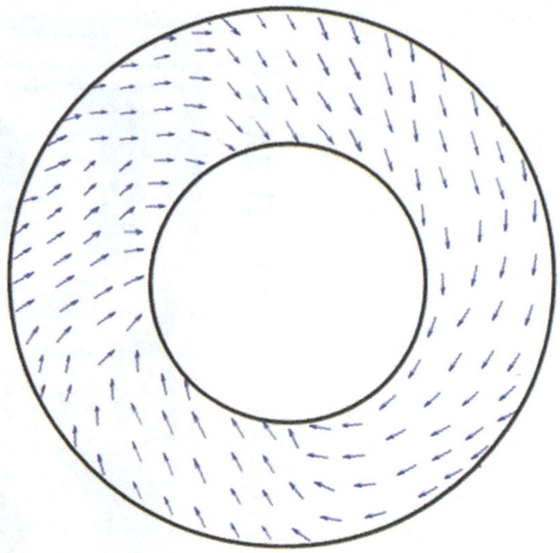

Fig. 3.29 Vector representation of the relative displacement field in the x_1x_2 plane defined by the 0° and 90° short axis. A certain amount of torsion can be observed

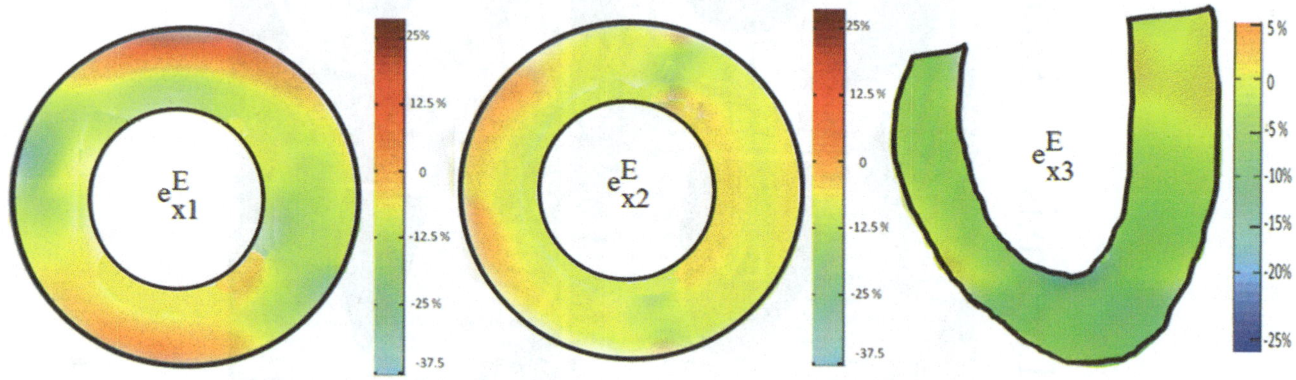

Fig. 3.30 Eulerian components of the strain tensor along coordinate axes

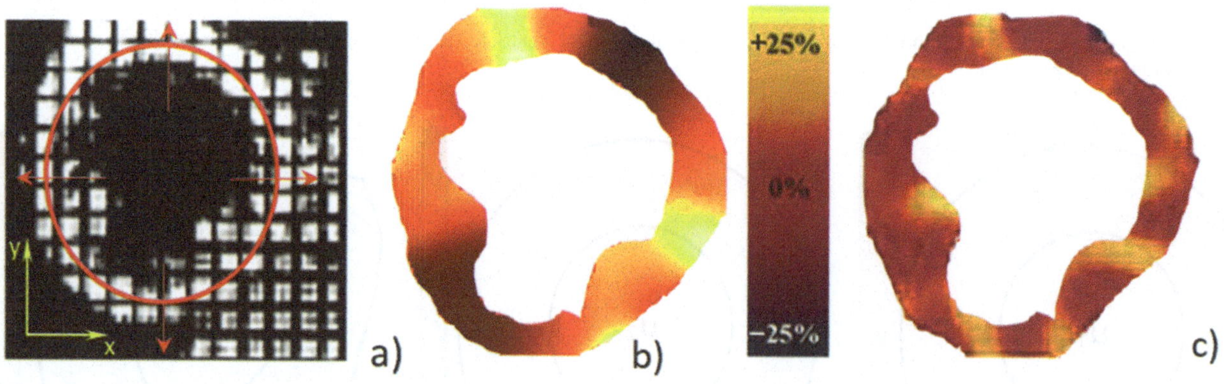

Fig. 3.31 (**a**) Tagged image of the lower ventricle short-axis [13] with indication of reference system and heart baseline utilized in the radial strain determinations, radial direction indicated by the red arrows. (**b**) Radial Eulerian strains HARP method. (**c**) Radial Eulerian strains carrier method

3.2.3 Discussion and Conclusions

In the initial sections of Part I of this paper, theoretical aspects of the coupling of Continuum Mechanics kinematics and optics through the general theory of optical signals analysis are presented. The purpose of this introduction is to look at the evolution of thought in approaching a question: How to experimentally measure displacement fields using optical methods as handy tools? The handy adjective is used here to indicate practical engineering tools, accurate and fast.

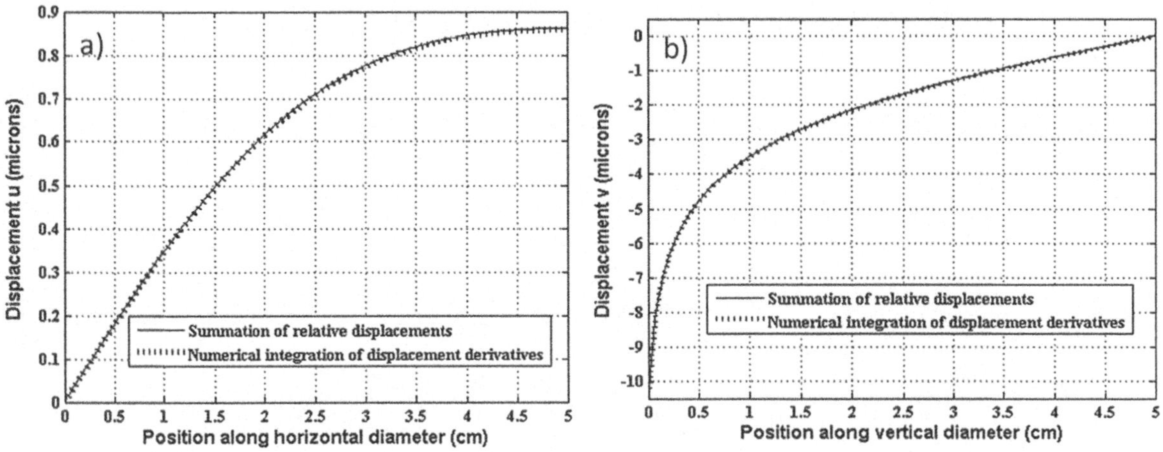

Fig. 3.32 Comparison of summation of relative displacements and numerical integration of displacement derivatives computed for the disk under diametrical compression: (**a**) u-displacement along the horizontal diameter; (**b**) v-displacement along the vertical diameter

To simplify the presentation of the relationship between scalar fields of intensities encoded in an image and the vectorial displacement field, it is chosen a 2D field. The first sections of Part I show the relationships among these two fields. The relationship is graphically represented in Fig. 3.12 of Part I and the added explanation in Part I-Sect. 3.1.8. The moiré method was the first method that evolved to measure displacements using a global model defined by Part I-Eq. (3.9) representing the intensity field by a phasor, a vector in the imaginary space representing displacements as phase angles. It was realized that this model needed to be modified by making each term in Part-I Eq. (3.10) a point function. This concept resulted in the multi-phase method to obtain point-wise values of the phase as is illustrated in Part I-Sect. 3.1.6.1. This approach created new problems, the need for matching local phases at the end of each 2π interval to secure correct displacement values, the process of phase unwrapping. This problem proved to be a very difficult one that presents multiple challenges and requires very complex software to solve many of the unwrapping cases encountered in actual patterns. In some cases, there are no automatic solutions and these cases require the intervention of the person doing the data processing. This initial model lacked two important variables, the angles $\theta_u(x)$, $\theta_v(x)$. In the initial model of moiré patterns (Part I-[4–6]), another approach to get the displacement information was presented, the in-quadrature model based in the Hilbert transform. For a long while this model was considered a formal representation but lacking practical implementation. This notion was dispelled when the theory of analytical functions was introduced and applied requiring the addition of carrier fringes (Part I-[4, 5]). Finally, the model was completed when two main improvements were plugged in the theory, the concept of local phase and the monogenic phasor (Part I-[5]). The monogenic phasor includes the angles $\theta_u(x)$, $\theta_v(x)$ enabling a general solution for closed fringes without the need of special solutions. These angles need to be determined by separate procedures from the application of Fourier or Hilbert transforms to get the components of the projected displacement vectors. This model has been the basic model that has been illustrated in Part II-Sect. 3.2.1. This model bypasses the problem of unwrapping as it is illustrated in Fig. 3.32 (Part II-[16]). Given the values of the displacement vectors [16], it is possible to add them, for example, along the horizontal diameter and vertical diameter and compute the accumulated relative displacements utilizing Eq. (3.8) of Part I. The summation operations can be performed along any selected trajectory besides the reference axis. The concept of local phase and the possibility of utilizing vectorial additions enables the analysis of patterns containing complex singularities, examples have been worked out.

In view of two important problems, the complexity of unwrapping methods and the loss of optical correlation between images utilizing rough wave fronts, alternatives paths have been developed to simplify the process of analyzing displacement fields captured in images. Digital image correlation solves these two main problems replacing the optical correlation by digital correlation and eliminating the unwrapping by removing the concept of phase as bearer of displacement information.

The choice of random patterns accomplishes another simplifying objective. It removes the technically difficult problem of tagging deterministic carriers to surfaces. To transform the scalar field of intensities into a vectorial field, it utilizes the property of the displacement field that has been summarized in Part I-Fig. 3.12, and that transforms the operation of getting the resultant displacement vector into a search symbolically illustrated in Part I-Fig. 3.16. We have already discussed the different concepts and procedures that are included under the name of DIC and the difficulty implied in evaluating a given DIC software package in view of the large number of alternatives possible to accomplish the different tasks. Going back to Part I-Fig. 3.16, that is illustrative of the example analyzed in Sect. 3.2.2.2, a specimen with parameters given in Table 3.2

Table 3.3 Comparison of experimentally measured strain for the diametrically compressed disk

	Aluminum disk					Polycarbonate disk			
	10^{-6}	MPa	Theor. ratio	Exp. ratio		10^{-6}	MPa	Theor. ratio	Exp. ratio
ε_{x1a}	188	–	–	–	ε_{x1p}	3250	–	–	–
ε_{y1a}	−211	–	–	–	ε_{y1p}	−4713	–	–	–
σ_{x1a}	–	5.6	–	–	σ_{x1p}	-	4.98	–	–
σ_{y1a}	–	−11.7	–	–	σ_{y1p}	–	−10	–	–
r_{1a}	–	–	1	1	r_{1p}	–	–	1	1
r_{2a}	–	–	2	2.09	r_{2p}	–	–	2	2
ε_{x2a}	220	–	–	–	ε_{x2p}	5143	–	–	–
ε_{y2a}	−375	–	–	–	ε_{y29}	−8525	–	–	–
σ_{x2a}	–	7.42	–	–	σ_{x2p}	–	6.62	–	–
σ_{y2a}	–	−23.76	–	–	σ_{y2p}	–	−21.21	–	–
r_{3a}	–	–	–	3.20	r_{3p}	–	–	–	3.20

is analyzed utilizing a single camera, the corresponding pixel size of the sensor is 98 μm, and a ROI of 1000x1234 is adopted. The selection of δ defines the correlation process based on Part I-Eq. (3.33) and the adopted linear shape function. The derivatives of the displacements are not provided in the paper where the example has been taken. In Part II-[14], for comparison purposes, it is added a theoretical solution for the disk under diametrical compression. Taking into consideration the fact that in the case of a plane stress solution for a simply-connected region the stress distribution is independent of the elastic constants and within the limitations that a color bar code provides, the strain and stresses of the aluminum disk and of the theoretical solution provided in Part II-[14] can be compared. The obtained results are presented in Table 3.3. The strains and stresses at two sections are compared. One section corresponds to the horizontal axis of the disk, the other section corresponds to the intersection of the vertical axis of symmetry of the disk with the upper side of the ROI. The notation utilized in Table 3.3 is as follows, the subscripts x1a and y1a correspond to the diameter of the aluminum disk, the subscripts x2a and y2a at the intersection of the vertical axis of the disk with the upper edge of the ROI. The ratio $r_{1a} = \sigma_{x1a}/\sigma_{avg}$, where $\sigma_{avg} = P/\pi Rt$, R disk radius t thickness of the disk, P the applied load. The theoretical solution gives $\sigma_{x1a} = \sigma_{avg}$ and $\sigma_{y1a} = 2\ \sigma_{avg}$, from the analysis of many experimental cases $\sigma_{y1a} \approx 2\ \sigma_{avg}$. In Part II-[14], P is not provided but from the preceding relationship P = 7719 N. The ratio $r_{3a} = \sigma_{y2a}/\sigma_{avg}$.

Similar notations are utilized for the polycarbonate disk. One can relate the strains in the two disks and the mean average is 21.425 microstrains with the standard deviation of the mean equal to (+,−) 1.39 about 5%. One can see through the comparison of the results of the aluminum disk with the polycarbonate disk along the horizontal diameter that the condition of independence of the distribution of stresses from the elastic constants is satisfied. The stresses are very close in value and they are far from the yield stresses of both materials. One also should notice that the strains in the polycarbonate disk are 21 microstrains larger than in the aluminum disk. If we analyze the condition that stresses values independent of E and ν at the intersection of the vertical axis of the disk with the upper edge of the ROI, it is required that $\varepsilon_{x2p} = 5143 \times 10^{-6}$, if one looks at Fig. 3.26 one finds from the color code bar that $\varepsilon_{x2p} = 4500 \times 10^{-6}$, the difference is −12.5%. Furthermore, if one utilizes the value $\varepsilon_{x2p} = 4500 \times 10^{-6}$ the corresponding stresses are $\sigma_{x2p} = 3.08$ MPa and $\sigma_{y2p} = 16.4$ MPa besides not satisfying the condition of stress distribution independent of the elastic constants the vertical stress σ_{y2p} does not satisfy the equilibrium condition since the value of σ_{y2p} does not increase enough to compensate the reduction of the disk cross-section. These considerations indicate that the DIC computations with the parameter values dictated by the geometry, camera position and pixel value of the sensor, in the analyzed region have not the spatial resolution required by the strain gradients present in the region approaching the load. The strain difference is $\Delta\varepsilon_{y2p} = 643 \times 10^{-6}$, the paper provides an estimation of $\Delta\varepsilon_{y2p} = 360 \times 10^{-6}$ which is 56% smaller than the value obtained from the condition of stress independence of elastic constants.

Having provided an idea of the strain field we can now look at the displacement values. Table 3.4 provides the displacements of the points of intersection of the ROI with the symmetry axis. Looking at the deformation of the disk under loading, the supported edge of the disk remains practically fix in the case of the polycarbonate disk while all the motion take place at the upper edge. For the horizontal axis and the *u*-displacement, the center of the disk remains immobile and the maximum movement corresponds to the intersection of the horizontal axis with the edge of the disk. Hence, in the table, the *u*-displacement corresponds to half of the horizontal dimension of the ROI and for the *v*-displacement the total vertical dimension of the ROI is considered. The standard deviation of the measurements in the aluminum disk are obtained from the values of the FT, the Hilbert transform and the FE. For the polycarbonate disk the paper contains estimates of the value. From the table, it is possible to see that the measured STd for the estimated horizontal values of the polycarbonate disk is almost 2 times larger than for the aluminum disk.

Table 3.4 Comparison of deformations measured/estimated for the diametrically compressed disk

	microns	STd μm	Relat. error%
U_{Tp}	147	2	1.3
V_{Tp}	490	2	1.3
U_{Ta}	2.428	0.0153	0.63
V_{Ta}	16	0.2	1.25

In Table 3.4, the subscript Tp indicate total displacements for the polycarbonate disk and Ta the corresponding quantities for the aluminum disk. The ratios of the displacements are: $r_{dv} = 490/16 = 30.6$ and $r_{du} = 147/2.428 = 60.5$. The ratio of the displacements in the vertical direction is close to the ratio of the Young's moduli E of the two materials, 28.5, difference 2.1%.

To compare the results of the polycarbonate disk with the aluminum disk, it is convenient to review briefly the involved procedures. Both patterns correspond to speckle methods. The aluminum disk image corresponds to speckle interferometry and the fringes are produced by optical correlation. The polycarbonate corresponds optically speaking an equivalent to white light speckle photography. In Fig. 3.15 of Part I we have an example of white light speckle photography with a total displacement of 830 μm, 70% larger than the one present in the polycarbonate disk. In DIC, the correlation of the signals loaded/unloaded is numerically done, and one could make an analogy assuming that the quantity δ is equivalent to the pitch of the fringes in this particular case the pitch will be 1.568 mm. Part I-Fig. 3.16 provides a visual input to clarify the DIC scheme in the case under discussion. Part I-Fig. 3.17 gives an enlarged view of the sub-elements, the vector displacements and for one element the pixel content that carries the information. The accuracy and precision of DIC depends of many different factors. To mention some of the fundamental ones, the speckle size, the subset size that defines the number of pixels as illustrated in Part I-Fig. 3.16. The shape function that is closely connected to the selection subset size and the algorithm that defines the registration of the undeformed and deformed sub-sets. The most utilized in the literature of DIC is the Newton-Raphson algorithm performed in the physical space, that is implemented with the zero-mean normalized cross-correlation norm. The intensity interpolation algorithms to get subpixel information. In the example chosen for comparison, the defining factor of the accuracy of the results is the selected sub-set and its connection with the actual field of displacements. A first observation is the signal input level respect to possible noise contamination is high, the strains of the plastic disk are 21 times higher than those of the metallic disk. The same observation applies to the displacements, in the vertical direction the displacements are 30 times higher for the plastic disk and in the horizontal direction 60 times. The influence of the intensity of the signal can also be detected by the displacements of the plastic disk itself. In Fig. 3.24 the distribution of the horizontal displacements is not well detected, while the vertical displacements of Fig. 3.25 are captured. The other observation comes from the analysis of the derivatives. Although in the paper no derivatives are directly provided, indirectly the authors present a formulation combining a theoretical solution of the disk with DIC to get the actual elastic constants of the material.

We have compared the results of this approach in two sections, the strain at the center of symmetry of the disk and at the interception of the vertical axis of symmetry of the disk with the upper boundary of the ROI; comparison based on the fact that the resultant stresses are independent of the elastic constants. In the center of symmetry, the results are satisfactory while in the other point are not, the obtained vertical strain has an error of 12.5%. The difference is a consequence of the sub-element size compared with the gradient of strains in the field. If the scale of the image is such that sub-element size cannot match the gradient present in the field, a direct consequence of the Nyquist condition, the output will average the actual local distribution. In the center of the disk the field of strains mimics better a uniform field than the upper point region. By its own formulation, results in DIC are dependent on the evaluation of a group of pixels as representative of a field. It is well known in DIC literature that the correlation results improve with the number of pixels in the sub-element, but as counterpart the spatial resolution is reduced when this number increases.

Summarizing, developments in the experimental determination of the continuum kinematics in the 2-D field are presented to facilitate the understanding of the most prevalent methods that have evolved to obtain displacement field information from images that contain scalar information of light intensity represented as level of gray levels. The aim of all the methods is the same, graphically illustrated in Fig. 3.12 of Part I of this paper: to obtain relative displacement vectors by converting scalar recorded gray levels into vectors. The carrier method works with projections of the displacement vectors and a mathematical scheme that is based in the classical approach in optics of representing the optical signal by an amplitude and a phase. The concept of local phase is introduced and the whole displacement information is contained on a pixel-based-phasor. Two key impediments of the classical so-called moiré method are overcome, the phase unwrapping is replaced by summations of local phases, the problem presented by the type of fringe's loci is removed.

DIC and HARP select the relative displacement vector as the unknown quantity to be directly determined. DIC utilizes numerical correlation of finite areas (local operation) to get the vector displacement thus eliminating the complex process of phase unwrapping. The numerical correlation requires the tessellation of the region of interest that results in a complex relationship between spatial resolution and pixel content of the area elements. The selection of levels of gray as a variable eliminates the conversion to phase but as a compensation needs gray level smoothing and interpolation. Finally, another consequence of the tessellation of the region of interest is the need of introducing regularization functions to connect the different local displacement vectors (global operation). The regularization function finally provides the displacements and the corresponding derivatives. The use of the regularization function implies a hybrid method where experimental information is complemented by numerical optimization of approximated models of the continuum.

The HARP method is closely connected to digital carrier method but similarly to DIC searches for the relative displacement vector directly utilizing the Newton-Raphson method. To get displacements, local relative displacements vectors are computed. Local sub-routines are developed to deal with discontinuities caused by noise or the end of the cycle of a selected spatial frequency. HARP has been designed to handle large deformations and large rotations. Reference points are tracked from one image to the following and differentiation is done by finite differences.

Experimentalists have a large variety of tools at their disposal to get information on the kinematics of the continuum. Are there selection criteria that can provide a clear-cut preference? This is a very difficult question to answer because the selection of a technique depends on many factors including efficiency, sensibility, accuracy considerations, feasibility of obtaining signals with a good signal to noise ratio. Furthermore, familiarity with a given method of analysis is a factor that cannot be neglected.

References

1. Bedrosian, E.: A Product Theorem for Hilbert Transforms. Memorandum RM-3439-PM. The Rand Corporation, Santa Monica, CA (1962)
2. Bedrosian, E.: Hilbert transform of bandpass functions. Proc. IEEE. **51**, 868–869 (1963)
3. Nuttall, A.H., Bedrosian, E.: On the quadrature approximation to the Hilbert transform of modulated signals. Proc. IEEE. **54**, 1458–1459 (1966)
4. Sciammarella, C.A., Lamberti, L.: Mathematical models utilized in the retrieval of displacement information encoded in fringe patterns. Opt. Lasers Eng. **77**, 100–111 (2016)
5. Sciammarella, C.A., Lamberti, L.: Generalization of the Poincare sphere to process 2D displacement signals. Opt. Lasers Eng. **93**, 114–127 (2017)
6. Sciammarella, C.A., Lamberti, L.: Determination of displacements and their derivatives from 3D fringe patterns via extended monogenic phasor method. Opt. Lasers Eng. **104**, 117–125 (2018)
7. Sutton, M.A., Orteu, J.J., Schreier, H.: Image Correlation for Shape, Motion and Deformation Measurements: Basic Concepts, Theory and Applications. Springer, New York (2009)
8. Sutton, M.A., Reu, P.L. (eds.): International Digital Imaging Correlation Society, Conference Proceedings of the Society for Experimental Mechanics Series, 2016. New York, Springer (2017)
9. Axel, L., Dougherty, L.: Heart wall motion: improved method of spatial modulation of magnetization for MR imaging. Radiology. **172**(2), 349–350 (1989)
10. McVeigh, E.: MRI of myocardial function: motion tracking techniques. Magn. Reson. Med. **14**(2), 137–150 (1997)
11. Fischer, S.E., McKinnon, G.C., Scheidegger, M.B., Prins, W., Meier, D., Boesiger, P.: True myocardial motion tracking. Magn. Reson. Med. **31**(4), 401–413 (1994)
12. Wyman, B.T., Hunter, W.C., Prinzen, F.W., McVeigh, E.R.: Mapping propagation of mechanical activation in the paced heart with MRI tagging. Am. J. Physiol. Heart Circ. Physiol. **276**, 881–891 (1999)
13. Garot, J., Bluemke, D.A., Osman, N.F., Rochitte, C.E., McVeigh, E.R., Zerhouni, E.A., Prince, J.L., Lima, J.A.C.: Fast determination of regional myocardial strain fields from tagged cardiac images using harmonic phase MRI. Circulation. **101**, 981–988 (2000)
14. Hild, F., Roux, S.: Digital image correlation: from displacement to identification of elastic properties. Strain. **42**(2), 69–80 (2006)
15. Roux, S., Hild, F., Pagano, S.: A stress scale in full-field identification procedures: a diffuse stress gauge. Eur. J. Mech. A Solids. **24(3**, 442–451 (2005)
16. Sciammarella, C.A., Lamberti, L.: Basic models supporting experimental mechanics of deformations, geometrical representations, connections among different techniques. Meccanica. **50**, 367–387 (2015)

Chapter 4
Measuring Spallation Strength of Epoxy by Laser Spallation Technique

Sarthak S. Singh and R. Kitey

Abstract The laser spallation technique in combination with Michelson interferometry is employed to estimate the spallation strength of epoxy film. The epoxy layer of thickness 160 μm is deposited onto the microscopic glass (MG) substrate. The specimens are subjected to high amplitude short duration stress waves, developed by the pulsed laser ablation of sacrificial absorbing layer, deposited at the back surface of the substrate. The complex interaction of rarefaction waves with the mode converted tensile pulse from the free surface of the film develops high magnitude tensile region. For sufficiently high stresses spallation is observed in the top part of the epoxy layer. Profilometric analysis confirms spallation in the epoxy layer while the MG/epoxy interface remains intact. Interferometric measurements exhibit distinctly spaced fringe patterns. The temporal span of the wave arrivals confirms that the fringes correspond to the longitudinal and the rarefaction waves. Based on the space-time wave travel analysis the spallation depth is predicted which is in excellent agreement with the profilometric observations. By employing dynamic wave propagation analysis in combination with the interferometric data, the spallation strength of epoxy is estimated to be 260 ± 15 MPa.

Keywords Thick film · Rarefaction waves · Spallation · Michelson interferometer · Micro-cracks

4.1 Introduction

Polymers need wide range of characterization because their mechanical and failure behavior depend strongly on time, temperature and strain rate. While standard test methods are available to determine quasi-static properties, performing transient measurements have always been challenging. The test strategies further complicate when the loading is extreme dynamic such as in case of the development of shock waves at and around the explosion sites. The materials' strength in such cases is obtained from their spallation characteristics. The flyer-plate-impact along with the VISAR technique is often used to measure the spallation strength [1–3]. In current investigation the *laser spallation technique* is proposed as an alternative for spallation characterization. This non-contact method, often used to measure thin film adhesion strength [4–6], has simpler experimental setup and involves easier data reduction.

4.2 Experimental Details

The test samples are prepared by depositing 160 μm thick epoxy layer onto microscopic glass (MG) substrate. A thin reflective coating is transferred onto the top of the film for performing in-situ interferometric measurements. Pulsed laser ablation of sacrificial metallic layer, deposited on the back surface of MG, develops high amplitude short duration compressive stress pulse (For experimental and measurement details refer [6]). The stress wave transmitted across the MG/epoxy interface is followed by rarefaction waves, propagating at a lower speed. The complex interaction of rarefaction waves with the mode converted tensile pulse from the free surface of the film develops high magnitude tensile region. For sufficiently high stresses spallation is observed in the epoxy layer. The spallation strength of the epoxy is inferred from a combination of interferometric data and computational wave propagation analysis.

S. S. Singh · R. Kitey (✉)
Department of Aerospace Engineering, Indian Institute of Technology, Kanpur, India
e-mail: kitey@iitk.ac.in

L. Lamberti et al. (eds.), *Advancement of Optical Methods & Digital Image Correlation in Experimental Mechanics, Volume 3*, Conference Proceedings of the Society for Experimental Mechanics Series, https://doi.org/10.1007/978-3-319-97481-1_4

Transient free surface displacement of the film is recorded by focusing the probe beam of Michelson Interferometer at the spallation spot. The variation of fringe intensity, I(t), at the point of interest is recorded by using a photodiode and an ultra-high frequency oscilloscope. The fringe order N(t) is calculated by using,

$$I(t) = \frac{I_{\max} + I_{\min}}{2} + \frac{I_{\max} - I_{\min}}{2} \sin 2\pi N(t) \tag{4.1}$$

where, $I_{\max}$ and $I_{\min}$ are the maximum and the minimum intensity of the fringe in consideration, respectively. The out-of-plane displacement history u(t) at the free surface of the film, is calculated next by applying the Doppler shift [7],

$$u(t) = \frac{\lambda}{2} N(t) \tag{4.2}$$

In the above equation λ (= 514 nm) represents the wavelength of the probe beam. The stress history ($\sigma(t)$) in the epoxy layer is evaluated by [5],

$$\sigma(t) = -\frac{1}{2} (\rho C_{\mathrm{L}}) \frac{\partial u}{\partial t} \tag{4.3}$$

where, ρ and C_{L} density and the longitudinal wave speed, respectively.

4.3 Results and Discussions

The representative optical micrograph illustrated in Fig. 4.1a shows the spallation observed in thick epoxy film specimen when a laser fluence of 210 mJ/mm^2 is impinged. The spallation region consists of a circular area with crumbled pattern. This indicates spallation induced micro-cracks developed in the material. The spallation zone encloses a rough dark area where the film appears to be completely ruptured. A halo around the circular region indicates the area of MG surface upon which the Nd:YAG energy was impinged. This is partially visible because of the transparent MG and epoxy layers. The profilometric scan across the spalled region (Fig. 4.1b) reveals the nature of spallation along with its depth. The epoxy film lift-off begins from the spallation boundary with a crater formation at the center is evident from the scan. The spallation depth of 25 μm confirms the failure in the epoxy layer and not at the MG/epoxy interface. Based on the optical and profilometric imaging the spallation geometry is sketched in Fig. 4.1c, illustrating the spallation depth and the film lift-off resulting into partial rupture of the spalled film.

From the space-time wave travel analysis and the measured failure depth it is inferred that the spallation in the epoxy film is due to the complex interplay of two or more stress waves. Prior to identifying the nature of these waves various waves speeds are characterized in the epoxy layer. The longitudinal (C_{L}) and the shear wave (C_{S}) speeds in epoxy are separately measured by employing ultrasonic pulse echo method. The two speeds are extracted to be, 2610 m/s and 1220 m/s, respectively. By using the equation given below the value of the bulk wave speed (C_0) in epoxy is calculated to be to be 2197 m/s.

$$C_0 = \sqrt{\left(C_L^2 - \frac{4}{3} C_s^2\right)} \tag{4.4}$$

For quantifying the spallation strength of epoxy interferometric measurements are performed on the MG substrate as well as on the MG/epoxy specimen. Representative interferometric data, recorded in MG/epoxy specimen is plotted in Fig. 4.2a. The fringes correspond to the 72 mJ/mm^2 laser fluence after which the failure has initiated. Two distinct set of fringes are evident from the figure. Corresponding wave arrivals are marked by arrows in the plot. The first wave reaches the free surface of epoxy at 377 ns. Based on the wave arrival time, the longitudinal wave speed in epoxy layer is calculated to be 2600 m/s which is very close to the measured longitudinal wave speed. This first set of fringes corresponds to the compressive longitudinal stress wave, typically observed in a laser spallation experiments [4–6]. Another set of fringes which is distinctly apart from the first set, indicates the arrival of second wave at 393 ns. From the wave speed calculations it is confirmed that the secondary wave is not developed due to the primary wave reflections or mode conversions.

It is well known that at the end of a compressive pulse rarefaction (release) waves are released in the form of a series of fan waves [8, 9]. These waves travel at a velocity of bulk wave speed. The spreading fringes correspond to the secondary

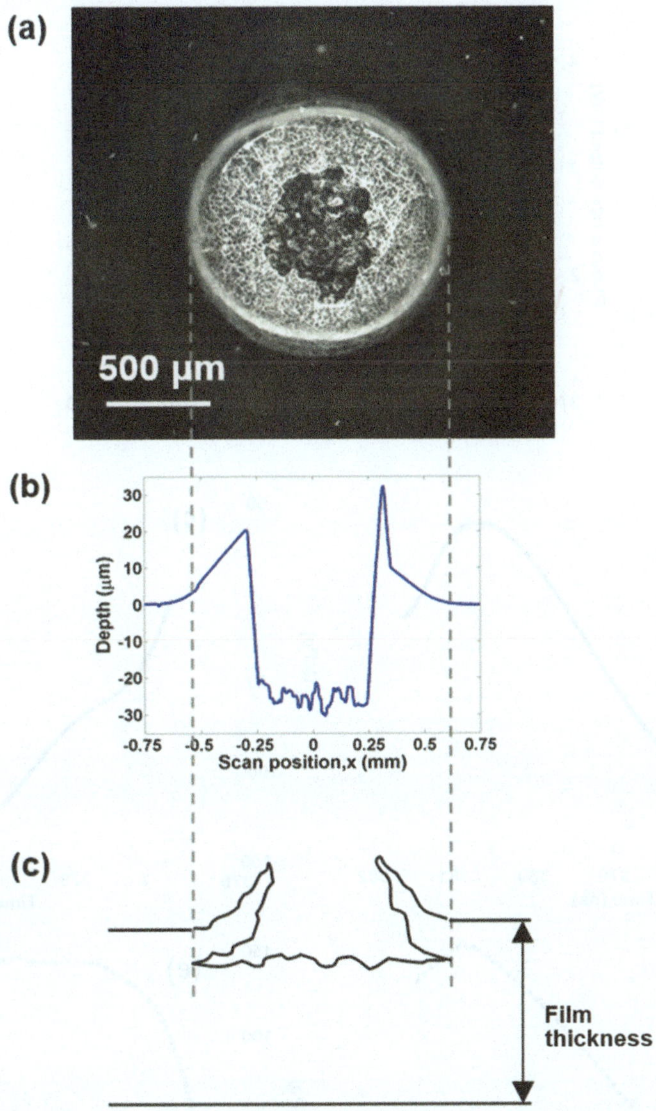

Fig. 4.1 (**a**) Spallation in thick epoxy film specimen, (**b**) Profilometric scan across the spalled region, (**c**) Sketch illustrating the spallation geometry

wave is the signature of release waves (see, Fig. 4.2a). Also, from the secondary wave arrival time its speed is calculated to be 2190 m/s which is very close to the earlier calculated bulk wave speed in the epoxy. For further ascertaining the nature of the waves, the two interferometric signals are analyzed by using Eqs. (4.1)–(4.3). Free surface displacement and the stress history associated to the primary wave are plotted in Fig. 4.2b, c, respectively. Corresponding data for the secondary wave counterpart are illustrated in Fig. 4.2d, e, respectively. While the primary stress wave subject the free surface to a maximum displacement of ~225 nm, corresponding stress history shows nearly triangular pulse with sharp rise time. On the contrary from the stress history corresponds to the secondary wave it appears that the stress wave temporally elongated and its magnitude remains nearly steady for ~6 ns.

It is worth noting that the magnitude of the secondary stress pulse is quite comparable to the primary one. This is attributed to the development of rarefaction (decompression) shock in MG. The analysis of interferometric data, separately obtained by performing experiments on MG specimens, indicates the delayed arrival of decompression shock to the MG surface. The development of rarefaction shock in fused quartz was first demonstrated by Barker and Hollenbach [10] and further reported by Wang et al. [11]. The wave speed calculations indicate that the mode converted tensile waves from this shock would interact with the rarefaction waves. Therefore, in view of the discussed observations, it is conjectured that the combination of two waves subject the bulk wave traveling into the epoxy with high amplitudes.

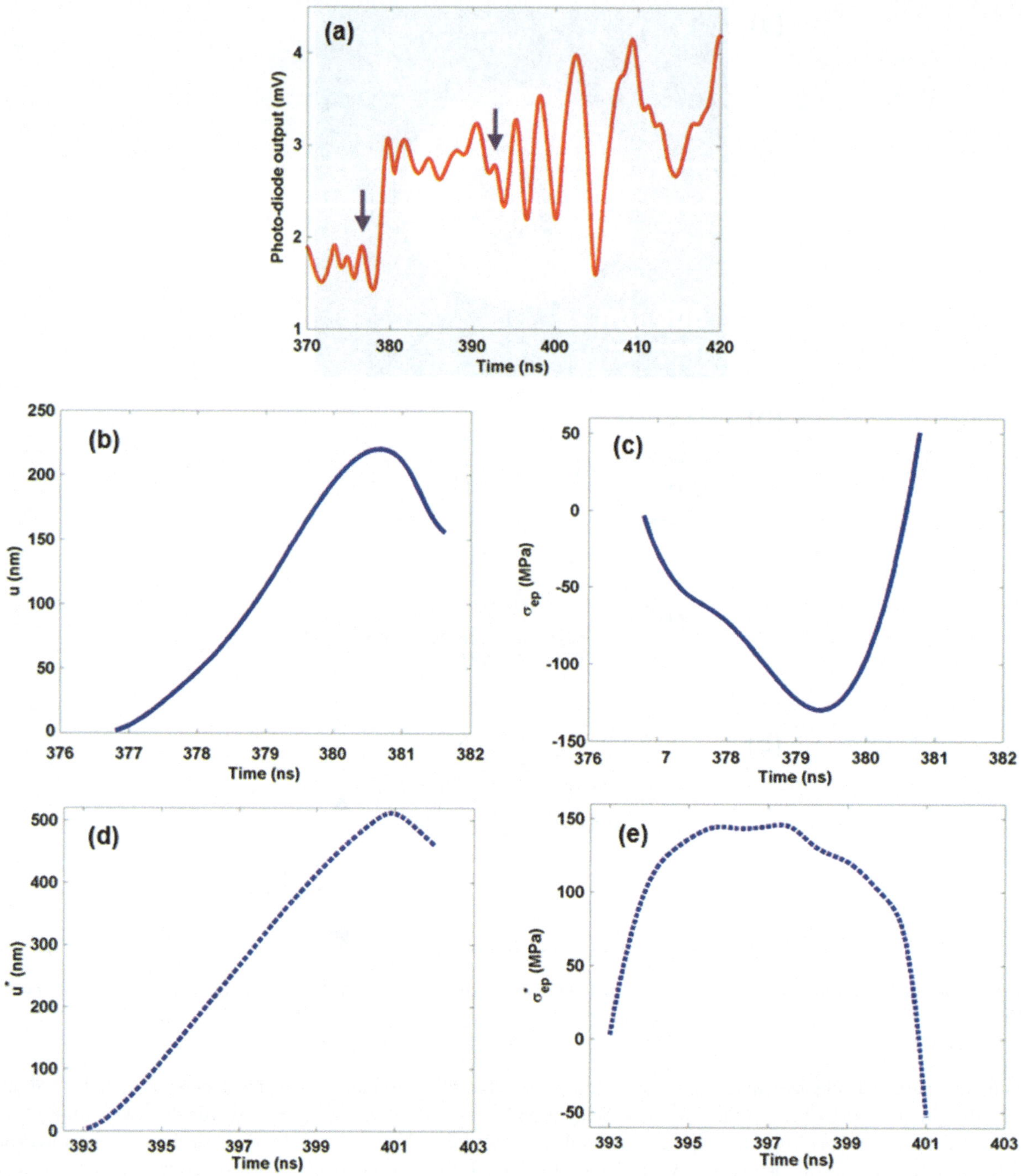

Fig. 4.2 (**a**) Representative variation in fringe intensity obtained from interferometric measurements for a 160 μm thick epoxy film (**b**) and (**c**) corresponds to the out-of-plane displacement and the stress history for the first compressive pulse (**d**) and (**e**) represent the out-of plane displacement and the stress history associated with the secondary pulse

The spallation strength is estimated by performing two dimensional planar wave propagation analyses in ABAQUS 6.12 where only the epoxy layer geometry is modeled instead of FQ/epoxy combination. Stress histories obtained from the experiments (Fig. 4.2c, e) are applied as the loading at the back surface. Simulations exhibit maximum stress developing at 20 μm from the top surface of the epoxy which is very close to the experimentally observed spallation depth. The spallation strength of the epoxy is estimated to be 260 ± 15 MPa.

References

1. Kanel, G.I., Razorenov, S.V., Bogatch, A., Utkin, A.V., Fortov, V.E., Grady, D.E.: Spall fracture properties of aluminum and magnesium at high temperatures. J. Appl. Phys. **79**(11), 8310–8317 (1996)
2. Field, J.E., Walley, S.M., Proud, W.G., Goldrein, H.T., Siviour, C.R.: Review of experimental techniques for high rate deformation and shock studies. Int J Impact Eng. **30**(7), 725–775 (2004)
3. Zaretsky, E., Perl, M.: The response of glass fibers reinforced epoxy composite to an impact loading. Int. J. Solids Struct. **41**(2), 569–584 (2004)
4. Gupta, V., Argon, A.S., Cornie, J.A., Parks, D.M.: Measurement of interface strength by laser-pulse-induced spallation. Mater. Sci. Eng. A. **126**, 105–117 (1990)
5. Wang, J., Weaver, R.L., Sottos, N.R.: A parametric study of laser induced thin film spallation. Exp. Mech. **42**(1), 74–83 (2002)
6. Kitey, R., Sottos, N.R., Geubelle, P.H.: A hybrid experimental/numerical approach to characterize interfacial adhesion in multilayer low-κ thin film specimens. Thin Solid Films. **519**(1), 337–334 (2010)
7. Barker, L.M.: Laser interferometry in shock wave research. Exp. Mech. **12**(5), 209–215 (1972)
8. Gilath, I., Eliezer, S., Dariel, M.P., Kornblit, L., Bar-Noy, T.: Laser induced spall in aluminum and copper. Le Journal de Physique Colloques. **49**(C3), C3–191 (1988)
9. Asay, J.R., Shahinpoor, M. (eds.): High-Pressure Shock Compression of Solids, Chapter-2. Springer Science & Business Media, New York, NY (2012)
10. Barker, L.M., Hollenbach, R.E.: Shock-wave studies of PMMA, fused silica, and sapphire. J. Appl. Phys. **41**(10), 4208–4422 (1970)
11. Wang, J., Weaver, R.L., Sottos, N.R.: Laser-induced decompression shock development in fused silica. J. Appl. Phys. **93**(12), 9529–9536 (2003)

Chapter 5
Uncertainty Quantifications for Multiviewcorrelation

F. Hild and S. Roux

Abstract The aim of this extended abstract is to introduce a general framework for the quantifications of uncertainties associated with displacement measurements via multiview correlation. The latter is an extension of stereocorrelation approaches to systems with more than two cameras.

Keywords Multiview systems · Noise · Registration · Uncertainty quantification

5.1 Introduction

Stereocorrelation is a noncontact technique to measure 3D surface shapes and their deformations under mechanical loads. The standard setup consists of two calibrated cameras [1]. Such configurations can be generalized to multiview systems in which more than two cameras acquire pictures of the same surface of interest [2]. In order to analyze the gain provided by the use of multiple cameras, the assessment of measurement resolutions is very helpful. Closed-form expressions were derived for Digital Image Correlation [3], and Digital Volume Correlation [4]. It is proposed to discuss their extension to 3D surface measurements. For the sake of conciseness, it is proposed to study 3D displacement uncertainties.

5.2 General Setting

Let us consider the homogeneous coordinates of 3D points $\{\mathbf{X}\} = \{X, Y, Z, 1\}^t$, and their corresponding displacement $\{\mathbf{U}\} = \{U_x, U_y, U_z, 0\}^t$. For each considered camera c, the projection matrix $[\mathbf{P}^c]$ enables the homogeneous coordinates of each pixel to be related to those of the considered surface point $s^c\{\mathbf{x}^c\} = [\mathbf{P}^c]\{\mathbf{X}\}$, where $\{\mathbf{x}^c\} = \{x^c, y^c, 1\}^t$ denotes the homogeneous coordinates in the detector plane of camera c, and s^c the scale parameter. Similarly, the displacement $\{\mathbf{U}\}$ induces motions $s^c\{\mathbf{u}^c\} = [\mathbf{P}^c]\{\mathbf{U}\}$ in the camera plane. The intrinsic surface brightness is defined as $f(\mathbf{X})$, whereas that captured by camera c is denoted by f^c. Let us consider a surface of area A subject to a uniform translation along $\mathbf{\Phi}$ and of magnitude U, $\mathbf{U} = \mathbf{\Phi} U$. A perturbation analysis of the gray level conservation equation [5] enables displacement increments δU to be derived

$$(\nabla f^c \cdot \varphi^c)\,\delta U = s^c \rho^c \tag{5.1}$$

where ρ^c is the local residual associated with an offset to gray level conservation, $[\mathbf{\Pi}^c]$ the 2×3 reduced projection matrix so that $\varphi^c = [\mathbf{\Pi}^c]\mathbf{\Phi}$. In this expression, no uncertainty is assumed for the projection matrix $[\mathbf{P}^c]$, nor for the initial shape of the surface of interest. White noise of the image η^c is considered with $\left\langle \eta^c\left(\mathbf{x}_i^c\right)\eta^c\left(\mathbf{x}_j^c\right)\right\rangle = I_{ij}(\sigma^c)^2$, where σ^c is the standard deviation of Gaussian white noise, and $[\mathbf{I}]$ the identity matrix. This noise is considered as the sole cause of registration

F. Hild (✉)
LMT (ENS Cachan/CNRS/University Paris-Saclay), Cachan, France
e-mail: hild@lmt.ens-cachan.fr

S. Roux
LMT, ENS Paris Saclay/CNRS UMR 8535/University Paris Saclay, Cachan, France

L. Lamberti et al. (eds.), *Advancement of Optical Methods & Digital Image Correlation in Experimental Mechanics, Volume 3*, Conference Proceedings of the Society for Experimental Mechanics Series, https://doi.org/10.1007/978-3-319-97481-1_5

residual. Stereo- (or multiview-)correlation consists in combining the partial information provided by each individual camera to extract the shape or displacements [1]. The standard practice is to give equal weight to each image. However, it can be shown that the optimal way of assembling those different channels of information is with a weight equal to the inverse of noise variance. The displacement variations then read

$$\sum_{c}(\sigma_c)^{-2}\sum_{S^c}(\nabla f^c \cdot \varphi^c)^2\delta U = \sum_{c} s^c(\sigma_c)^{-2}\sum_{S^c}\eta^c(\nabla f^c \cdot \varphi^c) \tag{5.2}$$

The presence of noise induces an uncertainty on the displacement U for multiview systems with a variance expressed as

$$\sigma_u^2 = \frac{\sum_{c}\left(\frac{s_c}{\sigma_c}\right)^2 \sum_{S_c}(\nabla f^c \cdot \varphi^c)^2}{\left(\sum_{c}\left(\frac{1}{\sigma_c}\right)^2 \sum_{S_c}(\nabla f^c \cdot \varphi^c)^2\right)^2} \tag{5.3}$$

This expression is the first key result. It accounts for the individual characteristics of each camera, and also the gradual degradation of the uncertainty as the observed surface is more inclined (with respect to a given camera).

5.3 Mean-Field Approximation

When the correlation length of the random pattern ξ is much smaller than any characteristic length scale L of the surface of interest, an order of magnitude of the matrix, which needs the spatial average of the tensor product of the picture gradient with itself, reads

$$\sum_{S^c}(\nabla f^c \otimes \nabla f^c) \approx S^c\langle\nabla f^c \otimes \nabla f^c\rangle \tag{5.4}$$

where S^c is the projected surface of interest. The 3D surface normal is denoted by **N**. Its projection onto the image plane of camera c is $s^c\{\mathbf{n}^c\} = [\mathbf{P}^c]\{\mathbf{N}\}$. Let **A** and **B** be two orthogonal tangent vectors to the surface (i.e., (**A**,**B**,**N**) is a basis, and their 4th homogeneous coordinate is set to 0). They are projected as $s^c\{\mathbf{a}^c\} = [\mathbf{P}^c]\{\mathbf{A}\}$ and $s^c\{\mathbf{b}^c\} = [\mathbf{P}^c]\{\mathbf{B}\}$, so that

$$\|\mathbf{a}^c\|^2 = (s^c)^2\{\mathbf{A}\}^t[\mathbf{P}^c]^t[\mathbf{P}^c]\{\mathbf{A}\} \tag{5.5}$$

A condensed matrix is constructed as

$$[\mathbf{Q}^c] = \begin{bmatrix}\mathbf{a}^c\cdot\mathbf{a}^c & \mathbf{a}^c\cdot\mathbf{b}^c\\ \mathbf{a}^c\cdot\mathbf{b}^c & \mathbf{b}^c\cdot\mathbf{b}^c\end{bmatrix} = (s^c)^2\begin{bmatrix}\{\mathbf{A}\}^t[\mathbf{P}^c]^t[\mathbf{P}^c]\{\mathbf{A}\} & \{\mathbf{A}\}^t[\mathbf{P}^c]^t[\mathbf{P}^c]\{\mathbf{B}\}\\ \{\mathbf{A}\}^t[\mathbf{P}^c]^t[\mathbf{P}^c]\{\mathbf{B}\} & \{\mathbf{B}\}^t[\mathbf{P}^c]^t[\mathbf{P}^c]\{\mathbf{B}\}\end{bmatrix} \tag{5.6}$$

which is 2 × 2 symmetric. The eigenvalues (q_1^c, q_2^c), and eigenvectors $(\mathbf{e}_1^c, \mathbf{e}_2^c)$ give the correspondence between physical distances in the 3D space to pixels in the camera c plane. More precisely, in the direction $\mathbf{e}_\alpha^c$ (in the camera c plane) one pixel corresponds to a physical length $a_\alpha^c = 1/\sqrt{q_\alpha^c}$. Equivalently

$$\mathbf{Q}^c = q_1^c\mathbf{e}_1^c \otimes \mathbf{e}_1^c + q_2^c\mathbf{e}_2^c \otimes \mathbf{e}_2^c \tag{5.7}$$

so that the projected surface area $A^c = A\sqrt{\det([\mathbf{Q}])} = A\sqrt{q_1^c q_2^c}$. When observed with a pixel size of a, the random pattern is modeled as $\widehat{f_a} = G_a{}^* f$, where G_a is a Gaussian of width a. To compute the needed pair correlation function, a move to Fourier space, denoted by F, is convenient $F\left(\widehat{f_a}\right) = F(G_a)\cdot F(f)$. The expectation of the quadratic norm of the gradient of $\widehat{f_a}$ is the second moment $g(a) = \int |\mathbf{k}|^2|F(G_a)|^2|F(f)|^2 d\mathbf{k}$. The expectation of the projector along the texture gradient then reads

$$\langle\nabla f^c \otimes \nabla f^c\rangle = g^c\left(1/\sqrt{q_1^c}\right)\mathbf{e}_1^c \otimes \mathbf{e}_1^c + g^c\left(1/\sqrt{q_2^c}\right)\mathbf{e}_2^c \otimes \mathbf{e}_2^c \tag{5.8}$$

Introducing the tensor $\mathbf{T}^c = \sqrt{q_1^c q_2^c}\left(g^c\left(1/\sqrt{q_1^c}\right)\mathbf{e}_1^c \otimes \mathbf{e}_1^c + g^c\left(1/\sqrt{q_2^c}\right)\mathbf{e}_2^c \otimes \mathbf{e}_2^c\right)$ gives access to a compact expression for the uncertainty due to noise

$$\sigma_u^2 = \frac{1}{A}\left(\sum_c \left(\frac{s_c}{\sigma_c}\right)^2 \varphi^c \cdot \mathbf{T}^c \cdot \varphi^c\right)\left(\sum_c \left(\frac{1}{\sigma_c}\right)^2 \varphi^c \cdot \mathbf{T}^c \cdot \varphi^c\right)^{-2} \tag{5.9}$$

This tensor accounts for the change of brightness as the surface becomes more inclined and in particular its anisotropic features. Similarly, the weight of a given surface that varies with the projected area is included. The optimal weighting of each camera mitigates the effect of unfavorable geometry of the observed surface.

5.4 Conclusions

The standard displacement resolution associated with the registration for a surface facet was derived within the framework of small perturbations due to Gaussian white noise. This type of result can be generalized to more complex surfaces, and to the effect of calibration errors and shape reconstruction uncertainties. Practical evaluations are to be performed in order to compare the theoretical predictions with practical cases.

References

1. Sutton, M.A., Orteu, J.-J., Schreier, H.W.: Image Correlation for Shape, Motion and Deformation Measurements: Basic Concepts, Theory and Applications. Springer, New York, NY (2009)
2. Dufour, J.-E., Hild, F., Roux, S.: Shape, displacement and mechanical properties from isogeometric multiview stereocorrelation. J. Strain Anal. Eng. Des. **50**(7), 470–448 (2015)
3. Hild, F., Roux, S.: Comparison of local and global approaches to digital image correlation. Exp. Mech. **52**(9), 1503–1519 (2012)
4. Leclerc, H., Périé, J.-N., Hild, F., Roux, S.: Digital volume correlation: what are the limits to the spatial resolution? Mech. Ind. **13**, 361–371 (2012)
5. Hild, F., Roux, S.: Digital image correlation. In: Rastogi, P., Hack, E. (eds.) Optical Methods for Solid Mechanics. A Full-Field Approach, pp. 183–228. Wiely-VCH Verlag, Weinheim (2012)

Chapter 6
Image Analysis of Curvature Using Classical Mechanics, the Elastica

Charles Wilson and James Dawson

Abstract The primary loading condition for implanted leads and other slender/compliant implantable medical devices is bending. Curvature, which is the inverse of radius of curvature, quantifies the magnitude of bending deformation and is used to motivate bench tests as well as used for computational modeling. The Association for the Advancement of Medical Instrumentation (AAMI) Transvenous Leads Working Group is currently studying how to best measure the curvature of leads under deformation, both in-vivo and on the bench.

Deformed lead segments bent to varying levels of severity, were imaged and the resultant curvature was analyzed using several optical analysis methods, including fitting circles, ellipses and splines to the deformed shape. Of all the methods studied, the one which showed the least amount of variability while also not over parameterizing was using the Elastica.

The Elastica, which was first formulated by James Bernoulli in 1691 and later solved by Leonhard Euler in 1744, is the closed form solution to large scale deformations of buckled structures. It is an example of bifurcation theory in the field of solid mechanics. The Elastica can take many different shapes while only having a few degrees of freedom. More importantly, the Elastica is the exact analytical solution to a long, slender beam under large deformation—which is precisely what is being studied. This approach has been implemented in an R script that allows the user to load an image, identify points in the image to which the Elastica should be fit, and optimizes the fit parameters using a least squares regression. Outputs from the script are the magnitude and location of the maximum curvature, the moment arm distance, and a goodness of fit metric.

Keywords Image analysis · Bending · Buckling · Curvature · Elastica

6.1 Introduction

The pacemaker (IPG) and implantable cardioverter defibrillator (ICD) are both devices that are used when a patient's heart is not functioning normally. For the IPG or ICD (from here on called the device) to be able to treat the heart condition, the device needs to be connected to the heart. Since the device is too big to be implanted in the heart itself, it is instead connected to the heart using an electrical wire. This type of wire is called a lead (see Fig. 6.1 below). At the proximal end closest to the device (located in the shoulder region of the patient), the lead is connected to the device. At the distal end furthest from the device (inside the heart), the lead is connected both electrically and mechanically to the heart wall.

An implanted lead is subjected to large scale deformation that cannot be analyzed by traditional theory of elasticity. Cardiac leads are slender compliant structures which are subjected to buckling and bending deformation inside the body. As shown in the Fig. 6.2 below, the tip of the lead is typically implanted into the right ventricle of the heart and is connected to the heart wall either with an active or passive fixation mechanism, both of which result in a pinned boundary condition at the tip. As the tricuspid valve closes during ventricular systole, the lead is pinched between the leaflets of the valve. As the heart contracts, the lead is forced into compression via ventricular ejection as the heart pumps blood from the heart to the lungs.

The Association for the Advancement of Medical Instrumentation (AAMI) Transvenous Leads Working Group is currently studying how to best measure the curvature of leads under deformation, both *in-vivo* and on the bench. In terms of measuring curvature on the bench, two approaches have been studied thus far—using circles and cubic splines to fit the deformed lead shapes. There are inherent limitations and challenges with each of these approaches, as neither circles nor cubic polynomials match the true analytical shape of a long, slender beam under large deformations.

C. Wilson (✉) · J. Dawson
Medtronic, Inc., Minneapolis, MN, USA
e-mail: charles.l.wilson@medtronic.com

L. Lamberti et al. (eds.), *Advancement of Optical Methods & Digital Image Correlation in Experimental Mechanics, Volume 3*, Conference Proceedings of the Society for Experimental Mechanics Series, https://doi.org/10.1007/978-3-319-97481-1_6

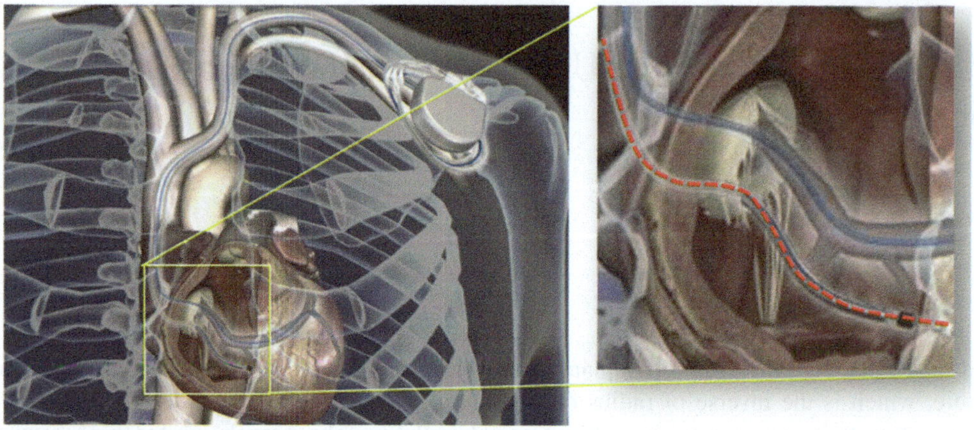

Fig. 6.1 Typical implant of a cardiac system consisting of a cardiac lead and device

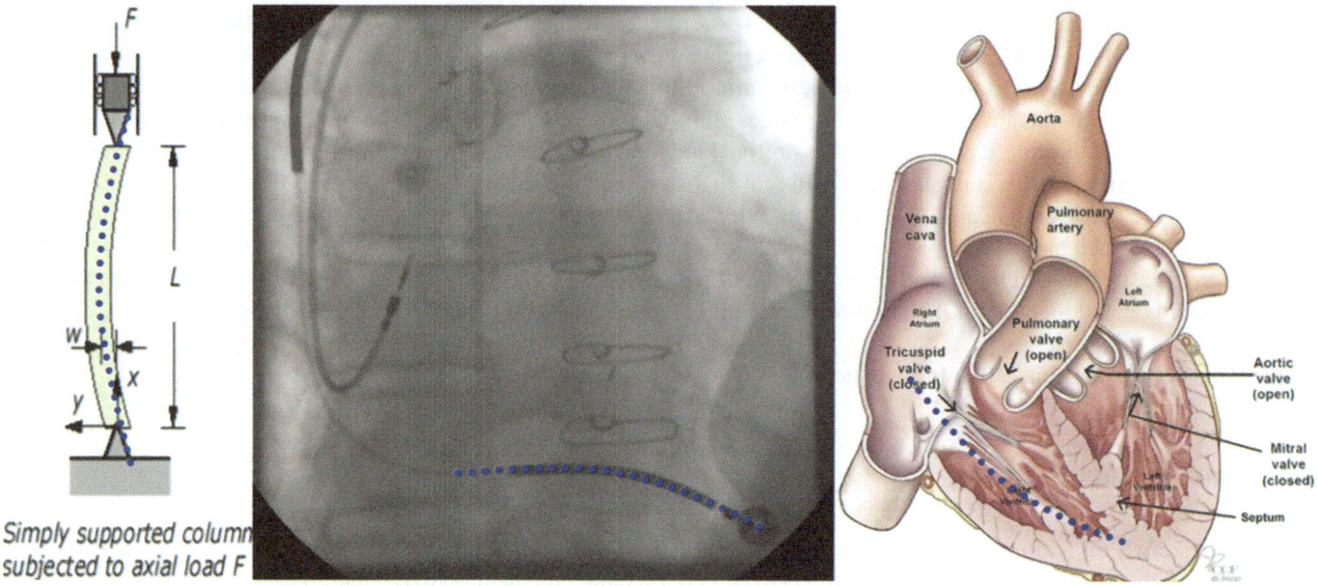

Fig. 6.2 Collage of images showing the correlation in shape between a buckled column (**a**), an actual X-ray image of a lead in patient (**b**), and the diagram of the heart with a lead path outlined (**c**)

The simpler of the two methods is fitting a circle, which has the characteristic of constant curvature. This works well when measuring curvature over very short arclengths. However, because shorter arclengths are needed to achieve consistent results, this can reduce the fidelity of the measurement. Additionally, depending on the arclength chosen, different values of curvature can be measured. The best success has been with trying to measure constant curvature over an included angle.

In contrast, fitting cubic splines can fit nearly any desired shape. However, this introduces a different problem than the circle—a spline ends up having too many degrees of freedom and over-fitting becomes problematic. Trying to balance goodness-of-fit and highly variable curvature measurement has proven to be a difficult task.

The Elastica, which was first formulated by James Bernoulli in 1691 and later solved by Leonhard Euler in 1744, is the closed form solution to large scale deformations of buckled structures. It is an example of bifurcation theory in the field of solid mechanics. The Elastica can take many different shapes while only having a few degrees of freedom. More importantly, the Elastica is the exact analytical solution to a long, slender beam under large deformation—which is precisely what is being studied. Examples of various Elastica shapes are shown in Fig. 6.3.

Fig. 6.3 Examples of various Elastica shapes

6.2 Methodology

The method for measuring curvature is to fit an Elastica curve to set of points along a deformed shape using a least squares regression. Once the best fitting Elastica curve is found, the magnitude and location of the maximum curvature of the shape can be computed directly.

Levien [1] showed that the equations governing the inflectional Elastica can be parameterized as:

$$x(s) = s - 2\, E\,(am\,(s, m)\,, m)$$

$$y(s) = -2\,\sqrt{m}\; cn\,(s, m)$$

$$\text{for} \quad 0 \le s \le 4\, K(m)$$

where m is the *elliptic modulus* and s is the arclength along the curve. The functions E, K, am and cn are elliptic integrals and Jacobi elliptic functions defined as:

$$E\,(\varphi, m) = \int_0^{\varphi} \sqrt{1 - m\,\sin^2(t)}dt$$

$$K(m) = \int_0^{\pi/2} \frac{1}{\sqrt{1 - m\,\sin^2(t)}}dt$$

$$am\,(u, m) = \text{the value of } \varphi \text{ such that}$$

$$u = \int_0^{\varphi} \frac{1}{\sqrt{1 - m\,\sin^2(t)}}dt$$

$$cn\,(u, m) = \cos\,(am\,(u, m))$$

Curvature can be defined as:

$$\kappa\,(s, m) = 2\sqrt{m}\; cn\,(s, m)$$

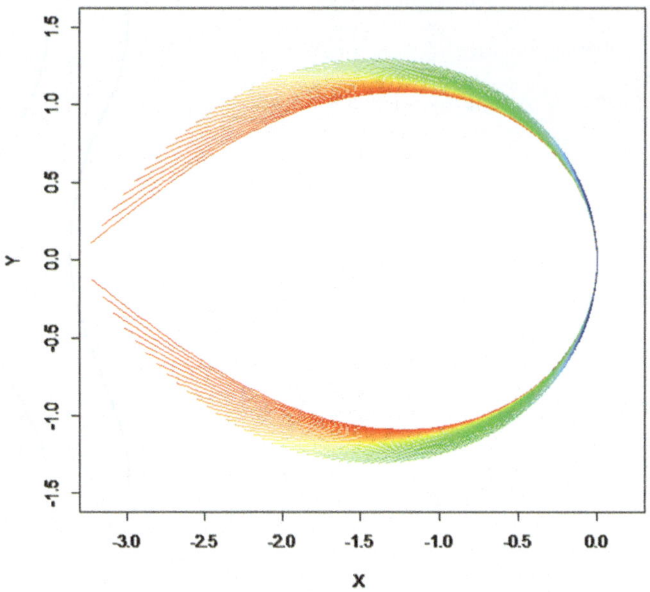

Fig. 6.4 Family of Elastica curves whose maximum curvature are equal to 1. The color of the curve corresponds with differing values of m, the red curves are m values closer to 0 and the blue curves are m values closer to 1

Table 6.1 Elastica curve fit parameters

Parameter	Description
curveID	An index identifying a single curve within the family of predefined Elastica curves computed using various *m* values
Scale	A value that multiplies the x and y values of the Elastica curve
theta	A value that rotates the Elastica curve about the apex
dx	A value that offsets the x values of the Elastica curve
dy	A value that offsets the y values of the Elastica curve

The maximum curvature occurs at the midpoint of the Elastica curve so that the maximum curvature for any value *m* is:

$$\kappa_{max}(m) = 2\sqrt{m}\, cn\,(2\; K(m), m) = 2\sqrt{m}$$

Because computation of the Elastica shape requires the use of Elliptic integrals, it is computationally advantageous to predefine a family of curves that will be used during optimization of the other fitting parameters. If a family of Elastica curves whose maximum curvature are all equal to 1 is created, then all that is necessary to calculate the curvature is a scaling parameter. An example family of Elastica curves is shown in Fig. 6.4.

Any Elastica curve can now be defined as a function of 5 parameters (Table 6.1):

If a set of XY coordinates along a deformed shape are known, then optimization techniques can be used to find the parameter values that minimize the sum of squared distances between each point and the optimal Elastica curve.

This approach has been implemented in an R script that allows the user to load an image, identify points in the image to which the Elastica should be fit, and optimizes the fit parameters using a least squares regression. Outputs from the script are the magnitude and location of the maximum curvature, the moment arm distance, and a goodness of fit metric.

The only setup required to run the script is to copy ElasticaFit_v1.1.R and ElasticaHelper_v1.1.R (see Appendix A) into a directory. Launch a new R session and change the working directory to the folder to which you copied the R scripts. Once the working directory has been set, start the script by typing:

```
source('ElasticaFit_v1.1.R')
```

Note: The first time the script is run, any libraries that need to be installed will install automatically. Additionally, if the csv file of Elastica curves is not present, it will need to be generated. This may take several minutes but will only occur the first time the script is run.

The script will begin by prompting the user to select which image or images they wish to analyze. The script will accept jpeg, bmp, and png image formats. Tiff images are currently experimentally supported and can be chosen by selecting "All files" from the file type dropdown. After the images are chosen, the script will begin to loop through each image for analysis. The graphics window must be selected for the interactive commands to work.

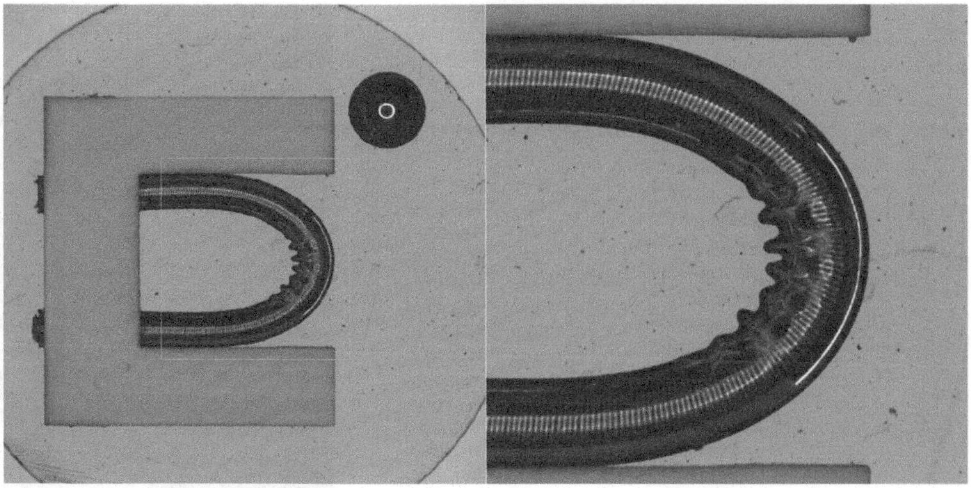

Fig. 6.5 Original image with (**a**) zoom box drawn and (**b**) zoomed image

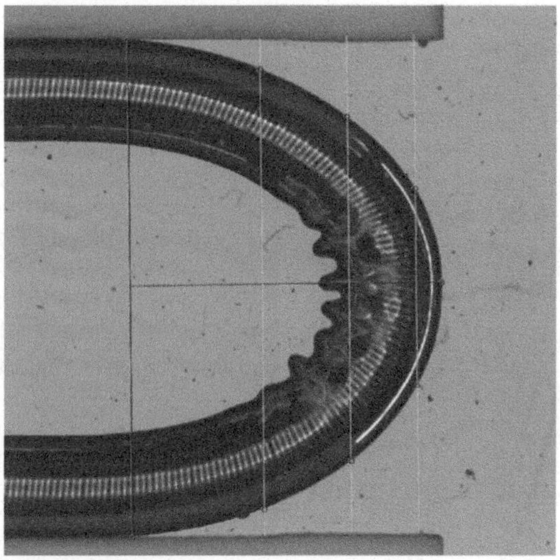

Fig. 6.6 Support and moment lines (red) and guidelines (green) drawn with selected lead body edge points

The first prompt for the image will ask the user if the shape is two-sided or not. The user may select yes or no by entering 'y' or 'n' on the keyboard. If yes is selected, then the user will select points on both sides of the apex of the bend, otherwise points will only be selected on one side of the apex. This option is useful for measuring curvature on one side of a transition point.

The next prompt will be for the user to draw a box around a region of interest. Ensure that the apex of the bend and the point at which the lead contacts the support are included in the zoom box. An example of zooming an image is shown in Fig. 6.5.

Next, the user will select the apex of the bend and two support points—which should be the location at which the deformed segment comes into contact with its support. A support line is then drawn between the support points and a moment line is drawn perpendicular to the support line and through the apex. Then, a guideline is drawn parallel to the support line. The user will then indicate where this guideline intersects the lead body edge. This step is then repeated two more times for a total of 7 selected edge points, 6 from guideline intersection and 1 from the apex, which are the red points shown in Fig. 6.6.

After the apex and 6 edge points have been selected, the Elastica curve parameters are then optimized and the best fit Elastica is overlaid on the image as shown in Fig. 6.7.

The final prompt in the script will ask the user if they are satisfied with the analysis. The user may select yes or no by entering 'y' or 'n' on the keyboard. If no is selected, analysis on the current image will be repeated, erasing the current analysis. If yes is selected, the script will advance to the next image. After all images have been analyzed, a summary of the analysis is printed to the R terminal and an output data file is created in the working directory.

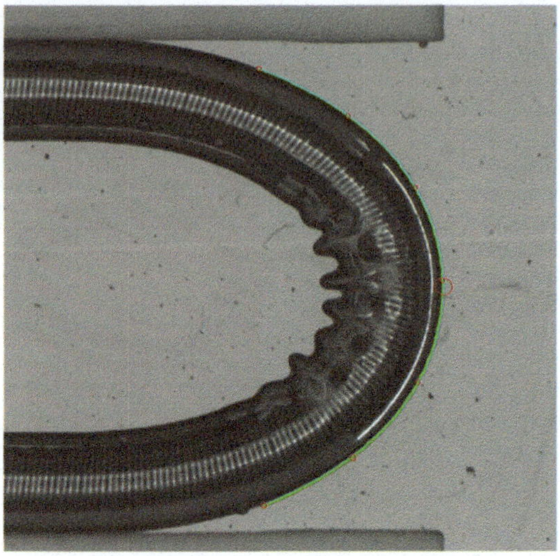

Fig. 6.7 Final Elastica curve fit, the circle indicates the location of maximum curvature

Table 6.2 Gage R&R results from the various curvature fitting methods

Source	Circle	Spline	Elastica
Repeatability	0.088	0.252	0.112
Reproducibility	0.276	0.234	0.122
Total Gage R&R	0.290	0.344	**0.166**

6.3 Results

Images of deformed leads were analyzed using circles, splines and the Elastica method by six operators. The resulting curvature data was then analyzed using a statistical software package to determine the repeatabiliy, reproducibility and total gage R&R. The results in Table 6.2 show that the total gage variation was lowest for the Elastica method.

6.4 Conclusion

There is challenges associated with determing the magnitude of deformation of deformed leads in vivo. The Elastica curvature fitting method has been developed to minimive the subjectivity associated with measuring curvature. Results show that the Elastica method, when compared to circle or spline fits, shows the least amount of variability (Total Gage R&R). Another advantage using the Elastica has over the other methods is that the basis function for the deformed shape. Conversely, the circle fit would be a good choice if the deformed shape were circular like that of a beam in pure bending. For instance, the deformed shape of a homogeneous beam subjected to 4 point bending between the inner supports would be circular.

Acknowledgements The authors of this publication would like to acknowledge these individuals for their contribution to this study.

J. Splett, D. McColskey, T. Luther, D. Smith, N. Duraiswamy, and S. Raymond for conducting the curvature measurements and A. Himes for his technical consult.

References

1. Levien, R.: The elastica: a mathematical history. EECS Department, University of California, Berkeley, Technical Report No. UCB/EECS-2008-103 (2008)

Chapter 7
Fast Adaptive Global Digital Image Correlation

Jin Yang and Kaushik Bhattacharya

Abstract Digital image correlation (DIC) is a powerful experimental technique to compute full-field displacements and strains. The basic idea of the method is to compare images of an object decorated with a speckle pattern before and after deformation, and thereby to compute displacements and strains. Since DIC is a non-contact method that gives the whole field deformation, it is widely used to measure complex deformation patterns. Finite element (FE)-based Global DIC with regularization is one of the commonly used algorithms and it can be combined with finite element numerical simulations at the same time (Besnard et al., J Strain Anal Eng Design 47(4):214–228, 2012). However, Global DIC algorithm is usually computationally expensive and converges slowly. Further, it is difficult to directly apply an adaptive finite element mesh to Global DIC because the stiffness matrix and the external force vector have to be rebuilt every time the mesh is changed.

In this paper, we report a new Global DIC algorithm that uses adaptive mesh. It builds on our recent work on the augmented Lagrangian digital image correlation (ALDIC) (Yang and Bhattacharya, Exp Mech, submitted). We consider the global compatibility condition as a constraint and formulate it using an augmented Lagrangian (AL) method. We solve the resulting problem using the alternating direction method of multipliers (ADMM) (Boyd et al., Mach Learn 3(1):1–122, 2010) where we separate the problem into two subproblems. The first subproblem is computed fast, locally and in parallel, and the second subproblem is computed globally without image grayscale value terms where nine point Gaussian quadrature works very well. Compared with current Global DIC algorithm, this new adaptive Global DIC algorithm decreases computation time significantly with no loss (and some gain) in accuracy.

Keywords Digital image correlation (DIC) · Adaptive mesh · Augmented Lagrangian · Alternating Direction Method of Multipliers (ADMM) · Heterogenous deformation

7.1 Introduction

Digital Image Correlation (DIC) is a popular optical experimental technique to measure deformations and strains in solids. In this method, we first take a sequence of grayscale digital images of a test specimen during deformation. Then by comparing images in the sequence, we determine displacements and strain fields of the specimen using image tracking algorithms [1]. Most DIC algorithms can be cast into two categories, Local Subset DIC method [2] and Global DIC method [3]. In Local Subset DIC method, as its name implies, we first discretize both reference image and deformed image into many subsets and then find the deformation of each subset independently. In the Global DIC method, we represent the global deformation using a basis set (often based on a finite element discretization), and then analyze the global image to obtain the coefficients relative to this basis set.

DIC is especially useful in the study of complex phenomena with very heterogeneous strain fields at various length scales. In heterogeneous strain fields, a uniform local subsets or global finite element mesh discretization leads very different accuracy in the different regions; however the precision only depends on the worst subset/element. In other words, a high precision requires a very fine mesh if it is uniform. We can overcome this by using an adaptive mesh in DIC to save computation time with no loss in accuracy. For example, in a fracture experiment, the near fracture tip region is much more heterogeneous than the far field. Thus we can apply denser mesh or higher order element near the crack tip. This idea is already commonly used in the adaptive finite element method [4].

J. Yang (✉) · K. Bhattacharya
Division of Engineering and Applied Science, California Institute of Technology, Pasadena, CA, USA

L. Lamberti et al. (eds.), *Advancement of Optical Methods & Digital Image Correlation in Experimental Mechanics, Volume 3*,
Conference Proceedings of the Society for Experimental Mechanics Series, https://doi.org/10.1007/978-3-319-97481-1_7

Currently, there are already several DIC methods, which adaptively choose local subset or finite element size. Yuan et al. [5] modifies Local Subset DIC method to automatically refine the subsets where the WZNSSD correlation function values are bigger than the setting tolerance. However, correlation function is not always a robust standard to refine the discretization. Further, in Local Subset DIC methods, smaller the DIC subset size can lead to larger DIC error [6]. Wittevrongel et al. [7] applies p-adaptive strategy in the global DIC and there is no need to re-compute the whole stiffness matrix after mesh refinement, which has computation time in the same order with current global DIC method. However, their method works well for very big subset/elements with size from 50 × 50 to 150 × 150 pixels, and the RMS error of solved displacement field is around 2%, which is larger than the typical DIC precision [1]. Besides p-adaptive strategy, h-adaptive strategy is also commonly used in adaptive FEM [4]. However, it is very expensive to directly apply h-adaptive finite element refinement strategy to global DIC because the stiffness matrix and the external force vector have to be rebuilt every time the mesh is changed. Recomputing the stiffness matrix and the external force vector is expensive since it involves a many evaluations of image grayscale value.

In this paper, we report a new Global DIC algorithm that uses h-adaptive mesh efficiently. It builds on our recent work on the augmented Lagrangian digital image correlation (ALDIC) and can be solved very fast using the alternating direction method of multipliers (ADMM). We introduce ALDIC formulation in the next section. We then introduce the new proposed adaptive mesh ALDIC algorithm. We demonstrate our algorithm using synthetic mode-I fracture deformation in the Numerical Demonstration section.

7.2 ALDIC Algorithm Formulation

The grayscale function of the image of the undeformed specimen or reference image is denoted as $f(\mathbf{X})$, while the grayscale function of the image of the deformed specimen is denoted $g(\mathbf{y})$. If we assume the deformation convects the grayscale, then we have

$$f(\mathbf{X}) = g(\mathbf{y}(\mathbf{X})) \tag{7.1}$$

The DIC problem is to solve the inverse problem: obtain the mapping $\mathbf{y}(\mathbf{X})$ from $f(\mathbf{X})$ and $g(\mathbf{y})$. This can be posed as a optimization problem, for example, which minimizes the summed square differences:

$$C = \int_{\Omega} |f(\mathbf{X}) - g(\mathbf{y}(\mathbf{X}))|^2 \quad d\mathbf{X} \rightarrow \text{minimize over } \mathbf{y} : \Omega \rightarrow \mathbb{R}^n. \tag{7.2}$$

In Local Subset DIC method, we decompose the whole domain into a finite number of subsets $\Omega = \mathrm{U}_i\Omega_i$, and make the ansatz that the deformation is piecewise affine $\mathbf{y}(\mathbf{X}) = \mathbf{X} + \mathbf{u}(\mathbf{X}) = \mathbf{X} + \Sigma_i (\mathbf{u}_i + \mathbf{F}_i (\mathbf{X} - \mathbf{X}_{i0}))\, \chi_i(\mathbf{X})$, where $\mathbf{u}_i$ is the translation vector and $\mathbf{F}_i$ is the affine deformation gradient tensor of each local subset, $\mathbf{X}_{i0}$ is the coordinates of the center of each local subset, χ_i is the characteristic or index function

$$\chi_i = \begin{cases} 1, & \mathbf{X} \in \Omega_i \\ 0, & \mathbf{X} \notin \Omega_i \end{cases} \tag{7.3}$$

We can solve the affine deformation by minimizing each local subset's correlation function C_i:

$$C_i = \int_{\Omega_i} |f(\mathbf{X}) - g(\mathbf{X} + \mathbf{u}_i + \mathbf{F}_i (\mathbf{X} - \mathbf{X}_{i0}))|^2 d\mathbf{X} \rightarrow \text{minimize over } \mathbf{F}_i, \mathbf{u}_i. \tag{7.4}$$

Since the optimization is conducted within each local subset and the number of unknown parameters is small, this optimization can be solved very fast using Inverse Compositional Gauss-Newton (IC-GN) method [8] or Inverse Compositional Levenberg-Marquadt (IC-LM) method [9].

But in a global picture, we notice that the unknown affine deformation parameters are not independent. They satisfy a global constraint instead: $\mathbf{F} = \mathbf{D}\mathbf{u}$, where $\mathbf{D}$ is the finite difference operator. We embed this global constraint onto the Local Subset DIC correlation function in the augmented Lagrangian form and introduce an auxiliary compatible displacement field $\widehat{\mathbf{u}}$ such that $\widehat{\mathbf{u}}(\mathbf{X}_{i0}) = \mathbf{u}_i$ and $\mathbf{D}\widehat{\mathbf{u}}(\mathbf{X}_{i0}) = \mathbf{F}_i$.

$$\mathcal{L} = \sum_i \int_{\Omega_i} \left(|f(\mathbf{X}) - g(\mathbf{X} + \mathbf{u}_i + \mathbf{F}_i(\mathbf{X} - \mathbf{X}_{i0}))|^2 + \frac{\beta}{2}\left|(\mathbf{D}\widehat{\mathbf{u}})_i - \mathbf{F}_i + \mathbf{W}_i\right|^2 + \frac{\mu}{2}|\widehat{\mathbf{u}}_i - \mathbf{u}_i + \mathbf{v}_i|^2 \right) d\mathbf{X} \tag{7.5}$$

where the coefficients β and μ are fixed, and Lagrangian multipliers $\mathbf{W}_i$ and $\mathbf{v}_i$ are updated during iterations. We use Alternating Direction Method of Multipliers (ADMM) Scheme [10] to decompose above minimization into two subproblems consisting of following iterations:

$$\left\{\mathbf{F}_i^{k+1}\right\}, \left\{\mathbf{u}_i^{k+1}\right\} = \underset{\{\mathbf{F}_i\},\{\mathbf{u}_i\}}{\arg\min} \mathcal{L}\left(\{\mathbf{F}_i\}, \{\mathbf{u}_i\}, \left\{\widehat{\mathbf{u}}_i^k\right\}, \left\{\mathbf{W}_i^k\right\}, \left\{\mathbf{v}_i^k\right\}\right) \quad \text{(Subproblem 1, solved locally)} \tag{7.6}$$

$$\left\{\widehat{\mathbf{u}}_i^{k+1}\right\} = \underset{\{\widehat{\mathbf{u}}_i\}}{\arg\min} \mathcal{L}\left(\left\{\mathbf{F}_i^{k+1}\right\}, \left\{\mathbf{u}_i^{k+1}\right\}, \{\widehat{\mathbf{u}}_i\}, \left\{\mathbf{W}_i^k\right\}, \left\{\mathbf{v}_i^k\right\}\right) \quad \text{(Subproblem 2, solved globally)} \tag{7.7}$$

$$\begin{cases} \mathbf{W}_i^{k+1} = \mathbf{W}_i^{k+1} + \left((\mathbf{D}\widehat{\mathbf{u}})_i^{k+1} - \mathbf{F}_i^{k+1}\right) \\ \mathbf{v}^{k+1} = \mathbf{v}^k + \left(\widehat{\mathbf{u}}^{k+1} - \mathbf{u}^{k+1}\right) \end{cases} \quad \text{(Lagrangian multipliers update)} \tag{7.8}$$

The functional $\mathcal{L}$ is not local because the discrete derivative $(\mathbf{D}\widehat{\mathbf{u}})_i$ depends on the value of $\widehat{\mathbf{u}}_i$ in neighboring subsets. We solve Subproblem 1 (7.6) using local subset method and it can be computed in parallel. We solve Subproblem 2 (7.7) globally. Crucially, this global problem does not involve any grayscale and involves a well-structured Laplace operator. Therefore, it can be solved very fast. In practice, the computational effort of the AL-DIC is at worst a factor two to three as expensive compared to Local Subset DIC, and significantly less expensive than the Global DIC.

7.3 Adaptive Mesh ALDIC Algorithm

Adaptive mesh (or h-adaptive) ALDIC solves the variational problem of (7.2) as loops of the form [4]:

$$Solve \to Estimate \to Mark \to Refine \tag{7.9}$$

- In the ***Solve*** part, we use the above ALDIC algorithm. Since the adaptive mesh can be irregular, we use finite element method to solve (7.7) and nine point Gaussian quadrature to approximate numerical integral.
- In the ***Estimate*** part, we use the same idea of adaptive mesh finite element method, and compute a posterior element error estimator $\varepsilon(T)^2$ as a weighted summation of elements' interior error r and jump error j.

$$\varepsilon(T)^2 = h_T^2 \int_T |r|^2 + h_T \int_{\partial T \backslash \partial\Omega} |j|^2 \tag{7.10}$$

- where T is the union of all the discretized elements with smallest element size h_T, and Ω denotes the whole domain.
- In the ***Mark*** part, we mark the elements with large *a posteriori* errors, which will further be refined in the next part. There are three popular marking strategies proposed in the literature, and we choose the Dorfler Strategy [4] in this paper.
- In the ***Refine*** part, we recursively refine the marked elements. Since this can differ from the integer pixels of the grayscale, it is necessary to interpolate them. Due to the fact that images always have grayscale values at integer pixels, Kuhn triangulation [4] and quad Octree square mesh [11] are two very good choices for the adaptive mesh. We have implemented and evaluated both Kuhn triangulation and quad Octree square mesh in 2D-DIC. In 3D-DVC, quad Octree box mesh is much easier to implement compared with 3D Kuhn triangulation.

7.4 Numerical Demonstration

We have implemented the proposed method using a series of case studies. These examples demonstrate the superior speed of the proposed algorithm and with good accuracy. We highlight it with an example where we use the real DIC experimental

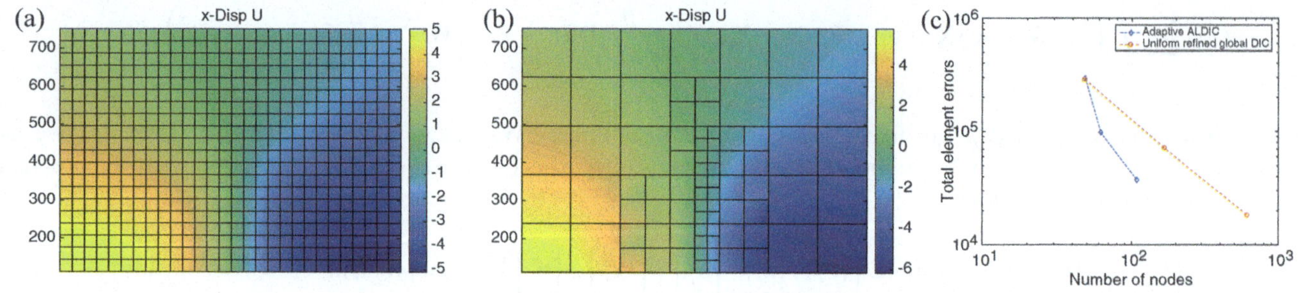

Fig. 7.1 The superior efficiency of the newly proposed ALDIC method. Mode-I fracture displacement obtained using different DIC algorithms. **(a)** Global DIC method. **(b)** Adaptive mesh ALDIC method. **(c)** Comparison of total a posterior error between Global DIC method and our adaptive mesh ALDIC method

Table 7.1 Comparison of computation time cost of Adaptive ALDIC and Global DIC algorithms for synthetic mode-I fracture experiment

Adaptive ALDIC							Global DIC	
Adaptive level	Min subset	Solve		Estimate	Mark	Refine	Element size	Computation cost
		Subproblem 1	Subproblem 2					
1	128 × 128	12.85 s	0.29 s	45.03 s	0.003 s	0.113 s	128 × 128	1634.5 s
2	64 × 64	27.40 s	0.28 s	51.41 s	0.162 s	0.229 s	64 × 64	3345.3 s
3	32 × 32	131.14 s	0.54 s	46.83 s	0.003 s	0.139 s	32 × 32	1651.2 s
Total	316.4 s							

image as the reference image, and synthesize deformed images using mode-I fracture deformation using Max Williams' exact solution

$$\begin{cases} u = \frac{K_I}{2E}\sqrt{\frac{r}{2\pi}}\,(1+\nu)\left[(2k-1)\cos\frac{\theta}{2} - \cos\frac{3\theta}{2}\right] \\ v = \frac{K_I}{2E}\sqrt{\frac{r}{2\pi}}\,(1+\nu)\left[(2k-1)\sin\frac{\theta}{2} - \sin\frac{3\theta}{2}\right] \end{cases} \tag{7.11}$$

where $E = 70$ GPa, $\nu = 0.3$, $K_I = 15$ GPa $\sqrt{\text{pixel}}$ are setting properties of the fracture deformation.

We start from uniform finite element mesh (element size: 128 × 128 pixels) and adaptively refine elements where *a posteriori* error is large. All the simulations are done using MATLAB on a personal laptop with 2.6 GHz Inter Core i5 Processor, 8GB 1600 MHz DDR3 Memory. Figure 7.1a plots the displacement field with the uniform finite element of size 32 × 32 pixels, shows the Fig. 7.1b shows the solution where the smallest finite element size is also 32 × 32 pixels. The displacement fields agree, but the computation time of Adaptive ALDIC is only 20% of the Global DIC. Figure 7.1c shows both in the Adaptive ALDIC and Global DIC, with more number of nodes, the results total error will decrease and the error of Adaptive ALDIC decreases faster than Global DIC. When we have the same number of nodes, Adaptive ALDIC has less total error compared with Global DIC. We compare the computation time of the new proposed adaptive mesh ALDIC method and uniform mesh Global DIC in Table 7.1, where we can find this new adaptive ALDIC method decreases computation time significantly compared with current Global DIC algorithm.

7.5 Conclusions

In this paper, we report a new Global DIC algorithm that uses an adaptive mesh. It builds on our recent work on the augmented Lagrangian digital image correlation (ALDIC) [12]. We solve the resulting problem using the alternating direction method of multipliers (ADMM). Compared with current Global DIC algorithm, this new adaptive Global DIC algorithm decreases computation time to 20% or even one order of magnitude with no loss (and some gain) in accuracy. Besides 2D-DIC, the early results of the application of Adaptive ALDIC to 3D-DVC are encouraging [13].

Acknowledgement We gratefully acknowledge the support of the US Air Force Office of Scientific Research through the MURI grant 'Managing the Mosaic of Microstructure' (FA9550-12-1-0458).

References

1. Sutton, M.A., et al.: Image Correlation for Shape, Motion and Deformation Measurements: Basic Concepts, Theory and Applications (2009)
2. Pan, B., Qian, K., Xie, H., et al.: Two-dimensional digital image correlation for in-plane displacement and strain measurement: a review. Meas. Sci. Technol. **20**(6), 062001 (2009)
3. Besnard, G., Leclerc, H., Hild, F., et al.: Analysis of image series through global digital image correlation. J. Strain Anal. Eng. Design. **47**(4), 214–228 (2012)
4. Nochetto, R.H., Siebert, K.G., Veeser, A.: Theory of adaptive finite element methods: an introduction. In: Multiscale, nonlinear and adaptive approximation, pp. 409–542. Springer, Berlin (2009)
5. Yuan, Y., Huang, J., Fang, J., et al.: A self-adaptive sampling digital image correlation algorithm for accurate displacement measurement. Opt. Lasers Eng. **65**, 57–63 (2015)
6. Hild, F., Roux, S.: Digital image correlation. Wiley-VCH, Weinheim (2012)
7. Wittevrongel, L., et al.: A self adaptive global digital image correlation algorithm. Exp. Mech. **55**(2), 361–378 (2015)
8. Baker, S., et al.: Lucas-Kanade 20 years on: a unifying framework. Int. J. Comput. Vis. **56**(3), 221–255 (2004)
9. Henn, S.: A Levenberg–Marquardt scheme for nonlinear image registration. BIT Numer. Math. **43**(4), 743–759 (2003)
10. Boyd, S., Parikh, N., Chu, E., et al.: Distributed optimization and statistical learning via the alternating direction method of multipliers. Mach. Learn. **3**(1), 1–122 (2010)
11. Haber, E., Heldmann, S., Modersitzki, J.: Adaptive mesh refinement for nonparametric image registration. SIAM J. Sci. Comput. **30**(6), 3012–3027 (2008)
12. Yang, J., Bhattacharya, K.: Augmented Lagrangian DIC. Submitted to Experimental Mechanics
13. Yang, J., Bhattacharya, K.: Fast Adaptive Global Digital Image Correlation. In preparation

Chapter 8
Speckle Image Rendering for DIC Performance Assessment

F. Sur, B. Blaysat, and M. Grédiac

Extended Abstract Assessing the metrological performance of the methods available for displacement field measurement is of primary importance in experimental mechanics. It often relies on synthetically rendered images of a specimen before and after deformation. In this context, the only varying quantity should be the displacement field in order to prevent any other phenomenon to bias the estimation of the metrological performance. Concerning digital image correlation (DIC), images of a random speckle pattern are needed. We propose a new algorithm to render such images with a classic model of stochastic geometry, namely the Boolean model.

Keywords DIC · Random speckle rendering · Boolean model

8.1 Introduction

The most popular methods for estimating the displacement field on the surface of a specimen subjected to a mechanical load are certainly the ones based on digital image correlation (DIC) applied on speckle images acquired before and after deformation. A crucial question, which numerous papers aim at addressing (see for instance [1]) is the characterization of the metrological performance of these methods. Using synthetically rendered speckle images allows comparing the estimated displacement field with the ground truth used for data generation. However, the resolution of the displacement field returned by any DIC software program is affected by both the DIC program itself and the method used for rendering synthetic data. We propose a new rendering algorithm which mimics the acquisition of an image with a digital sensor, from physical deformation of the specimen to smoothing caused by the lens point spread function (PSF), spatial discretization, and gray-level quantization. In particular, it does not require any interpolation scheme likely to bias the accuracy estimation, as discussed in [1]. The aim is to complement datasets as those given in [2] which are provided by software programs such as TexGen [3]. Datasets and Matlab codes are freely provided to the community [4], and more details are available in a companion paper [5].

8.2 Rendering Speckle Images

The proposed algorithm is based on the Boolean model, a classic model of stochastic geometry having applications ranging from material analysis [6] to film grain modeling [7]. In the Boolean model, a random set $\mathbf{P}$ is defined as the union of a set of N random disks $\mathbf{D}_i$:

$$\mathbf{P} = \cup_{1 \leq i \leq N} \mathbf{D}_i \tag{8.1}$$

F. Sur (✉)
Université de Lorraine, CNRS, INRIA projet Magrit, Vandoeuvre-lès-Nancy Cedex, France
e-mail: frederic.sur@loria.fr

B. Blaysat · M. Grédiac
Université Clermont Auvergne, SIGMA, Institut Pascal, UMR CNRS 6602, Clermont-Ferrand, France

L. Lamberti et al. (eds.), *Advancement of Optical Methods & Digital Image Correlation in Experimental Mechanics, Volume 3*, Conference Proceedings of the Society for Experimental Mechanics Series, https://doi.org/10.1007/978-3-319-97481-1_8

where the $\mathbf{D}_i$'s have random radii. These disks are spatially distributed following a Poisson point process. This approach models the formation of real speckle patterns in the sense that the surface of a tested specimen is often prepared by spraying ink, the disks representing randomly distributed ink droplets.

An infinite-resolution image $\mathcal{I}_0$ of the reference state is obtained by considering the indicator function of the random set $\mathbf{P}$, which takes value 1 at $\boldsymbol{\xi}$ on the surface under investigation if $\boldsymbol{\xi} \in \mathbf{P}$, and 0 otherwise.

A ground truth deformation map $\boldsymbol{\phi}$ being given, an infinite-resolution deformed image is given by:

$$\mathcal{I}'_0(\boldsymbol{\xi}) = \mathcal{I}_0(\boldsymbol{\phi}(\boldsymbol{\xi})). \tag{8.2}$$

A gray-level digital image is eventually rendered by quantizing over b bits (typically, $b = 8$ or 12 bits), at any discrete pixel $\boldsymbol{x}$, the value of the following integral:

$$I(\boldsymbol{x}) = \gamma \iint G_\sigma(\boldsymbol{x} - \boldsymbol{\xi})\, \widetilde{\mathcal{I}_0}(\boldsymbol{\xi})\, \mathrm{d}\boldsymbol{\xi} \tag{8.3}$$

where $\widetilde{\mathcal{I}_0}(\boldsymbol{\xi}) = \mathcal{I}_0(\boldsymbol{\xi})$ (reference image) or $\widetilde{\mathcal{I}_0}(\boldsymbol{\xi}) = \mathcal{I}'_0(\boldsymbol{\xi}) = \mathcal{I}_0(\boldsymbol{\phi}(\boldsymbol{\xi}))$ (deformed image), $0 < \gamma \leq 1$ is the contrast, and G_σ is the PSF of the optical device (here, an isotropic Gaussian function of variance σ^2, with $\sigma \simeq 1$ pixel).

Since no-closed form solution of this integral is available, we make use of Monte Carlo integration. We have proved in [5] that, if the size N_{MC} of the Monte Carlo sample satisfies:

$$N_{\text{MC}} \geq \frac{2^{2b-1}\gamma^2}{\pi\alpha^2} \tag{8.4}$$

then α is an upper bound of the quantization error probability in the rendered image (a finer bound is actually available in [5]). This proves that the proposed algorithm gives an exact gray-level value at any pixel, except for a proportion α of them. A typical value is $\alpha = 5\%$. In practice, the quantization error is always less than one gray value at these pixels. Such an error is well below the noise floor of any digital sensor.

8.3 Results

The interested reader is kindly referred to [5] for additional information on calculation time, computational speed-ups, rounding errors caused by floating point arithmetic, and a discussion of aliasing. Figure 8.1 shows typical speckle images rendered by the proposed algorithm.

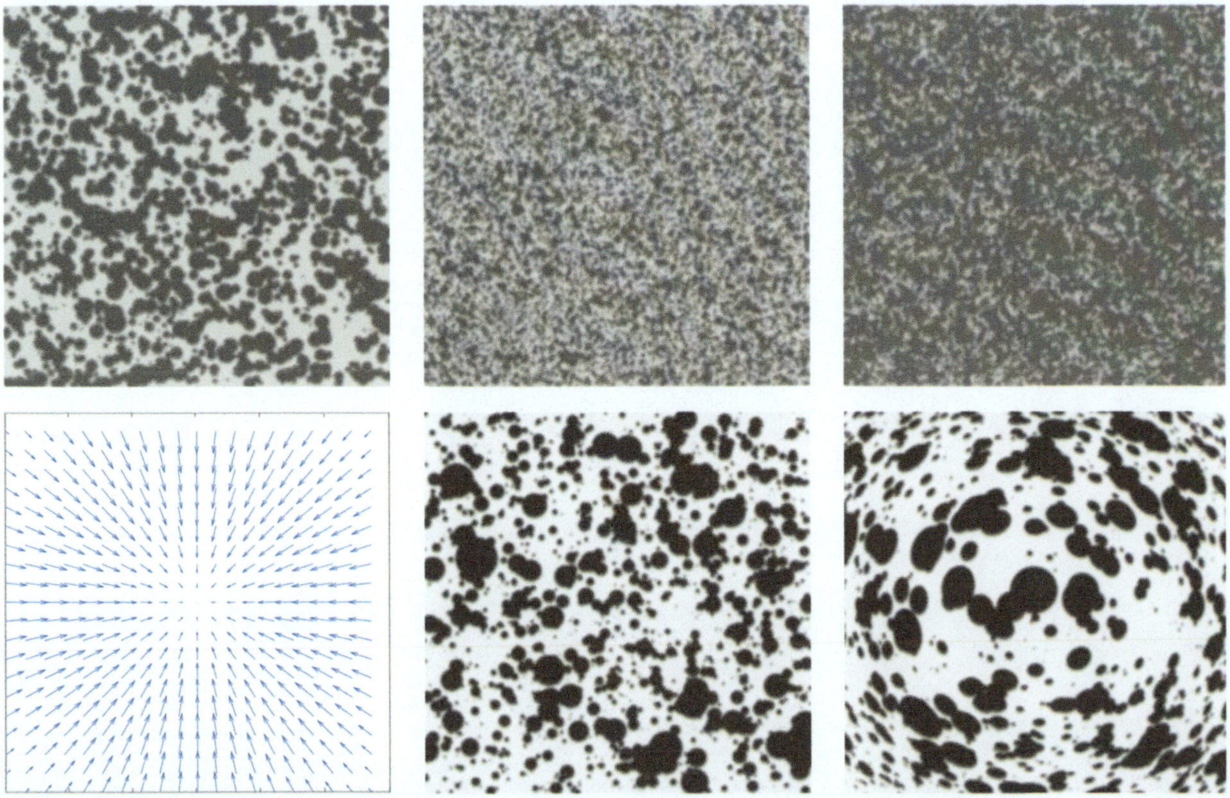

Fig. 8.1 Top: Speckle images rendered with the proposed algorithm. Bottom, from left to right: inverse displacement field (here, the "punch" function of [3]), reference (non-deformed), and deformed speckle images. A non-realistic displacement field of large amplitude is deliberately used here to facilitate the visualization of both rendered images

References

1. Bornert, M., Doumalin, P., Dupré, J.C., Poilâne, C., Robert, L., Toussaint, E., Wattrisse, B.: Shortcut in DIC error assessment induced by image interpolation used for subpixel shifting. Opt. Lasers Eng. **91**, 124–133 (2017)
2. Society for Experimental Mechanics: DIC challenge. http://sem.org/dic-challenge/
3. Orteu, J.J., Garcia, D., Robert, L., Bugarin, F.: A speckle texture image generator. Proc. SPIE **6341**, 63410H (2006)
4. BSpeckleRender: (2018). https://members.loria.fr/FSur/software/BSpeckleRender/
5. Sur, F., Blaysat, B., Grédiac, M.: Rendering deformed speckle images with a Boolean model. J. Math. Imaging Vis. **60**, 634–650 (2018)
6. Serra, J.: The Boolean model and random sets. Comput. Graphics Image Process. **12**(2), 99–126 (1980)
7. Newson, A., Delon, J., Galerne, B.: A stochastic film grain model for resolution-independent rendering. Comput. Graphics Forum **36**(8), 684–699 (2017)

Chapter 9
Speckles and DIC or Checkerboards and LSA?

M. Grédiac, B. Blaysat, and F. Sur

Abstract Two measurement techniques, namely DIC and LSA, are compared in this study. Both techniques consist in minimizing the optical residual. DIC performs the minimization in the real domain on speckles. LSA performs this minimization in the Fourier domain on periodic patterns such as checkerboards. It is shown here that for the same systematic error and the same spatial resolution, the noise level in displacement and strain maps is lower with LSA than with DIC.

Keywords Bias · Checkerboard · DIC · LSA · Random error · Speckle optimization

9.1 Extended Abstract

The principle of 2D subset-based DIC consists in finding the displacement field that occurs on the surface of a deformed specimen by minimizing the optical residual constructed over subsets Ω defined in the reference and current images. This residual writes as follows in its simplest and continuous form:

$$\iint_{(x,y)\in\Omega} \left(s^{ref}(x, y) - s^{cur}\left(x + u_x, y + u_y\right)\right)^2 dxdx \tag{9.1}$$

With DIC, this minimization is performed iteratively with respect to the sought quantities ux and uy. In [1], it is shown that under mild assumptions, this minimization has a unique and straightforward solution if the pattern is periodic. Indeed, the minimum of this residual is characterized by the fact that the exact solution for ux and uy is proportional to the phase change between reference and current images of this periodic pattern. Thus,

$$u_i = -\frac{p}{2\pi}\left(\Phi_i^{cur} - \Phi_i^{ref}\right), \quad i \in \{x, y\} \tag{9.2}$$

where p is the pitch of the periodic pattern. Φ_i^{cur} and Φ_i^{ref} are the phases of the current and reference images, respectively. Finding this phase change, and thus the displacement at the sub-pixel scale, can be made by using one of the Fourier-based methods available to process periodic patterns such as the Localized Spectrum Analysis (LSA) generally employed to extract displacement fields from grid images [2].

This link between DIC and LSA is interesting when considering the problem of the definition and use of optimal patterns for DIC. Indeed, as discussed in recent papers, [3] for instance, the optimal pattern for DIC in terms of sensor noise propagation to the displacement and strain maps is the checkerboard (if the images are suitably sampled) because this patterns maximizes the sum of the square of the subset intensity gradient (SSSIG). Indeed, this quantity directly governs the random error impairing the results provided by DIC [4]. In general, such a pattern cannot be successfully processed by DIC because of its periodicity, which induces convergence issues. This leads to deviate from this optimal pattern by considering

M. Grédiac (✉) · B. Blaysat
Université Clermont Auvergne, SIGMA, Institut Pascal, UMR CNRS 6602, Clermont-Ferrand, France
e-mail: michel.grediac@uca.fr

F. Sur
Université de Lorraine, CNRS, INRIA projet Magrit, Vandoeuvre-lès-Nancy Cedex, France

L. Lamberti et al. (eds.), *Advancement of Optical Methods & Digital Image Correlation in Experimental Mechanics, Volume 3*,
Conference Proceedings of the Society for Experimental Mechanics Series, https://doi.org/10.1007/978-3-319-97481-1_9

random ones which account for additional constraints specific to DIC, and thus to deliberately process non-optimal patterns in terms of sensor noise propagation, [3] for instance. The objective of the present study work is to show that the periodicity of checkerboard patterns is not a drawback, but a feature that enables us to process such patterns efficiently with LSA.

As other similar methods based on Fourier analysis, LSA has been mainly used so far to process grid images and not checkerboards [2], so it is first necessary to examine how to process checkerboards images instead of grid images with this technique. Two methods are investigated in this study. The first one consists in pre-processing checkerboard images in such a way that they are transformed into two perpendicular 1D grids, which are then processed individually by LSA to obtain phase fields along both the x- and y-directions. The sought phases encoded in the checkerboard image are unaffected by this pre-processing step while the modulus of the windowed Fourier transform used in LSA is maximized, which positively impacts the noise level in final displacement and strain maps since sensor noise in the final maps is inversely proportional to this modulus [2]. The second method consists in considering that a checkerboard is made of interlocked rows of diamond-shaped patterns along the $\pm 45°$ directions. Such a periodic pattern can be successfully processed by LSA, and the results show that both techniques provide displacement and strain maps affected by the same noise level. The second technique is however faster since no pre-processing step is needed, so only this technique is used below for comparison purposes with classic DIC.

LSA applied on checkerboard images and DIC applied on random speckles were compared. An intermediate situation is to process checkerboard images with DIC, which can be made by carefully choosing the initial values employed to initiate the iterative minimization process performed by DIC, and therefore avoid convergence issues classically observed with this periodic pattern when it is processed by DIC. This last case provides the lower bound for the noise level that can be observed in displacement and strain maps obtained by DIC because the SSSIG is maximum in this case.

The noise level observed in results provided by DIC and LSA depends on the size of the subset for DIC, as well as on the size of the Gaussian window used in LSA since the latter acts like the subset in DIC [2]. Indeed, the higher these quantities, the lower the noise observed with each technique. Since they are set separately for each technique and since their apparent sizes are not directly comparable (a Gaussian distribution has no compact support), special care must be taken to perform a fair comparison between results obtained with DIC and LSA. Two other parameters were therefore considered and kept constant in this comparison, namely the bias and the spatial resolution. The first one is the systematic error due the use of polynomial matching functions in DIC. By performing DIC pixelwise, it can be shown that this bias is equivalent to the bias observed when convolving the real and unknown displacement by a Savitzky-Golay filter [5]. This bias is therefore directly comparable with the one observed with LSA since the latter is also estimated with a convolution by a kernel, which is in this case directly the Gaussian window used in LSA [6]. The spatial resolution is as defined and used in [1]. Fixing a subset size for DIC and a Gaussian window for LSA in such a way that the same bias and the same spatial resolution are obtained for both techniques is therefore possible, and the resulting noise levels observed in displacement and strain maps are then fairly comparable. Other sources of systematic errors, such as that due to interpolation, were not considered here because they are specific to DIC and do not really concern LSA.

Experiments were carried out on real checkerboard patterns and on classic random patterns subjected to translation tests. Obtained results show that the noise level in displacement and strain maps obtained at iso-bias and iso-spatial resolution is lower with LSA than with classic subset-based DIC, thus experimentally confirming that checkerboard patterns are optimal in terms of sensor noise propagation. In addition, calculations are much faster with LSA than with classic DIC because they are performed in the Fourier domain thanks to the fast Fourier transform. Full details can be found in [7].

References

1. Grédiac, M., Blaysat, B., Sur, F.: A critical comparison of some metrological parameters characterizing local digital image correlation and grid method. Exp. Mech. **57**(3), 871–903 (2017)
2. Grédiac, M., Sur, F., Blaysat, B.: The grid method for in-plane displacement and strain measurement: a review and analysis. Strain. **52**(3), 205–243 (2016)
3. Bomarito, G.F., Hochhalter, J.D., Ruggles, T.J., Cannon, A.H.: Increasing accuracy and precision of digital image correlation through pattern optimization. Opt. Lasers Eng. **91**, 73–85 (2017)
4. Pan, B., Xie, H., Wang, Z., Qian, K., Wang, Z.: Study of subset size selection in digital image correlation for speckle patterns. Opt. Express. **16**(10), 7078 (2008)
5. Schreier, H.W., Sutton, M., Michael, A.: Systematic errors in digital image correlation due to undermatched subset shape functions. Exp. Mech. **42**(3), 303–310 (2002)
6. Sur, F., Grédiac, M.: Towards deconvolution to enhance the grid method for in-plane strain measurement. Inv Prob Imaging. **8**(1), 259–291 (2014)
7. Grédiac, M., Blaysat, B., Sur, F.: Beyond speckles and DIC: checkerboards and LSA. (2017)

Chapter 10
Update on the 2D-DIC Challenge: Results and Conclusions

P. L. Reu, E. Toussaint, E. Jones, H. Bruck, M. Iadicola, R. Balcaen, D. Turner, T. Siebert, P. Lava, M. Simonsen, and M. Grewer

Abstract The 2D-DIC Challenge is organized by an international committee working to understand the accuracy of digital image correlation (DIC) through standardized image sets. The DIC Challenge is run under the auspices of the Society for Experimental Mechanics (SEM) and the International DIC Society (iDICs). The 2D-Challenge incorporates 19 image sets that can be used in evaluating 2D-DIC algorithms. The full results of the study and description of the image sets may be found in Reu et al. (Exp Mech, 2017). A new round of the 2D Challenge is being launched at SEM 2018 and will seek to probe the concept of spatial resolution.

Keywords Digital image correlation (DIC) · Full-field · Optical methods · Uncertainty quantification (UQ) · Aliasing

10.1 Introduction

The 2D-DIC challenge seeks to provide images that may be used for comparing DIC algorithms and determining the ultimate resolution of DIC. To this end, 19 image sets were developed that challenge various aspects of DIC including contrast change, image noise, image rotation, etc. The images listed in Table 10.1 are all freely available at the SEM website, and researchers are encouraged to use these images in publications, particularly when improvements in DIC algorithms are claimed. The images include synthetically generated images with known answers that can be used for DIC code verification, i.e. ensuring that the DIC code is correctly implemented and for probing the ultimate error regime of DIC. Experimental images with known displacements are also available and can be used for code validation (Sample 16) or for comparing different DIC implementations (Sample 12, 13, and 17). These images have been widely vetted and a detailed discussion of their generation and use are described in [1].

P. L. Reu (✉) · E. Jones · D. Turner
Sandia National Laboratories, Albuquerque, NM, USA
e-mail: plreu@sandia.gov

E. Toussaint
Université Clermont Auvergne, Clermont-Ferrand, France

H. Bruck
University of Maryland, College Park, MD, USA

M. Iadicola
National Institute of Standards, Gaithersburg, MD, USA

R. Balcaen
KU Leuven, Ghent, Belgium

T. Siebert
Dantec Dynamics GmbH, Ulm, Germany

P. Lava
MatchID, Ghent, Belgium

M. Simonsen
Correlated Solutions Inc., Columbia, SC, USA

M. Grewer
LaVision GmbH, Goettingen, Germany

L. Lamberti et al. (eds.), *Advancement of Optical Methods & Digital Image Correlation in Experimental Mechanics, Volume 3*, Conference Proceedings of the Society for Experimental Mechanics Series, https://doi.org/10.1007/978-3-319-97481-1_10

Table 10.1 Sample image set numbering and information

Description	Set name	Method [2, 3]	Contrast	Noise σ (GL)	Shift (pixels)	# Images
TexGen Shift X,Y	Sample 1	TexGen	Varying	1.5	$X = Y = 0.05$	20
TexGen Shift X,Y	Sample 2	TexGen	0 to 50	8	$X = Y = 0.05$	20
FFT Shift X,Y	Sample 3	FFT Shift	0 to 200	1.5	$X = Y = 0.1$	10
FFT Step Shift	Sample 3b	FFT Shift	0 to 200	1.5	0.05 to 0.5	5
FFT Shift X,Y	Sample 4	FFT Shift	0 to 50	8	$X = Y = 0.1$	10
FFT Shift X,Y	Sample 5	FFT Shift	Varying	1.5	$X = Y = 0.1$	10
Prosilica Bin	Sample 6	Binning	0 to 200	0.26	$X = Y = 0.1$	10
Prosilica Bin	Sample 7	Binning	0 to 50	0.66	$X = Y = 0.1$	10
Rotation TexGen	Sample 8	TexGen	0 to 100	2	Θ by 1	10
Rotation FFT	Sample 9	FFT	0 to 100	2	Θ by 1	10
Strain Gradient	Sample 10	TexGen	0 to 200	2	Sinusoid	10
Strain Gradient	Sample 11	TexGen	60 to 130	2	Sinusoid	10
Strain Gradient	Sample 11b	FFT	0 to 200	1.5	Tri. .01 to 1	6
Ex1—Plate Hole	Sample 12	Exper.	Good	Low	N/A	12
Ex2—Weld	Sample 13	Exper.	Poor	Low	N/A	52
Varying Strain	Sample 14	FFT	0 to 200	5	N/A	4
Varying Strain	Sample 15	TexGen	80 to 180	2	N/A	9
Rigid Motion	Sample 16	Exper.	0 to 254	0.26	≈ 0.1	11
Interpolant	Sample 17	Exper.				

Image sets are available at: https://sem.org/dic-challenge/

10.2 Results

Two of the sample sets (Sample 14 and 15) had DIC results supplied from 9 participants with 12 different code implementations including both global and local implementations. Both samples supplied a realistic speckle image with known displacement and strain gradients that the participants were to determine. Only one result could be submitted, and then the bias and noise errors of the results were investigated. This required a compromise between spatial resolution and noise when doing the analysis. The "error" for all 12 codes plotted in Fig. 10.1 is the difference between the submitted result and the known answer for Sample 14. Both bias and noise errors were observed. It is interesting to note that some codes (Code A) applied more filtering, but had lower noise, while other codes (Code E) had less filtering and bias error, but more noise in the results. These results illustrate the fundamental compromise between noise and spatial resolution.

10.3 Conclusions

The complete results of the challenge are available in a journal article already published [1], but in brief, nearly all the codes supplied excellent results. The largest difference amongst the codes was in the trade-off between spatial resolution and noise. Unfortunately, the current challenge images did not have a high enough spatial gradient to truly challenge the DIC codes. To remedy this, a second 2D challenge has been initiated with images more suited to determination of spatial resolution [4]. New images will be available at www.sem.org. Researchers are strongly encouraged to use these images for better understanding of their commercial codes, improvements in their own DIC codes, and for publications that claim improvements in DIC algorithms. The current state of journal publications makes it impossible for readers to verify improvement claims made in journal articles. Instead, using the DIC Challenge images as a benchmark allows a clear path to an improved understanding of 2D DIC.

10.4 Future Work

The DIC Challenge format is continuing forward by developing images for Stereo-DIC, Scanning Electron Microscope DIC, and Digital Volume Correlation. Interested participants should contact the author for inclusion in future challenge rounds.

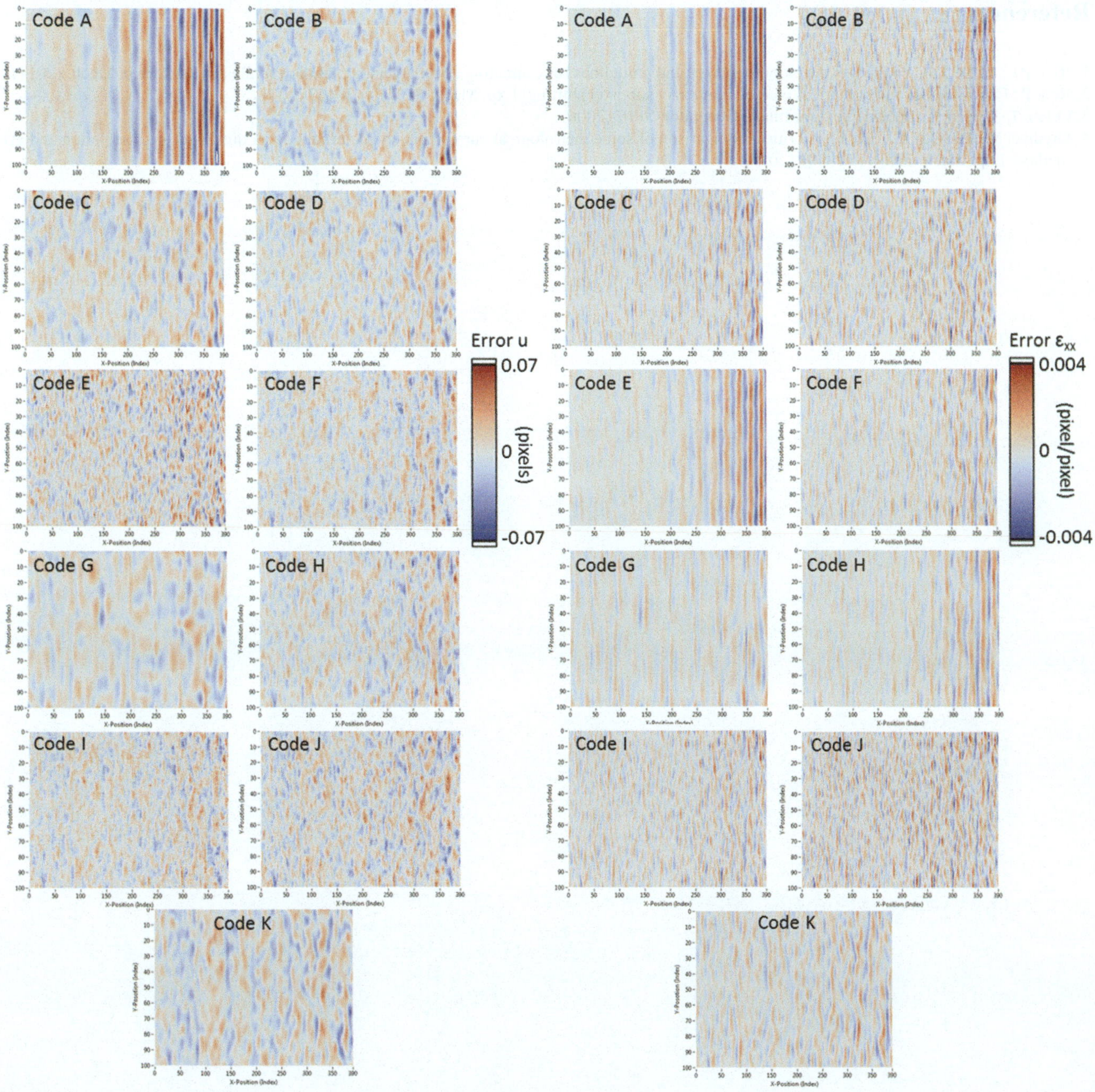

Fig. 10.1 Full-field error plots for all 12 codes. Displacement (left) and strain (right) with all scales identical [1]

Acknowledgements I would like to thank the DIC Challenge board for their help in completing this work, as well as the DIC Challenge participants. Without broad collaboration, making progress on these important DIC measurement problems would be difficult or impossible.

Sandia National Laboratories is a multimission laboratory managed and operated by National Technology and Engineering Solutions of Sandia, LLC., a wholly owned subsidiary of Honeywell International, Inc., for the U.S. Department of Energy's National Nuclear Security Administration under contract DE-NA-0003525.

References

1. Reu, P.L., et al.: DIC challenge: developing images and guidelines for evaluating accuracy and resolution of 2D analyses. Exp. Mech. (2017)
2. Reu, P.: Experimental and numerical methods for exact subpixel shifting. Exp. Mech. **51**(4), 443–452 (2011)
3. Orteu, J.-J., et al.: A Speckle Texture Image Generator. SPIE (2006)
4. Grediac, M., Blaysat, B., Sur, F.: A critical comparison of some metrological parameters characterizing local digital image correlation and grid method. Exp. Mech. **57**(6), 871–903 (2017)

Chapter 11
Eliminating Air Refraction Issues in DIC by Conducting Experiments in Vacuum

P. L. Reu and E. M. C. Jones

Abstract A major and often unrecognized error source in digital image correlation (DIC) is the influence of the intervening air between the cameras and sample. Minute differences in air temperature, composition, or both can cause index of refraction changes that act as a lens and cause distortions in the DIC displacement and strain results (Jones and Reu, Exp Mech, 2017). There are limited options to correct this problem as it is both spatial and temporal in nature. One method is to use X-rays for imaging that are not affected by air refraction, but this requires costly equipment. A second method uses a vacuum chamber to minimize the intervening air to remove the distortions, but unfortunately this requires inconvenient setups.

Keywords Digital image correlation (DIC) · Full-field · Optical methods · Uncertainty quantification (UQ)

11.1 Introduction

Stereo-DIC measurements are subject to many error sources that can create either bias errors or variance errors in the results. There is some discussion as to what the largest remaining error sources are for a typical DIC setup. Aliasing, heat waves and camera motion are all possibilities. Of course, this will be very setup dependent. This study looks at index of refraction changes caused by heat waves that are an often-ignored source of temporal and spatial errors [1]. To help elucidate their effects on DIC, an experiment was setup inside a vacuum chamber with a stationary plate to determine a possible lower bound on the noise floor.

11.2 Experimental Setup

To determine the influence of the air on the DIC results we used a stereo-rig observing a stationary speckle pattern with a field-of-view of approximately 100-mm (See Table 11.1 for setup details and Fig. 11.1 for a photo of the setup inside the vacuum chamber). The entire system was placed in a large vacuum chamber that was pumped down to pressures between 3 Torr and atmosphere. The minimum pressure was dictated by the chamber size and leak rate. After shutting off the pumps to minimize vibrations, we acquired images at 5 fps for 3 min at a variety of pressures. Images were also acquired at 140 Hz to look at the frequency response of the system. To better illustrate the influence of heat waves, a hotplate was placed under the speckle pattern at a temperature of 60 °C, which simulates the temperature of LED lights. Note that this position of the heat source is a worst-case scenario. Other tests positioned the hotplate either behind the cameras to better simulate a typical experimental lighting setup, or with no hot plate for a best-case scenario.

P. L. Reu (✉) · E. Jones
Sandia National Laboratories, Albuquerque, NM, USA
e-mail: plreu@sandia.gov

L. Lamberti et al. (eds.), *Advancement of Optical Methods & Digital Image Correlation in Experimental Mechanics, Volume 3*,
Conference Proceedings of the Society for Experimental Mechanics Series, https://doi.org/10.1007/978-3-319-97481-1_11

Table 11.1 Stereo-rig setup parameters

Cameras	FLIR 2.3 mega-pixel
Lens	Edmund optics 35-mm
Software	Vic3D 7.2.6
Image scale	≈20 pixel/mm
Stereo-angle	22.7°
Correlation	ZNSSD
Subset	35 × 35 pixels
Step	15 × 15 pixels
Strain window	15 × 15 data points
Shape function	Affine
Normalization	ZNSSD
Pre-processing	Low pass filter

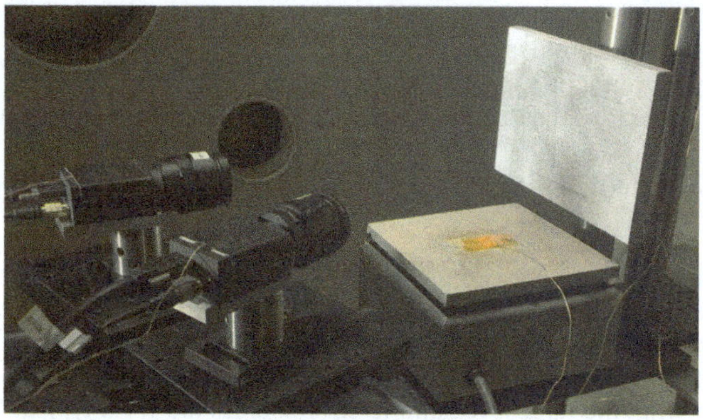

Fig. 11.1 Setup photo of cameras inside the large vacuum chamber. A hotplate was placed immediately under the target to create heat waves

11.3 Results

The stereo-images were analyzed using DIC software with the settings in Table 11.1. Approximately 900 images were analyzed (5 fps for 3 min) to look at both the spatial distribution of noise and the temporal distribution of noise. We noted that after removal of the air from the chamber, other error sources started to become important. For example, sample or camera motion, on the order of 200 nm and small temperature changes of 2 °C were measurable, particularly for the setup with no hotplate. We attributed the improved measurement results, even at atmospheric temperature, to the fact that the lights and other heat sources, along with disturbances from building HVAC systems were isolated outside the measurement chamber. This gave the system, even at atmospheric pressure, a better noise floor than is typically observed in a standard laboratory setting. At this point it becomes difficult to separate the different error sources. For simplicity, only the horizontal displacement (U) from the experiment with the heat source are shown. Temporal errors are measured by taking the standard deviation of the center subset through the 3-min record time, approximately 900 image pairs. Spatial errors are the standard deviation of the U-displacement at frame 125 at 25 s after the start of the experiment. Figure 11.2 shows the results at vacuum on the left and atmospheric pressure on the right. An order of magnitude improvement is seen in the error. Figure 11.3 shows the steady improvement of the displacement errors, both spatial and temporal, as the air is removed. Plots are in engineering units, but the corresponding errors in terms of pixels are 0.046 pixels at atmosphere and 0.0032 pixels in vacuum. The atmospheric noise floor is a little larger than the typical rule of thumb of 0.01 pixels, mainly due to the positioning of the hotplate immediately below the sample. However, in vacuum, the value is now approaching the theoretical noise floor for DIC—even in a very poor experimental setup with a heat source immediately below the sample. Results for the hotplate behind the cameras and without the hotplate are similar in trends, but typically with lower errors. However, once below 200 Torr, vibrational motion of either the camera or the sample corrupted the results. To improve on these results, a floating vacuum chamber will likely be required as we are approaching the resolution of interferometry.

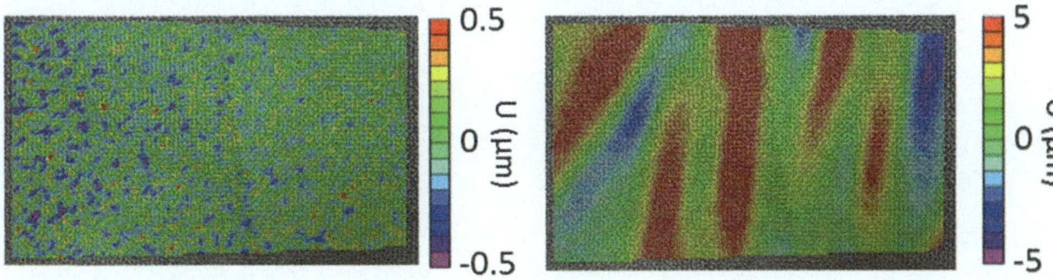

Fig. 11.2 Noise floor characteristics between (left) vacuum measurement with heat source and (right) atmospheric with heat source both for frame 125. Notice an order of magnitude improvement in vacuum

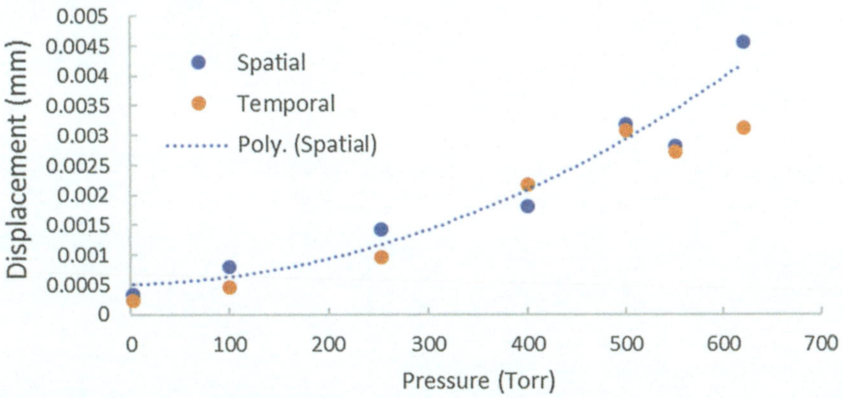

Fig. 11.3 Spatial and temporal strain resolution at various pressures

11.4 Conclusions

The results clearly show that index of refraction changes in the air can be an important error source for DIC. Removal of this effect via a vacuum chamber is effective, but with a more complicated experimental setup. It will be important to judge whether this improvement is important for any given experiment. The results show that typical rules of thumb of 0.01 pixels or 50 $\mu\varepsilon$ errors in DIC results are more influenced by the air than the fundamentals of the DIC algorithms. Placing the cameras outside the chamber unfortunately did not solve the problem. Even the small amount of air within the lens and between the lens and the viewport was enough to cause measurable index of refraction errors. This helps explain why simulated image errors are typically an order of magnitude lower than experimental errors and are extremely difficult to replicate.

Acknowledgements We would like to thank Rusty Spillers, Seth Spitzer and Paul Farias for help in setting up and conducting these experiments.

Sandia National Laboratories is a multimission laboratory managed and operated by National Technology and Engineering Solutions of Sandia, LLC., a wholly owned subsidiary of Honeywell International, Inc., for the U.S. Department of Energy's National Nuclear Security Administration under contract DE-NA-0003525.

Reference

1. Jones, E., Reu. P.: Distortion of Digital Image Correlation (DIC) Displacements and strains from heat waves. Experimental Mechanics: 1–24 (2017)

Chapter 12
Identification of Deformation Mechanisms in Biomaterials Through AFM and Digital Image Correlation

Horacio D. Espinosa

Abstract Most biological composite materials achieve higher toughness without sacrificing stiffness and strength. Interrogating how Nature employs these strategies and decoding the structure-function relationship of these materials is a challenging task that requires knowledge about the actual loading and environmental conditions of the material in their natural habitat, as well as a complete characterization of their constituents and hierarchical ultrastructure. In this work, we present an experimental framework that combines in situ and ex situ fracture testing with digital image correlation to allow the identification and quantification of toughening mechanisms involved during fracture of natural systems. We present this methodology in two case studies: (1) pangolin scales, and (2) nacre from seashells. We envision that the outcome of this research will pave the way for more bio-inspired design systems that can subsequently shed light on how Nature has evolved materials to optimize mechanical properties.

Keywords Biomaterials · Bioinspired materials · Digital image correlation · Atomic force microscope · Nacre · Pangolin

12.1 Introduction

There is a strong demand for new paradigms in the design and development of advanced structural materials to enable the progress of high-performance applications where materials that are capable of fulfilling combinations of conflicting properties are required [1–3]. For instance, some applications require materials that are stiff and strong for structural purposes and that exhibit high toughness to avoid catastrophic failure under extreme events. Unfortunately, as shown in the red dashed line on Fig. 12.1 most engineering materials have an inverse relation between these mechanical properties [1, 3–5]. In contrast, the blue dash-dot line represents the map of the same properties for biological materials. Although the order of magnitude of the stiffness and toughness are lower than those observed from engineering materials, the trade-off between these competing properties is different in biological materials. In this regard, Nature has evolved efficient strategies to achieve remarkable combinations of conflicting properties from relative weak constituents [6–8].

The outstanding mechanical properties exhibited by biological materials harvest from the synergic interaction between distinct structural features that range from nanometers to the macroscopic scale. Devising how Nature employs these strategies, represents a critical step toward realizing strong and tough bio-inspired materials. In this work, we present the details about the experimental investigation into the toughening mechanisms observed in two natural systems: (1) Pangolin scales, and (2) Nacre. In situ and ex situ three point bending (TPB) fracture tests were performed and complemented with digital image correlation (DIC) in order to characterize the fracture resistance in these biological materials. The comprehensive methodology presented in this work provides a framework for broad experimental investigations into the failure mechanisms of bio- and bio-inspired materials.

H. D. Espinosa (✉)
Department of Mechanical Engineering, Northwestern University Technological Institute, Evanston, IL, USA
e-mail: espinosa@northwestern.edu

L. Lamberti et al. (eds.), *Advancement of Optical Methods & Digital Image Correlation in Experimental Mechanics, Volume 3*,
Conference Proceedings of the Society for Experimental Mechanics Series, https://doi.org/10.1007/978-3-319-97481-1_12

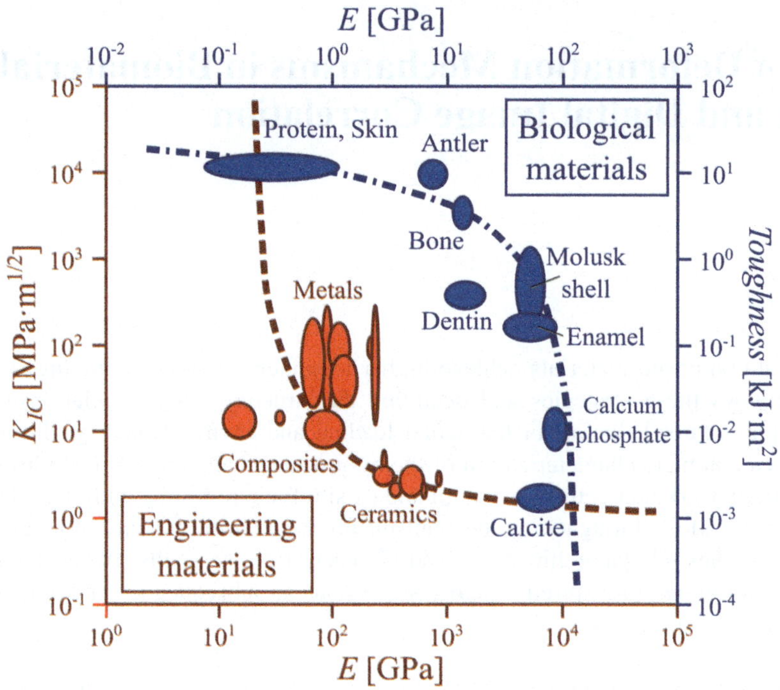

Fig. 12.1 Trends on mechanical performance of Man-made vs. Biological materials. Relative plot: Normalized Toughness vs. Stiffness (Adapted from [6–8])

12.2 Fracture Behavior and Toughness of Pangolin Scales

The pangolin is a nocturnal insectivore found primarily in parts of Africa and Asia that possesses a protective armor constructed from a network of overlapping scales. These scales are made from keratin lamellae and are able to resists penetration from predators such as: lions, tigers, and leopards [9]. In order to understand and quantify the toughening mechanisms involved during the fracture of pangolin scales we performed three point bending (TPB) fracture tests [10]. A photograph of a typical pangolin scale along with the relevant material directions is presented in Fig. 12.2a, and a schematic representation of the orientation of the keratin lamella along the thickness of the plates is shown in Fig. 12.2b(i). Samples for the fracture tests were laser cut to the appropriate shape and size, and a single edge notch (~1 mm) was introduced using a thin diamond-tipped saw (see Fig. 12.2b(ii)). In order to study the effects of hydration, dried and hydrated samples (relative humidity 20–30%) were considered.

The fracture resistance of the pangolin scales was evaluated following the ASTM E1820 standard [11] which permits a direct assessment of the *J*-integral using TPB testing of single edge cracked specimens. Digital image correlation (DIC) was implemented to determine the position of the crack-tip, using a procedure reported in [12]. In this technique, the vertical displacements were extracted from isosurfaces along a path in the vicinity of the crack tip (See Fig. 12.2c(i)). The extracted displacement profile is then fit to a parabolic function to determine the position of the crack-tip. The calculated crack extensions are then used to determine the components of *J*. The measured fracture resistance (J-R) curves from TPB tests of the pangolin scales are reported in Fig. 12.2c(ii). The *J* values presented in this figure represent the measurement window for each sample condition, which is limited by the initiation of unstable crack propagation in the pangolin scale. As shown in the figure, the *J* values for the scales tested at an ambient hydration level have peak toughness values of 4.3 and 6.3 kJ/m^2 for the longitudinal and transverse orientations, respectively. Significantly, in the hydrated condition, a large increase in the fracture resistance is measured with *J* reaching values as high ~25 kJ/m^2 in both material orientations, which represents an order of magnitude increase in material toughness. A key observation of the fracture behavior of pangolin scales is the non-uniform crack profile that develops during crack extension. Figure 12.2(d)(i)–(iii) shows different orthographic views of a 3D model of the crack profile taken from microCT images of a representative sample. The blue voxels represent the free surface of the crack, which exhibits a slanted and discontinuous crack front and the gray voxels contain the pre-notched region. The deflection of the crack in the inter region is best illustrated in Fig. 12.2d(iii) which images the fracture relief along an axis parallel to the propagation direction. These scans support the idea that the crack prefers to propagate between

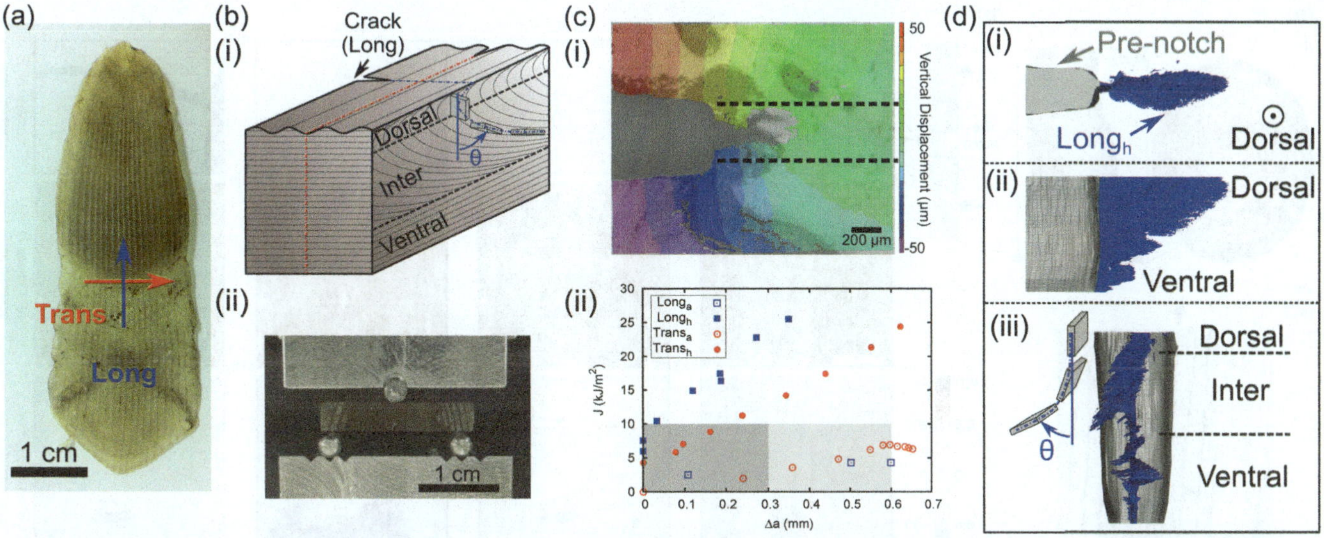

Fig. 12.2 Fracture behavior and toughness of pangolin scales [10]. (**a**) Photograph of a typical scale of the African tree pangolin. The transverse (Trans) and longitudinal (Long) material directions are indicated. (**b**) (i)A perspective cross-section schematic of the pangolin scale revealing the layering patterns of keratinous lamellae across the thickness of the scales (ii) A typical single edge cracked TPB specimen positioned within the loading frame. (**c**) (i) Digital image correlation map revealing isosurfaces of equivalent vertical displacement. These isosurfaces are used to generate a vertical displacement profile. (ii) The measured fracture resistance curves of the pangolin scales. (**d**) 3D reconstruction of microCT images of the crack profile from three orthogonal projections showing fracture surfaces consistent with the orientation of the lamellae structures. Blue and gray voxels represent the crack and pre-notched features, respectively

layers of the keratinous lamellae and that crack kinking and twisting contributes to the measured increase in toughness during crack extension [10]. This mechanism is illustrated by Fig. 12.2b(i).

12.3 Toughening Mechanisms in Nacre

Nacre, also known as mother-of-pearl, is a hard biological composite found in the inside layer of many shells such as oyster or abalone. It is composed of microscopic ceramic tablets arranged in layers and tightly stacked to form a three-dimensional brick wall structure, where the mortar is a thin layer of biopolymers (20–30 nm). Although mostly made of a brittle ceramic, the structure of nacre is so well designed that its toughness is several order of magnitudes larger that the ceramic it is made of. This extraordinarily tough material is part of a two-layer armor system (Fig. 12.3a(i)) deployed by seashells to protect themselves from predators. A hard calcite outer layer prevents penetration of the shell but is prone to brittle fracture. The tough nacreous inner layer is more compliant and arrests propagating cracks initiating in the brittle outer layer.

Using a similar methodology to the one described in Sect. 2, pre-notched fracture samples of natural nacre cut from Red Abalone shells were tested in three-point bending and imaged simultaneously in an atomic force microscope (AFM) [13]. Figure 12.3b(iv)–(viii) shows a representative series of AFM images captured in the scanned area during loading and after sample failure. Post-processing of the images using DIC enabled the quantification of tablet sliding, as well as identification of transverse dilation of tablets during crack propagation, all with sub-nanometer resolution (~1/10 pixel or 0.4 nm). As the crack initiates and approaches the scanned area (Fig. 12.3b(x)–(xi)), progressive sliding is seen through increasing color intensity in the DIC displacement fields. After the crack passes above the scanned area (Fig. 12.3b(xii)), a partial reversal of the displacement field occurs, resulting in partial closure of the gap between tablet ends. The sliding of microscale wavy tablets relative to each other results in dilation-induced hardening of the interface between tablets, which results in an increase in toughness.

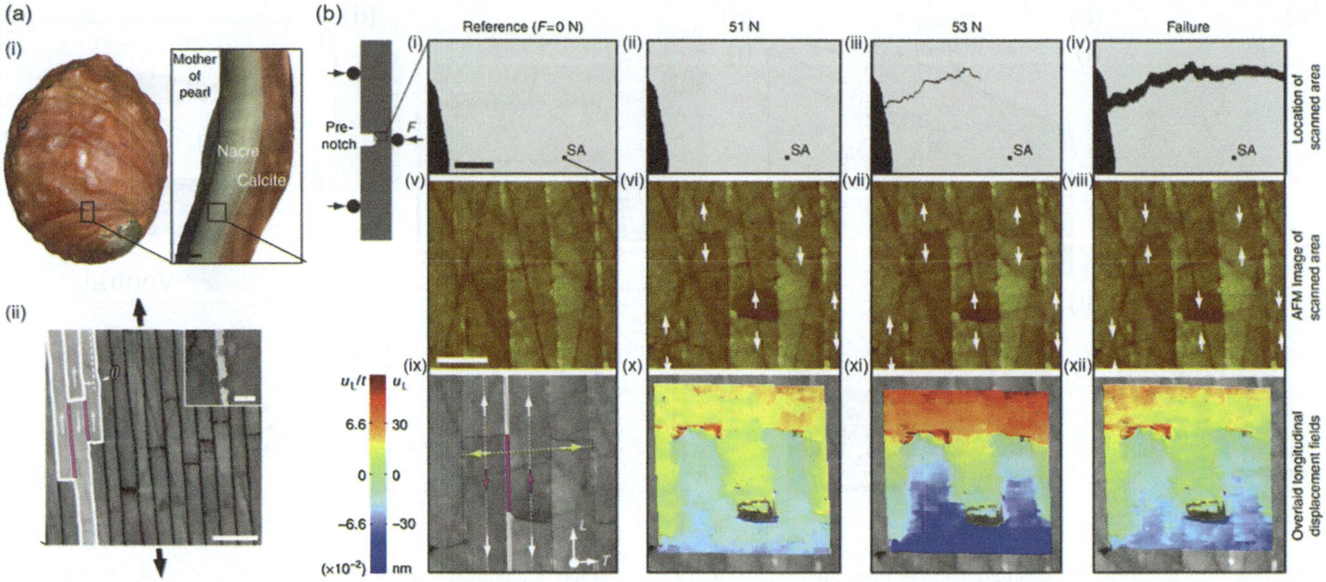

Fig. 12.3 Toughening mechanisms in nacre [13]. (**a**) (i) Red abalone shell. Inset shows a cross-section cut from the shell. (ii) Scanning electron micrograph showing micro/nanostructure of natural nacre. Scale bar, 1 μm. (**b**) (i)–(iv) images showing the position of the scanned area (SA) relative to the pre-notch and propagating crack. Scale bar, 100 μm. (v)–(viii) Series of AFM images of the SA in natural nacre during loading in three-point bending (left of a). Scale bar, 500 nm, height scale is 50 nm. (ix)–(xii) Results of the DIC performed on images

12.4 Conclusions

In contrast to man-made materials, Nature can produce materials with remarkable mechanical properties from relative weak constituents. This feature is achieved by establishing controlled synthesis and hierarchical assembly of nano- to micro-scales building blocks. Devising methods to allow a comprehensive understanding of the multi-scale mechanisms observed in high performance biological materials is a critical step towards realizing strong, tough, and multifunctional materials. In this work, we advanced in the understanding of these mechanisms by detailing an experimental methodology based on in-situ and ex-situ three point bending fracture tests in combination with digital image correlation. This experimental framework is aimed to allow the quantification and identification of toughening mechanisms observed in biological materials. Two case studies have been presented: the first case corresponds to the analysis of the pangolin scales where it was possible to identify a crack deflection mechanism triggered by the changing orientation of keratin lamellae across the thickness of the scales. The second case corresponds to nacre, where combining in situ AFM fracture tests with digital image correlation techniques allowed the identification of a new toughening mechanisms involving the sliding of the tablets that form the microstructure of this system. We envision that the outcome of this research will pave the way for more bio-inspired design systems that can subsequently shed light on how Nature has evolved materials to optimize mechanical properties.

References

1. NSF.: NSF-EC workshop on nanomaterials and nanotechnology (2002).
2. Materials Genome Initiatives for Global Competitiveness. In: Council, National Science and Technology, 211AD.
3. Ashby, M.: Designing architectured materials. Scr. Mater. **68**(1), 4–7 (2013)
4. Brechet, Y., Embury, J.D.: Architectured materials: expanding materials space. Scr. Mater. **68**(1), 1–3 (2013)
5. Dunlop, J.W.C., Brechet, Y.J.M.: Architectured structural materials: a parallel between nature and engineering. In: MRS Proceedings, vol. 1188, no. 15, pp. 1188-NaN-4 (2009)
6. Fratzl, P., Gupta, H.S., Paschalis, E.P., Roschger, P.: Structure and mechanical quality of the collagen–mineral nano-composite in bone. J. Mater. Chem. **14**(14), 2115–2123 (2004)
7. Espinosa, H.D., Rim, J.E., Barthelat, F., Buehler, M.J.: Merger of structure and material in nacre and bone—perspectives on de novo biomimetic materials. Prog. Mater. Sci. **54**(8), 1059–1100 (2009)
8. Ashby, M.F., Gibson, L.J., Wegst, U., Olive, R.: The Mechanical Properties of Natural Materials. I. Material Property Charts. Proc. R. Soc. London A Math. Phys. Eng. Sci. **450**(1938), 123–140 (1995)

9. Wang, B., Yang, W., Sherman, V.R., Meyers, M.A.: Pangolin armor: overlapping, structure, and mechanical properties of the keratinous scales. Acta Biomater. **41**, 60–74 (2016)
10. Chon, M.J., et al.: Lamellae spatial distribution modulates fracture behavior and toughness of african pangolin scales. J. Mech. Behav. Biomed. Mater. **76**, 30–37 (2017)
11. 2015 ASTM(E1820-15a).: Standard test method for measurement of fracture toughness (2015)
12. Barthelat, F., Espinosa, H.D.: An experimental investigation of deformation and fracture of nacre-mother of pearl. Exp. Mech. **47**(3), 311–324 (2007)
13. Espinosa, H., Juster, A., Latourte, F., Loh, O., Gregoire, D., Zavattieri, P.: Tablet-level origin of toughening in abalone shells and translation to synthetic composite materials. Nature. **2**, 173 (2011)

Chapter 13
Fast, Sub-pixel Accurate Digital Image Correlation Algorithm Powered by Heterogeneous (CPU-GPU) Framework

Mullai Thiagu, Sankara J. Subramanian, and Rupesh Nasre

Abstract Digital Image Correlation (DIC) is a popular non-contact image-based full-field deformation measurement tool widely used in mechanics. In spite of its significant advantages, it is still primarily used as a post-processing tool due to its computational cost. In recent years, parallel computing platforms such as multi-core processors and Graphics Processing Units (GPUs) have been used to improve the speed of the DIC algorithm, with GPUs being well-suited for implementing data-parallel operations. Previous works have performed GPU-based DIC wherein each sub-image (*i.e. a collection of a few pixels in the local neighborhood of a point of interest*) is allocated to a single thread on the GPU, thus achieving parallelism across sub-images. However, this is not the only type of parallelism that is possible: one can also achieve parallelism within a sub-image as well as across whole images. The aim of this work is to efficiently implement 2D-DIC such that parallelism within a sub-image as well as across sub-images leads to considerable reduction in computation time. We use a heterogeneous framework consisting of an Intel Xeon octa-core CPU and an Nvidia Tesla K20C GPU card in this work. The CPU is used to handle image pre-processing, whereas the GPU is used to process four compute-intensive tasks: affine shape function computation, B-Spline interpolation, residual vector calculation and deformation vector update. Parallelization within and across sub-images is achieved in this work by efficient thread handling and use of pre-compiled BLAS libraries. In order to estimate the speedup provided by the GPU, the same four tasks were also evaluated on the octa-core CPU; a speedup of approximately 7 to 5 times was observed for a single sub-image whose size varies from 21×21 to 61×61 respectively. However, it is expected that for a larger number of sub-images, the GPU speedup will be higher and this is indeed the case: when the affine shape function computation and B-Spline interpolation steps were evaluated on 1869 21×21 pixel sub-images, the speedup was around a more impressive 453 times. Further GPU optimization as well as parallelization across image pairs is currently underway and even faster GPU-assisted DIC seems achievable.

Keywords Full-field displacement · Sub-image · Parallel computing · Heterogeneous framework · Compute Unified Device Architecture (CUDA) · Thread · Kernel

13.1 Introduction

Digital Image Correlation (DIC) is a well-established optical technique used to measure full-field displacements with sub-pixel accuracy [3, 5, 6, 10, 12]. Since DIC is computationally expensive [2, 15], it continues to be used as a post-processing tool. In general, a computation pattern that comes under the category of Single Instruction Multiple Data (SIMD) [9] such as DIC is well-suited for implementation on a GPU. We take motivation from recent advancements in GPU-based parallel computing [1, 9] to address the computational efficiency issue of conventional DIC and present in this paper our recent work on developing a fast, sub-pixel accurate 2D-DIC powered by a heterogeneous (CPU-GPU) framework.

Recent works [7, 8, 17] have demonstrated that GPUs can be effectively used for DIC. Le Besnerais et al. [8] implemented a fast, iterative Lucas Kanade-based (LK-based) DIC towards dense estimation of 2D and 3D displacements. The increase in computation speed in this work is due to a local translational motion model derived from Particle Image Velocimetry (PIV). An alternative approach to achieving high-performance was proposed by Gates et al. [7] for Digital Volume Correlation (DVC) tested on Keeneland, a cluster with 120 nodes, each with two 6-core Intel X5660 CPUs and three NVIDIA M2090 GPUs. In this study, the authors present two forms of parallelism: coarse-grained and fine-grained. The coarse-grained step

M. Thiagu · S. J. Subramanian (✉) · R. Nasre
Department of Engineering Design, Indian Institute of Technology Madras, Chennai, India
e-mail: shankar_sj@iitm.ac.in

L. Lamberti et al. (eds.), *Advancement of Optical Methods & Digital Image Correlation in Experimental Mechanics, Volume 3*, Conference Proceedings of the Society for Experimental Mechanics Series, https://doi.org/10.1007/978-3-319-97481-1_13

decomposes a pair of images to simultaneously calculate initial deformation values by using Message Passing Interface (MPI). The fine-grained step is a standard optimization algorithm using GPU to achieve sub-pixel accuracy. The results of sub-pixel registration based on inverse compositional Gauss-Newton (IC-GN) algorithm using GPU was illustrated by Zhang et al. [17]. A block with 128 threads was assigned to carry out single-precision computations of one sub-image. Although their work showed sub-images invoked in parallel, the thread handling scheme leads to a considerable number of idle threads. A few studies have also investigated the impact of realizing 2D-DIC in parallel using multi-core processors within a CPU. Improvement in performance by using multi-threading concepts offered by Intel hyper-threading technology and OpenMP to reduce computation time was investigated [4, 11, 13, 16]. A limitation of multi-core CPU is that the maximum speedup gained is dependent on the number of threads, which is relatively small (a few tens).

Interestingly, very few studies have concentrated on the nature of computationally demanding operations in data processing flow of the DIC algorithm. Earlier approaches evaluated the performance gained due to GPU usage by invoking parallel threads, with one thread per sub-image. We refer to this technique in which parallelism is achieved across sub-images as *inter-parallelism.*

However, DIC has a rich parallelism profile; one can parallelize computations within a sub-image as well as across all images or across the whole set of images. The former forms the topic of this work; the latter will be taken up in the future. Within a sub-image, an iteration technique such as Newton-Raphson (NR) is commonly used to incrementally update deformation shape function parameters until convergence. In conventional DIC this minimization of a sub-image is processed sequentially although these computations can be executed in parallel. In this paper, we address this performance hurdle by parallelizing minimization within a sub-image, which we call *intra-parallelism.* In particular, we propose and implement a heterogeneous framework on CPU and GPU for improving the performance of 2D-DIC. For this work, we use an Intel Xeon octa-core CPU and an NVIDIA Tesla K20C GPU card, with the CPU used for image pre-processing and the GPU for compute-intensive tasks. We achieve parallelization within and across sub-images through efficient thread handling and usage of cuBLAS libraries and show that our proposed framework achieves considerable speedup on a typical DIC dataset.

The remainder of the paper is organized as follows: in the next section, we discuss details of the heterogeneous framework, followed by the presentation of results in Sect. 13.3 and concluding remarks in Sect. 13.4.

13.2 Heterogeneous Framework

In this section, we briefly explain the sequential 2D-DIC algorithm and then motivate the need for heterogeneous parallel processing.

13.2.1 2-D DIC Algorithm: Sequential Version

The 2-D DIC algorithm performs region-based template matching derived from the Lucas-Kanade (LK) optical flow method to compute displacements between a pair of input images, the undeformed image $\mathbf{I}^R$ and the deformed image $\mathbf{I}^D$. First, a *shape function* that takes a sub-image in $\mathbf{I}^R$ to its map in $\mathbf{I}^D$ is assumed; in 2-D DIC, this shape function is commonly chosen to be the affine warp, characterized by 6 parameters $p_0 \ldots p_5$. Details of iteratively computing these parameters for a given subset in $\mathbf{I}^R$ are shown in Algorithm 1. The resulting shape function parameters are used to calculate the displacement vector $\mathbf{u}$ for this sub-image and are further propagated as initial estimates to a neighbouring sub-image and so on. Similar processing is repeated for other deformed images to compute displacements at each time step.

Nomenclature

$\mathbf{u}$	Displacement vector
$\mathbf{I}^R, \mathbf{I}^D$	Reference, deformed image
t, g	Number of sub-images, deformed images
w	Sub-image size (pixels)
i, j, k	Image, sub-image, iteration indices
m	Maximum number of iterations

r, c	Row, column indices of a matrix
$\mathbf{p}', \mathbf{p}, \Delta\mathbf{p}$	Initial, converged, incremental shape function parameter vector
$\mathbf{H}$	Hessian matrix
$\mathbf{q}$	Residual vector
ϵ	Norm value to terminate iterations

In the present work, an efficient inverse compositional approach is adopted to implement the parallel version of DIC and hence a comparison is drawn with its CPU version. There are two `while` loops in Algorithm 1. In the inner `while` loop (lines 11 through 17), the following are the compute-intensive tasks carried out within each sub-image: affine shape function computation (T_1), B-Spline interpolation (T_2), residual vector calculation (T_3) and deformation vector update (T_4). These tasks are executed in sequential order until the convergence condition ($\sum \Delta p_n/p_n \leq \epsilon$) is satisfied.[1] The converged shape function parameters p_n for a sub-image are used to find its respective displacement vector **u**. In the outer `while` loop (lines 6 through 19), the above steps are repeated across sub-images. It is to be noted that for the first sub-image, the estimation of the initial parameter vector $\mathbf{p}'$ is usually a heuristic, where the user selects a point of zero displacement. For the rest of the sub-images ($j = 1 \ldots t$), the converged parameters p_n are propagated in a serial manner to the neighbouring sub-image as initial estimates p_n'. Finally, the `for` loop (lines 4 through 20) is used to find the displacements across the consecutive pairs of undeformed and deformed images. Computation of the Hessian matrix **H** and the image gradients $\frac{\partial \mathbf{I}^R}{\partial r}$ and $\frac{\partial \mathbf{I}^R}{\partial c}$ for $\mathbf{I}^R$ in line 3 is done outside the loops since they are one-time computations in the inverse compositional iterative update scheme [14].

1 **Input:** $w, \mathbf{I}^R, \mathbf{I}_1^D, \ldots, \mathbf{I}_g^D$
2 **Output:** **u**.
3 Pre-compute: **H**, image gradients $\frac{\partial \mathbf{I}^R}{\partial r}, \frac{\partial \mathbf{I}^R}{\partial c}$
4 **for** $i \leftarrow (1\ to\ g)$ **do**
5   Read deformed image ($\mathbf{I}_i^D$)
6   **while** $(j \leq t)$ **do**
7   **if** $j = 1$ **then**
8   Initialize: $\mathbf{p}'$
9   **else**
10   Propagate: ($\mathbf{p}' \leftarrow \mathbf{p}$)
11   **while** $((k \leq m)$ **and** $(\sum \frac{\Delta p_n}{p_n} \leq \epsilon))$ **do**
12   Affine shape function computation
13   B-Spline interpolation
14   Residual vector calculation
15   Deformation parameter vector update
16   Calculate $\sum \frac{\Delta p_n}{p_n}$
17   k++
           end
18   Displacement vector (**u**) per sub-image
19   j++
       end
20   i++
end

Algorithm 1: Pseudo-code for the 2D-DIC algorithm realized using a sequential CPU

13.2.2 *Two-Dimensional DIC Algorithm: Heterogeneous Version*

It is evident from Algorithm 1 that these computations exhibit massive data parallelism across images (`for` loop), sub-images (outer `while` loop) and NR iterations for a sub-image (inner `while` loop). The compute-intensive tasks T_1 through T_4 in the inner `while` loop pose performance bottlenecks, which preclude the realization of real-time DIC measurement. Since these

[1] $n = 0 \ldots 5$ for affine shape function.

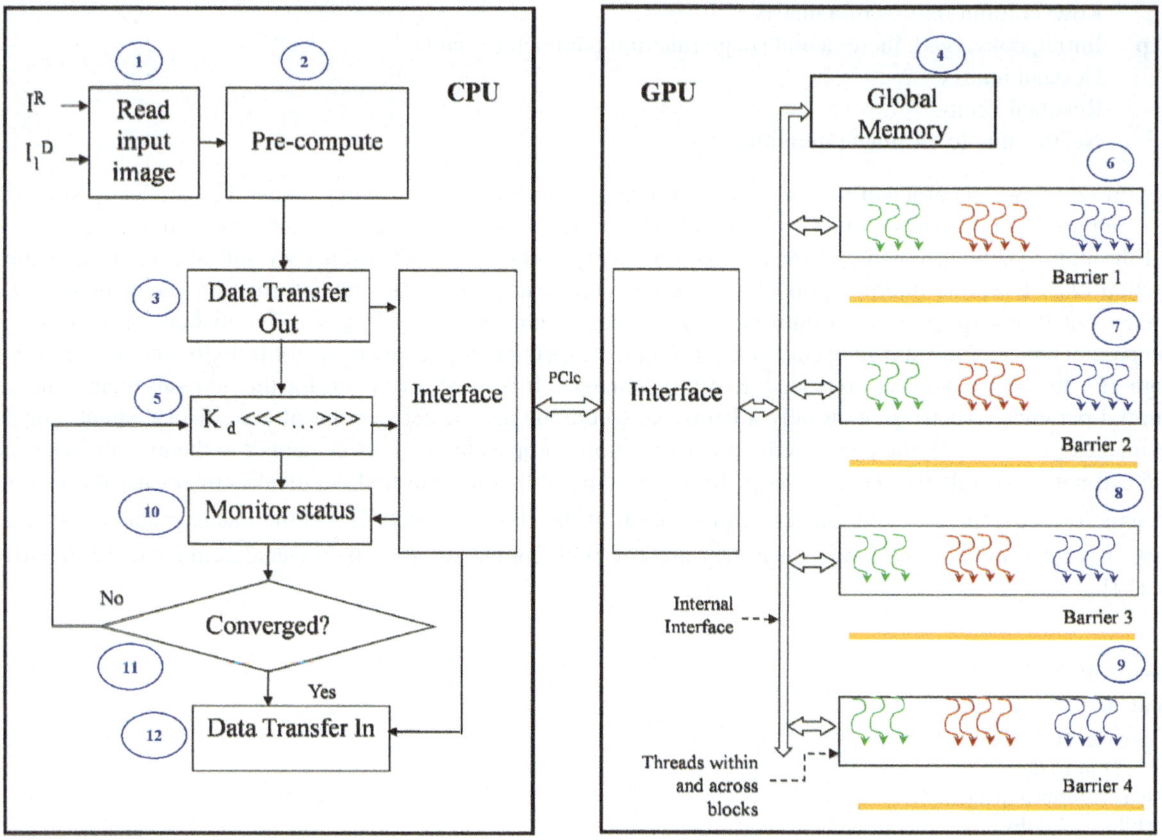

Fig. 13.1 Heterogeneous framework with CPU and GPU to implement the 2D-DIC algorithm

Table 13.1 Dependency analysis of compute-intensive tasks of 2D-DIC algorithm in a sequential version

Computation task	Intra-parallelism	Inter-parallelism
T_1: affine shape function	✓	✓
T_2: B-spline interpolation	✓	✓
T_3: residual vector update	✓	✓
T_4: deformation vector update	×	✓

are sequential processing steps, a GPU-only (without CPU) setup is not a preferred option. Therefore, in the present work, we implement the tasks T_1 through T_4 in a GPU through a generic heterogeneous computing framework with one CPU and GPU as depicted in Fig. 13.1. We assume that the initial affine parameter vector $\mathbf{p}'$ for t sub-images is known, from which DIC optimization is carried out in parallel. The goal of our framework is to exploit the individual strengths of CPU and GPU and to assign tasks suitable for their respective architecture. By combining their merits through inter-parallelism across sub-images and intra-parallelism within a sub-image, our framework is capable of significant performance gains; implementation details for tasks T_1 through T_4 are summarized in Table 13.1.

The reason for improved performance is that a GPU is analogous to a co-processor, which is used to perform SIMD (single instruction multiple data) computations carried out in an asynchronous manner with less CPU supervision. While the GPU is busy, resources of CPU are allocated for other high priority tasks. Meanwhile, if the GPU completes its execution, the final set of data is transferred back to the CPU. Therefore, a GPU is suited to execute compute-intensive operations in parallel and a CPU is the best fit for ordered realization of tasks. Aim of the former is to maximize throughput, while that of the latter is to minimize the execution latency. Thus, a heterogeneous computing framework with a CPU and GPU combination is a good choice for performance-oriented applications such as real-time DIC.

The CPU is responsible for one-time operations such as image-handling operations, pre-computations and transfer of data from CPU to GPU memory. In Fig. 13.1, a pair of images $\mathbf{I}^R$ and $\mathbf{I}_1^D$ are read and loaded into an on-board CPU memory through in-built routines from a standard image processing toolbox. Further, the pre-computed data from the CPU RAM are transferred as a one-dimensional array to the GPU global memory through PCIe interface. This is followed by launching a set of *kernels*, which define data operations handled within a thread to asynchronously execute tasks T_1 through T_4 on the GPU. During a kernel call, the total number of threads and blocks that implement intra- and inter-parallelism are specified as arguments (e.g. $K_d <<< \ldots >>>$), where d=1…4). A barrier (control point) is used to synchronize the threads and blocks before moving to the next computation step. The status of the task execution is continuously monitored by the CPU. After task T_4, convergence condition ($\sum \Delta p_n / p_n \leq \epsilon$) is evaluated. Upon convergence, the computed displacement vector **u** for t sub-images is transferred from the GPU global memory to the CPU RAM. The commonly used programming model, *CUDA* (Compute Unified Device Architecture), is used to develop source codes and validate them on the processor cores (threads) in the GPU.

13.3 Results and Discussion

For a pair of grayscale images, the NVIDIA Tesla K20C GPU card in heterogeneous version is programmed using CUDA C language (Toolkit version 9.0). The GPU has 5 GB onboard memory (RAM) and 2496 cores spread across 13 Streaming Multiprocessors (SMXs). Subsequently, the values computed using GPU are benchmarked with the serial version of DIC implemented in Matlab R2017. To evaluate performance, Sample 14 image pairs from the DIC Challenge dataset with a resolution of 588 × 2048 pixels are used with an ROI of 472 × 1916 pixels (Fig. 13.2). Sub-images are chosen to be non-overlapping and initial values are set to $\mathbf{p}' = (0, 0, 0, 0, 0, 0)^T$ for all of them. The tolerance value ϵ and the maximum number of iterations m are set to 5×10^{-3} and 20 respectively. The average computation time (usually in milliseconds) is measured during the execution of a kernel by repeating it three times. This computation time refers to amount of time taken to execute a kernel and does not include time taken to transfer data between the CPU and GPU. Standalone testing of tasks T_1 through T_4 is done for both sequential and heterogeneous versions.

13.3.1 Speedup Per Sub-image

A quantitative analysis of the computation time measured for the four compute-intensive tasks T_1 through T_4 is shown in Fig. 13.3. This is done for a selected sub-image in one NR iteration. The x-axis in Fig. 13.3 runs over various sub-image sizes ($21 \times 21 \ldots 61 \times 61$), while the y-axis shows the speedup factor, defined as the ratio of the time taken by the CPU to the time taken by the GPU to perform a computation. It is seen that the speedup of task T_2 is above 400, while that of the other three tasks is below 7. As explained in Sect. 13.2, improvement in performance per iteration is due to intra-parallelism and inter-parallelism.

Subsequently, we measure the execution time taken by a single sub-image for multiple iterations until convergence is reached. The results in Table 13.2 obtained by varying the size of the sub-image (w) show a speedup ranging between 6–7 times. It is observed that for varying w, the average number of iterations to reach convergence condition is four for the Sample 14 images studied in this work. Within the computation of task T_4, cuBLAS (CUDA Basic Linear Algebra Subroutines) solver routines, which are optimized for GPU hardware, are used to calculate the increment in deformation parameter vector, $\Delta\mathbf{p}$.

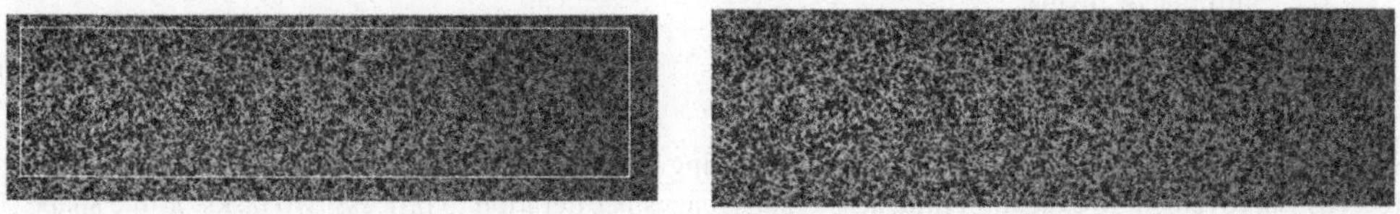

Fig. 13.2 Image pair from sample 14 - DIC challenge repository. (**a**) Reference image ($\mathbf{I}^R$) with a chosen region-of-interest. (**b**) Deformed image ($\mathbf{I}_1^D$)

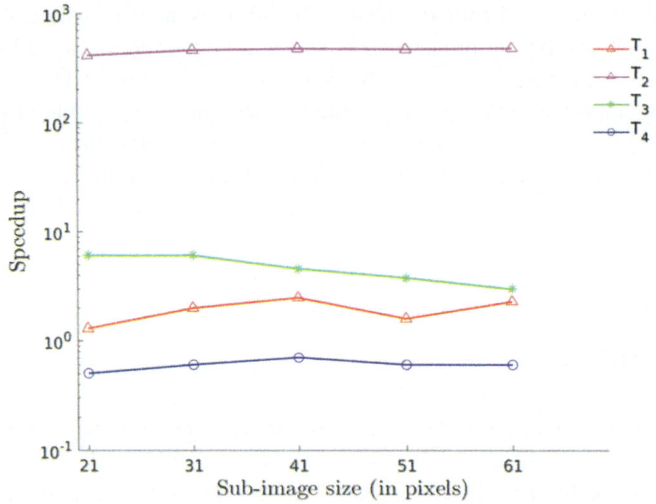

Fig. 13.3 GPU speedup of four compute-intensive tasks per sub-image and one iteration (across different sub-image sizes) - Breakup of step-wise improvement in speedup factor

Table 13.2 Speedup factor for repeated execution of one sub-image until convergence across different sub-image sizes; average computation time in (milliseconds) is shown for (a) sequential version implemented using MATLAB; (b) heterogeneous version implemented using MATLAB (for CPU processing) and CUDA C (for GPU processing)

Sub-image size (in pixels)	Sequential version	Heterogeneous version	Speedup factor	Iterations
21×21	30.1	4.0	7.5	4
31×31	30.7	4.8	6.4	4
41×41	31.4	5.2	6.0	4
51×51	32.6	5.2	6.3	4
61×61	36.7	6.2	5.9	4

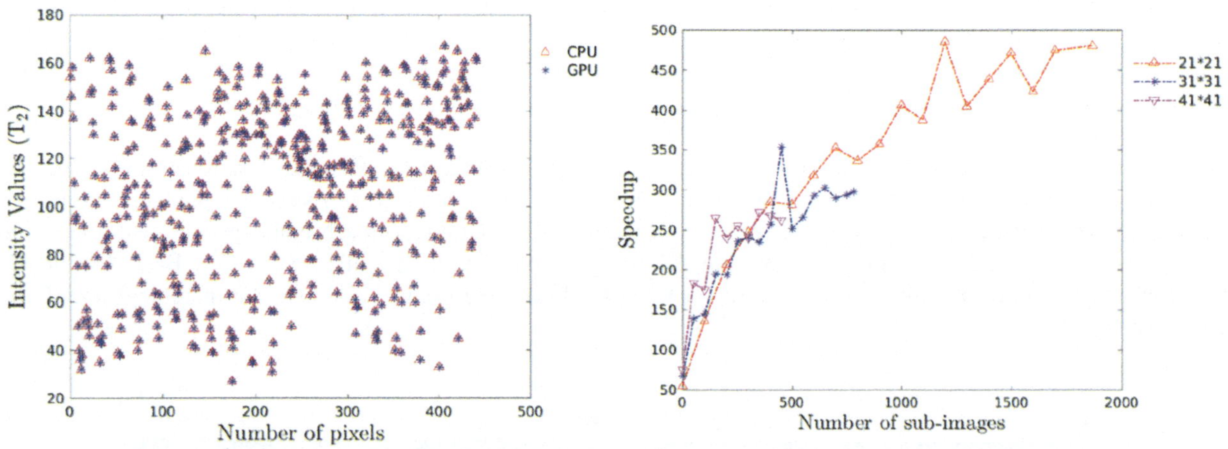

(a) Interpolation values of a sub-image **(b)** Speedup across sub-images per iteration

Fig. 13.4 Comparison of interpolation values numerically computed using both CPU and GPU with double precision arithmetic. (**a**) Interpolation values of a sub-image. (**b**) Improvement in execution time across sub-images per iteration to process two tasks: affine shape function computation and B-Spline interpolation

13.3.2 Speedup Across Sub-images

We also evaluate the performance obtained from parallelism within as well as across sub-images (Fig. 13.4b). The results are reported for two compute-intensive tasks: affine shape function and B-Spline interpolation on sequential and parallel

Table 13.3 Average computation time (in milliseconds) across sub-images per iteration

Sub-image size (in pixels)	Sequential version	Heterogeneous version	Speedup factor
21×21	417.50	0.92	453
31×31	301.26	0.96	323
41×41	245.36	0.91	269
51×51	198.93	0.83	239
61×61	184.76	0.79	233

Measured time is for computation of affine shape function and B-Spline interpolation of sub-images within selected region-of-interest

computing platforms. Comparing these results, it is evident that a significant improvement in execution time is obtained for varying sub-image sizes (21 × 21 . . . 61 × 61) in one iteration. Figure 13.4a shows numerically interpolated intensity values for a sub-image of size 21 × 21 with double-precision. It can be seen that the error between the two results for a given set of sub-images is of the order of machine precision. Implementing parallelism across all sub-images for tasks T_3 and T_4 is under development.

In general, we observe that our method scales well and the overall speedup improves with increase in number of sub-images; indeed, from Fig. 13.4b it is observed that the performance can further improve with increased number of image-pairs and sub-images. Thus, we remark that the performance of DIC computations can be further enhanced by exploiting parallelism at all levels (images, sub-images, NR iterations within a sub-image). While the sub-image sizes in our experiments were fixed (21×21 . . . 61×61), our heterogeneous framework is generic to work with other sub-image sizes (Table 13.3).

13.4 Conclusion

In this work, we have implemented a heterogeneous framework for 2D-DIC computations to exploit parallelism within and across sub-images synergistically on CPU and GPU. We have illustrated the effectiveness of our framework with various sub-image sizes.

In particular, the following are our contributions:

- The time taken per sub-image per iteration to execute tasks T_1 through T_4 for varying sub-image sizes (21 × 21 . . . 61 × 61) was evaluated. The task T_2 implemented using intra-parallelism contributes to increase in speedup.
- The convergence time for a sub-image with a variety of sizes (21 × 21 . . . 61 × 61) was studied and a speedup of around 5 to 7 times obtained.
- Further, for 1869 sub-images of 21 × 21 pixels, we have reported a more impressive speedup of approximately 453 per iteration.
- Work to parallelize across images is underway and is expected to result in significant improvement in computational speed.

References

1. Anderson, R.F., Kirtzic, J.S., Daescu, O.: Applying parallel design techniques to template matching with GPUs. In: International Conference on High Performance Computing for Computational Science, pp. 456–468 (2010)
2. Baker, S., Matthews, I.: Lucas-Kanade 20 years on: a unifying framework. Int. J. Comput. Vis. **56**, 221–255 (2004)
3. Bing, P., Hui-Min, X., Bo-Qin, X., Fu-Long, D.: Performance of sub-pixel registration algorithms in digital image correlation. Meas. Sci. Technol. **17**, 1615 (2006)
4. Blaber, J., Adair, B., Antoniou, A.: Ncorr: open-source 2D digital image correlation matlab software. Exp. Mech. **55**, 1105–1122 (2015)
5. Bornert, M., Brémand, F., Doumalin, P., Dupré, J.C., Fazzini, M., Grediac, M., Hild, F., Mistou, S., Molimard, J., Orteu, J.J., Robert, L.: Assessment of digital image correlation measurement errors: methodology and results. Exp. Mech. **49**, 353–370 (2009)
6. Cofaru, C., Philips, W., Van Paepegem, W.: Improved Newton–Raphson digital image correlation method for full-field displacement and strain calculation. Appl. Opt. **49**, 6472–6484 (2010)
7. Gates, M., Heath, M.T., Lambros, J.: High-performance hybrid CPU and GPU parallel algorithm for digital volume correlation. Int. J. High Perform. Comput. Appl. **29**, 92–106 (2015)
8. Le Besnerais, G., Le Sant, Y., Lévêque, D.: Fast and dense 2D and 3D displacement field estimation by a highly parallel image correlation algorithm. Strain **52**, 286–306 (2016)

9. Owens, J.D., Houston, M., Luebke, D., Green, S., Stone, J.E., Phillips, J.C.: GPU computing. Proc. IEEE **96**, 879–899 (2008)
10. Pan, B., Li, K.: A fast digital image correlation method for deformation measurement. Opt. Lasers Eng. **49**, 841–847 (2011)
11. Pan, B., Tian, L.: Superfast robust digital image correlation analysis with parallel computing. Opt. Eng. **54**, 034106 (2015)
12. Pan, B., Wang, B.: Digital image correlation with enhanced accuracy and efficiency: a comparison of two subpixel registration algorithms. Exp. Mech. **56**, 1395–1409 (2016)
13. Shao, X., Dai, X., He, X.: Noise robustness and parallel computation of the inverse compositional Gauss–Newton algorithm in digital image correlation. Opt. Lasers Eng. **71**, 9–19 (2015)
14. Sutton, M.A., Orteu, J.J., Schreier, H.: Image Correlation for Shape, Motion and Deformation Measurements: Basic Concepts, Theory and Applications. Springer, Berlin (2009)
15. Wang, Z., Wang, S., Wang, Z.: An analysis on computational load of DIC based on Newton–Raphson scheme. Opt. Lasers Eng. **52**, 61–65 (2014)
16. Wu, R., Kong, C., Li, K., Zhang, D.: Real-time digital image correlation for dynamic strain measurement. Exp. Mech. **56**, 833–843 (2016)
17. Zhang, L., Wang, T., Jiang, Z., Kemao, Q., Liu, Y., Tang, L., Dong, S.: High accuracy digital image correlation powered by GPU-based parallel computing. Opt. Lasers Eng. **69**, 7–12 (2015)

Chapter 14
Vibration Modal Analysis by High-Speed and Accurate Shape Measurement Using One-Pitch Phase Analysis Method

Yoshiharu Morimoto, Akifumi Takagi, Masaki Ueki, and Lionel Pirsig

Abstract To protect vibration of structures, vibration modal analysis is important. Conventional experimental vibration analysis method uses accelerometers or laser displacement meters. The analysis takes much time and troublesome work. Especially, by contacting accelerometers to objects the results of frequency mode have some errors because of contact of the accelerometer weight. The authors have previously proposed a new method for shape measurement, i.e. the one-pitch phase analysis (OPPA) method using moiré topography. In moiré topography, the number of pixels for one pitch is constant regardless of the object height. The OPPA method analyzes the phase of each pixel in a single-shot image by using brightness data from one pitch of a projected grating. A motion capture system has been previously developed using OPPA method and measured the 3D coordinates at every pixel of the surface of a human body. In this study, the OPPA method is applied to vibration modal analysis of a cantilever plate using a high-speed camera. This method is useful for inspection of automobiles, machines, buildings, etc. without contact, in a short time.

Keywords Shape measurement · Phase analysis · Vibration modal analysis · Frequency analysis · One-pitch phase analysis

14.1 Introduction

High-speed shape measurement is requested in many disciplines. Examples include motion capture for human body, inspection during factory automation and applications in vibration analysis. The authors have previously proposed a new method for shape measurement; the one-pitch phase analysis (OPPA) method, which uses moiré topography and analyze the phase of projected grating from a single shot image [1]. The OPPA method can measure three-dimensional coordinates at each pixel point of a moving object. Using the OPPA method, a motion capture system has been previously developed and measured the 3D coordinates of a human body at every pixel.

To protect vibration of structures, vibration modal analysis is important. Conventional experimental vibration analysis method uses accelerometers or laser displacement meters. The modal analysis takes much time and troublesome work. Especially, by contacting accelerometers to objects, the results of frequency mode have some errors.

In this study, the OPPA method is also extended to perform vibration modal analysis of a vibrating object without contact by using a high-speed camera and a projector using optical fiber.

14.2 Optical System and Theory for Obtaining Height by the OPPA Method

Figure 14.1 shows the optical system for moiré topography; the grating plane, reference plane, camera lens plane, and image recording plane of the camera are all parallel to each other, and the heights of the light source and the center of camera lens from the grating plane are equal. In this case, regardless of the height of the object, the image of the grating projection on

Y. Morimoto (✉)
Wakayama University, Wakayama, Japan

4D Sensor Inc., Wakayama, Japan
e-mail: morimoto@4d-sensor.com

A. Takagi · M. Ueki · L. Pirsig
4D Sensor Inc., Wakayama, Japan

L. Lamberti et al. (eds.), *Advancement of Optical Methods & Digital Image Correlation in Experimental Mechanics, Volume 3*, Conference Proceedings of the Society for Experimental Mechanics Series, https://doi.org/10.1007/978-3-319-97481-1_14

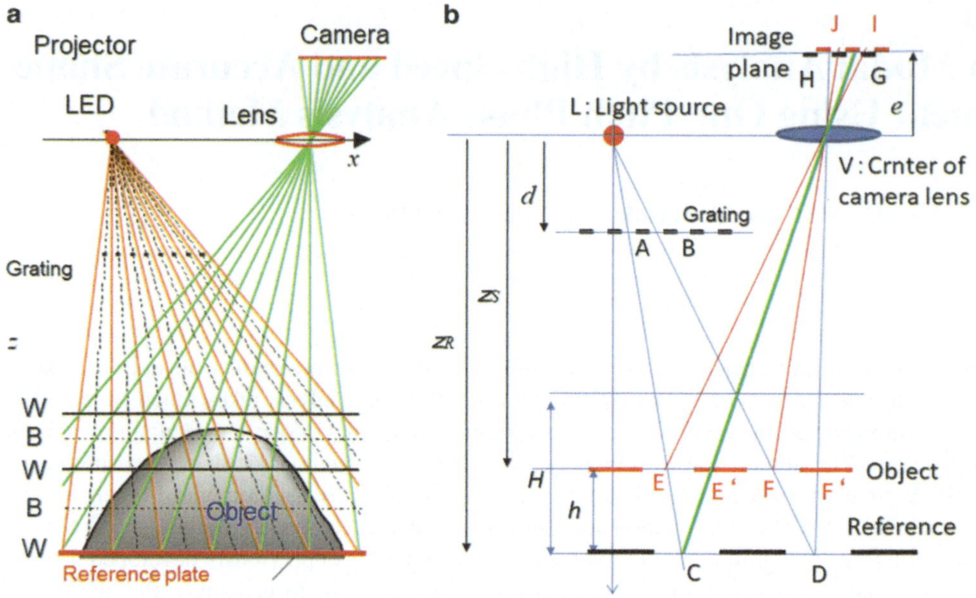

Fig. 14.1 Optical system of moiré topography for the OPPA method. (**a**) Optical system of moiré topography (**b**) Schematic for explanation

the object has a fixed grating pitch on the camera imaging plane, and only the phase of the grating shadow differs depending on the height of the object. It is possible to analyze the phase from the brightness data in one pitch of the grating projection image. First, before recording the grating projected on the reference plane, the grating projection is adjusted so that the pitch of the grating shadow image on the reference plane matches *N* (integer) pixels of the camera image sensor. Then, even if the grating is projected on an object with different heights, the one-pitch length of the grating projection image in the camera is always *N* pixels. However, the phase of the grating changes depending on the height of the object. By calculating the phase for each pixel, it is possible to obtain the phase distribution on the object.

In Fig. 14.1b, L is the location of a light source, and a grating plane is located horizontally at a distance *d* from the location L of the light source. The pitch of the grating is *p*.

In Fig. 14.1b, the *x*, *y* and *z* coordinates are set with the light source L being the origin, and the direction perpendicular to the grating plane is set as the *z* direction. The downward direction is the positive direction in the *z* direction. The direction perpendicular to the grating lines drawn on the surface of the grating plane is the *x* direction, and the direction parallel to the grating lines is the *y* direction. Location V is the center of a camera lens. The *x* directional component of the distance between the locations L and V is *v*.

In this optical system, the center V of the camera lens is at height *d* from the grating plane, and the distance from the light source to the reference plane is z_R. The height to be measured is $h = z_R - z_S$, and the camera imaging plane and the reference plane are parallel to the grating plane. The distance between the camera lens center and the image plane is *e*.

In Fig. 14.1b, AB is one pitch wide on the grating, CD is one pitch of the grating shadow on the reference plane, and EF is one pitch of the grating shadow on the object.

Θ_M is the phase difference between the phase θ_S of the grating shadow on the object and the phase θ_R of the grating shadow on the reference plane. That is, the phase of the moiré fringe obtained by superposing the grating shadows on the object and the grating on the reference plane. The height from the reference plane is obtained from the phase difference Θ_M of the grating as follows [1].

$$h = z_R - z_S = z_R - \frac{2\pi v d z_R}{p z_R \Theta_M + 2\pi v d} = \frac{p \Theta_M {z_R}^2}{p z_R \Theta_M + 2\pi v d} = \frac{p' \Theta_M z_R}{p' \Theta_M + 2\pi v} \tag{14.1}$$

where p' is the pitch of the projected grating on the reference plane.

$$p' = \frac{p z_R}{d} \tag{14.2}$$

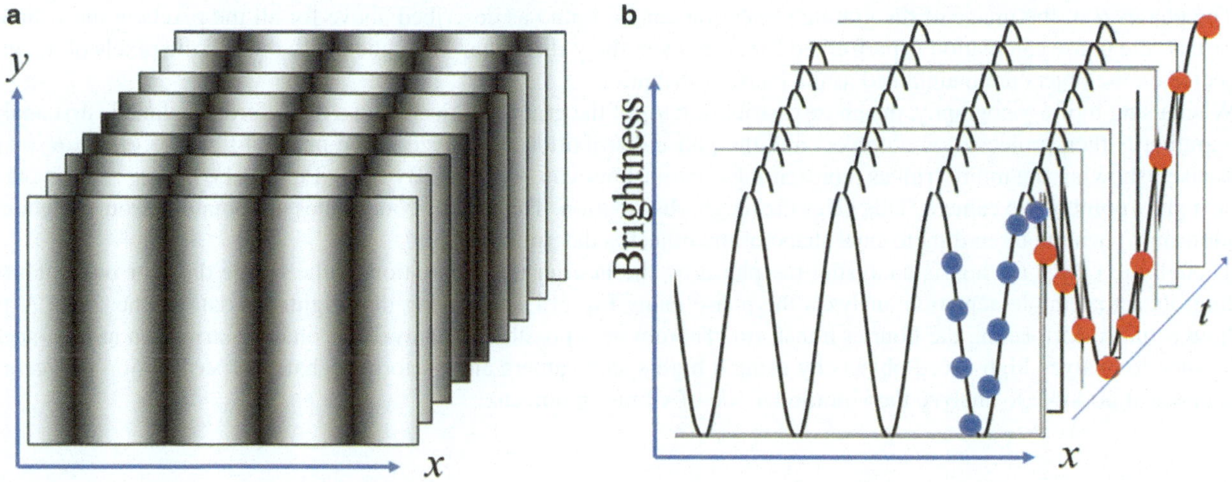

Fig. 14.2 Phase-shifted images for phase-shifting method and for OPPA method. Phase-shifted brightness (red points) for phase-shifting method and one-pitch brightness values (blue points) for OPPA method. (**a**) Photos during one cycle phase shifting (**b**) Brightness distribution along center line (blue points) and at edge (red points)

The dynamic range H is obtained as h by setting $\Theta_{\mathrm{M}} = 2\pi$ in Eq. (14.1).

$$H = \frac{pz_R^2}{pz_R + vd} = \frac{p' z_R}{p' + v} \tag{14.3}$$

Equation (14.1) does not depend on the coordinates x and y, or the camera pixel coordinates i and j on the image sensor.

14.3 Phase Analysis Via the OPPA Method

Figure 14.2 shows the schematic explanation of the phase analysis methods. The phase shifted images in Fig. 14.2a show one-cycle of the phase-shifting method. The brightness distributions in Fig. 14.2b are shown along the horizontal center lines in Fig. 14.2a. The brightness changes cosinusoidally. The brightness values of the right ends of the horizontal center lines during one-cycle of phase-shifting are shown in red points.

To analyze the phase, several methods were proposed [1–13].

The phase shifting method [2] uses these brightness values to obtain the initial phase. Since the phase-shifting method requires that the object not move during one cycle's shifting time, it is not useful for moving objects.

The Fourier transform method [3, 4] uses all brightness values along the centerline. It requires a lot of time to calculate the Fourier transforms of all the values along the centerline.

In the sampling moiré method [5], to analyze the phase at a pixel point, several interpolated images derived from the pixel data along two pitches are used. It requires a lot of memory and the calculation speed is slow.

In the OPPA method [1], the phase is obtained from the brightness values for one-pitch as shown in blue points in Fig. 14.2. A one-dimensional grating is projected onto an object or a reference plane. A digital camera then takes an image of the projected grating. The magnification is adjusted so that one pitch of the grating corresponds to N pixels e.g. $N = 9$ on the camera-imaging sensor. The phase $\theta(-\pi < \theta \leq \pi)$ is obtained at the center pixel using Eq. (14.4).

$$\tan\theta = -\frac{\sum_{n=-(N-1)/2}^{(N-1)/2} I_n \sin\left(n\frac{2\pi}{N}\right)}{\sum_{n=-(N-1)/2}^{(N-1)/2} I_n \cos\left(n\frac{2\pi}{N}\right)} \tag{14.4}$$

The phases θ of the image of the grating projection can be found as described above for all the pixels in the x direction. After that, the phase calculation is performed for all rows in the y-direction. Then, the phase θ of all the pixels of the grating projected on the object are maintained as the phase distribution.

When using moiré topography, the phase distribution θ_R of the grating projected on a flat reference plane is first analyzed. The grating is then projected on an object and the phase distribution is analyzed. The phase difference $\Theta_M = \theta_s - \theta_R$ at each pixel, showing the moiré fringes representing contour lines can be found. By using Eq. (14.1), $h = z_R - z_S$ is obtained at each pixel point of the camera. This gives the height distribution. The x and y coordinates are obtained from the geometric relationship. Then the three-dimensional shape of the object is determined.

Though the OPPA method is analyzing the phase by the Fourier transformation of the image data for one pitch of the grating, its actual implementation analyzes the phase using Eq. (14.4) by using the brightness data of the direct N pixels without actually performing the Fourier transform. Thereby it is possible to analyze the phase distribution at high-speed. It is possible to analyze high-speed objects by using a high-speed camera and performing batch processing on the recording. This makes it possible to analyze the vibration mode of vibrating objects.

14.4 Vibration Modal Analysis System

To protect vibration of structures, vibration modal analysis is important. Conventional experimental vibration analysis method uses accelerometers or laser displacement meters. The modal analysis takes much time and troublesome work. Especially, by contacting accelerometers to objects, the results of frequency mode have some errors because of the contact of the weight of the accelerometer.

In this study, the OPPA method is applied to frequency modal analysis. Figure 14.3 shows the vibration analysis system implemented using a high-speed camera. The light source is an LED that projects light through fiber-optic lines. The projection goes through a glass plate with grating lines and projects grating lines on an object as shown in Fig. 14.1. The projected grating on the object is recorded by the high-speed camera of 2000 fps. A cantilever aluminum plate with 200 mm × 100 mm × 2 mm is used as the specimen. The free edge of the cantilever is hit with a hammer, and 10 s of vibration are analyzed to obtain frequency modes. Figure 14.4 shows the display of the system showing the result of modal analysis. The power (left) and phase (right) distributions are shown with pseudo color. And amplitude and phase according to frequency at any pixel indicated by cross cursor are shown under the display of the power and phase. Figure 14.4a–d show the display at local peak frequencies; 40.0 Hz, 166.8 Hz, 249.6 Hz and 543.6 Hz, respectively. The vibration displays

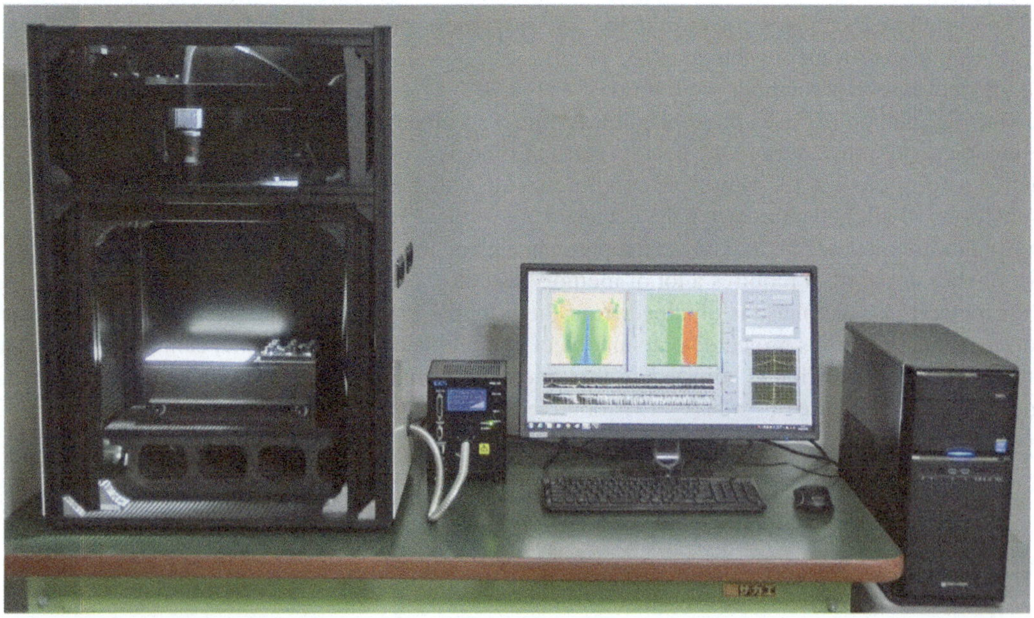

Fig. 14.3 Vibration analysis system using OPPA1

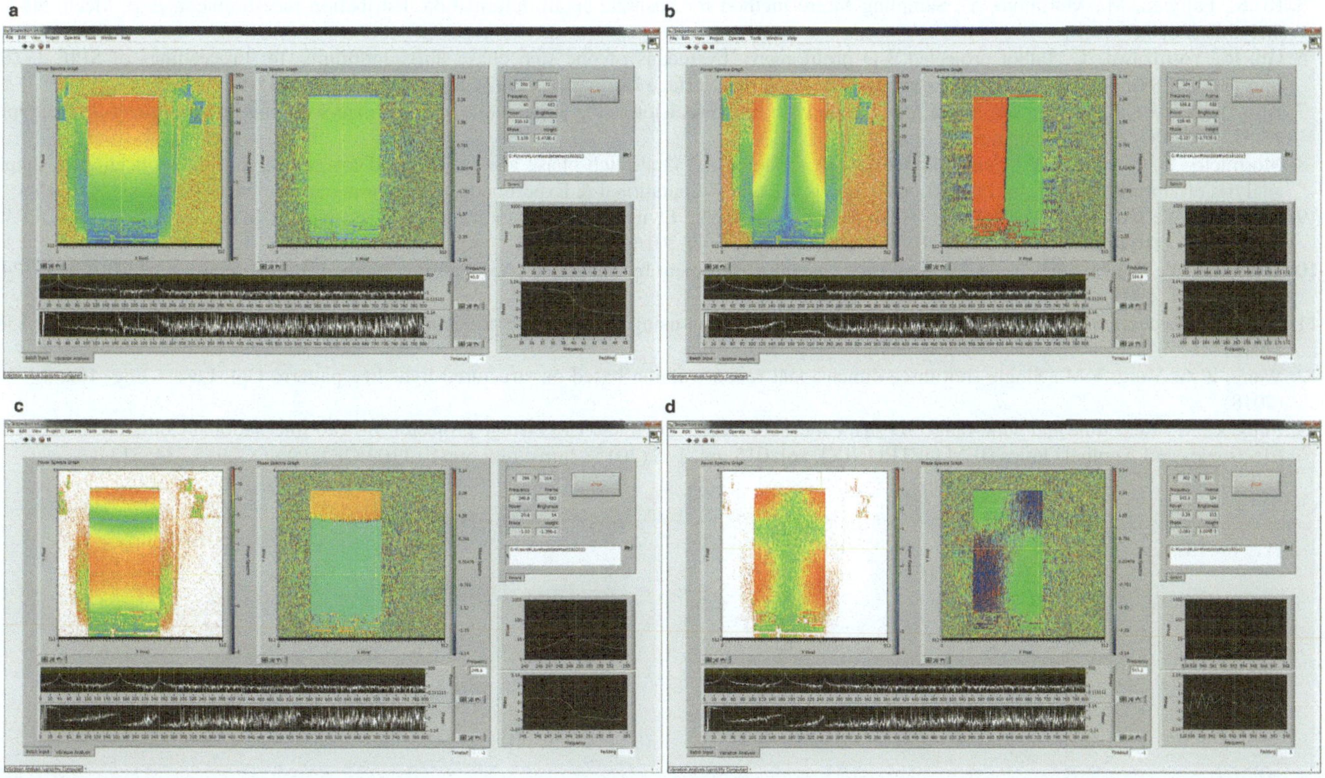

Fig. 14.4 Results of vibration mode analysis for a cantilever plate. (**a**) 1st mode (40.0 Hz); (**b**) 2nd mode (166.8 Hz); (**c**) 3rd mode (249.6 Hz); (**d**) 4th mode (543.6 Hz)

are corresponding to 1st, 2nd, 3rd and 4th mode, respectively. Like this, vibration modal analysis can be performed quickly without influencing analysis results through physical contact.

14.5 Conclusions

The OPPA method proposed by the authors can analyze the phase from a single image of a projected grating. Using this method, frequency modal analysis method is proposed. The mode change of cantilever vibration is displayed. The 1st, 2nd, 3rd and 4th mode are clearly observed. This method is useful for inspection of automobiles, machines, buildings, etc.

Acknowledgements The work in this paper was supported by many members of 4D Sensor Inc. The author appreciates Dr. A. Masaya, Mr. Y. Irino, Mr. Q. Quarles, Mr. A. Horn, Mr. I. Harker and, Mr. S. Seles and to their contributions, and Prof. H. Koike of Okayama University to his advice.

References

1. Morimoto, Y., Masaya, A., Ueki, M.: High-speed 3D shape measurement by one pitch phase analysis method using brightness values in small square area of one-shot image. Opt. Lasers Eng. (2018) (submitted)
2. Bruning, J.H., Herriott, D.R., Gallagher, J.E., Rosenfeld, D.P., White, A.D., Brangaccio, D.J.: Digital wave front measuring interferometer for testing optical surface and lenses. Appl. Opt. **13**(11), 2693–2703 (1974)
3. Takeda, M., Mutoh, K.: Fourier transform profilometry for the automatic measurement of 3-D object shapes. Appl. Opt. **22**(24), 3977–3982 (1983)
4. Huang, L., Kemao, Q., Pan, B., Asundi, A.K.: Comparison of Fourier transform, windowed Fourier transform, and wavelet transform methods for phase extraction from a single fringe pattern in fringe projection profilometry. Opt. Lasers Eng. **48**(2), 141–148 (2010)

5. Ri, S., Fujigaki, M., Morimoto, Y.: Sampling Moiré method for accurate small deformation distribution measurement. Exp. Mech. **50**(4), 501–508 (2010)
6. Gu, R., Yoshizawa, T.: Talbot projected 3-D profilometry by means of one step phase-shift algorithms. In: Proceedings of SPIE 1720, International Symposium on Optical Fabrication, Testing, and Surface Evaluation (1992). https://doi.org/10.1117/12.132157
7. Arai, Y., Kurata, T.: High-speed Moiré topography with high precision using a technique with a fringe scanning interferometer. Jpn. J. Opt. **15**(5), 402–406 (1986). in Japanese
8. Morimoto, Y., Kusunoki, Y., Ueki, M., Masaya, A., Takagi, A.: One-pitch phase analysis method for grating and its application to shape analysis of vibrating object. In: International Conference on Computational & Experimental Engineering and Sciences (ICCES'15), 2015
9. Morimoto, Y., Kusunoki, Y., Ueki, M., Takagi, A., Masaya, A.: Dynamic deformation measurement by one-pitch phase analysis (OPPA) method. In: 11th ISEM'16-Ho Chi Minh, Vietnam, 2-4 November 2016
10. Morimoto, Y., Fujigaki, M., Toda, H.: Real-time shape measurement by integrated phase-shifting method. In: Proceedings of SPIE. vol. 374, pp. 118–125 (1999)
11. Zhang, S.: Recent progresses on real-time 3D shape measurement using digital fringe projection techniques. Opt. Lasers Eng. **48**(2), 149–158 (2010)
12. Wang, Z.Z., Yang, Y.M.: Single-shot three-dimensional reconstruction based on structured light line pattern. Opt. Lasers Eng. **106**, 10–16 (2018)
13. Fujigaki, M.: Real-time and wide-range 3D shape measurement using linear LED fringe projector. In: 2nd International Conference on Opto-Electronics and Applied Optics (IEM OPTRONIX), vol. 015, pp. 1–5 (2015). https://doi.org/10.1109/OPTRONIX.2015.7345513

Chapter 15
DIC Image on FIB Ring-Core Analysis of Depth Sensing Residual Stress Measurement of Thin Films

Wen Chieh Pan, Ang-Ting Tsai, Fa-Yen Cheng, Terry Yuan-Fang Chen, and Ming-Tzer Lin

Abstract This work demonstrates a technique to measure the residual stress of thin films that involves a combination of focused ion beam (FIB) milling, SEM imaging and digital image correlation (DIC) analysis. Use focused ion beam (FIB) to surface milling different depth of the ring core on thin film and capture the clear images with sub-micron resolution SEM. The residual stress measurement of 1.5 μm thick silver film through PVD Sputter and E-gun deposition was discussed. By the use of digital image correlation (DIC) for full-field strain analysis and calculate the displacement-strain relationship of the silver film surface and SEM high-resolution surface deformation images. The residual stress state of the two kinds of silver films was evaluated, which indicated that FIB-DIC was well implemented to measure the residual stress.

Keywords Focused Ion Beam (FIB) · Digital Image Correlation (DIC) · Residual stress · Hole drilling · Ring-core

15.1 Abstract

Reliable measurement and modeling of residual stresses at the micron-scale is a big challenging task, as demonstrated by efforts that have been recently made in the Micro-Nanotechnology. The specific micro scale evaluation of residual stress gradients is a very critical issue in the mechanical design and reliability of small scale systems. The analysis of residual strain depth profiles requires detailed knowledge of in-depth lattice strain function in order to calculate the residual stress profile in a manner that considers both the mechanical anisotropy and the texture of the materials [1]. Also, the application of these techniques to the analysis of residual stress in amorphous materials is often not possible. The development of an independent procedure for depth-resolved measurement of residual stress in a microstructure is an issue of strategic interest. Full-field measurement of the deformation behaviour of small scale structures and nanostructured materials has recently been made possible by combining Scanning Electron Microscopy (SEM) and Digital Image Correlation (DIC) techniques [2]. A major problem is the application of a suitable pattern on the test sample surface for tracking from one image to another. Although a high-quality pattern can be artificially applied for better measurement, the mechanical behaviour of sample structures may become affected by the patterning process. The high magnification of SEM-DIC leads to complex image distortions. Long image scanning times can capture high noise sensitivity and drift from SEM parameters; therefore the development of a proper scheme for digital correlation of SEM images with natural surface features is needed. Moreover, correct implementation of these techniques may encounter complex problems than those encountered in macro-scale DIC. Here, we perform a digital correlation (DIC) of the specimen images acquired by incremental focused ion beam (FIB) ring-core drilling with various depth steps with each depth of 300 nm. A translation test was performed first to study the applicability of DIC to FIB images, and a proper procedure was established to obtain more accurate results. Sputtered deposited Ag and TiN films were used for this study. An example of the residual stress measurement of 1.5 μm Ag thin film was measured by a combination of focus ion beam (FIB) ring core milling, SEM imaging and DIC analysis. The total measurement of the relaxation strain at each drilling depth was obtained through the calculation of Kabiri [3] and Schajer [4] methods after surface deformation/strain revealed by DIC. A schematic experimental procedure is shown in Fig. 15.1. Figure 15.2 show the SEM images of the sample on the pattern area after FIB ring core drilling. Three high magnified 8000×

W. C. Pan · A.-T. Tsai · M.-T. Lin (✉)
Graduate Institute of Precision Engineering, National Chung Hsing University, Taichung, Taiwan
e-mail: mingtlin@nchu.edu.tw

F.-Y. Cheng · T. Y.-F. Chen
Department of Mechanical Engineering, National Cheng Kung University, Tainan, Taiwan

L. Lamberti et al. (eds.), *Advancement of Optical Methods & Digital Image Correlation in Experimental Mechanics, Volume 3*, Conference Proceedings of the Society for Experimental Mechanics Series, https://doi.org/10.1007/978-3-319-97481-1_15

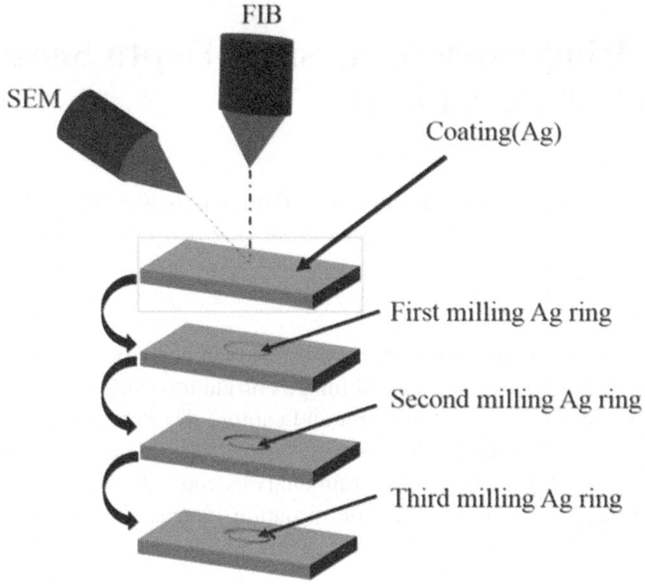

Fig. 15.1 Experimental setup of SEM and FIB

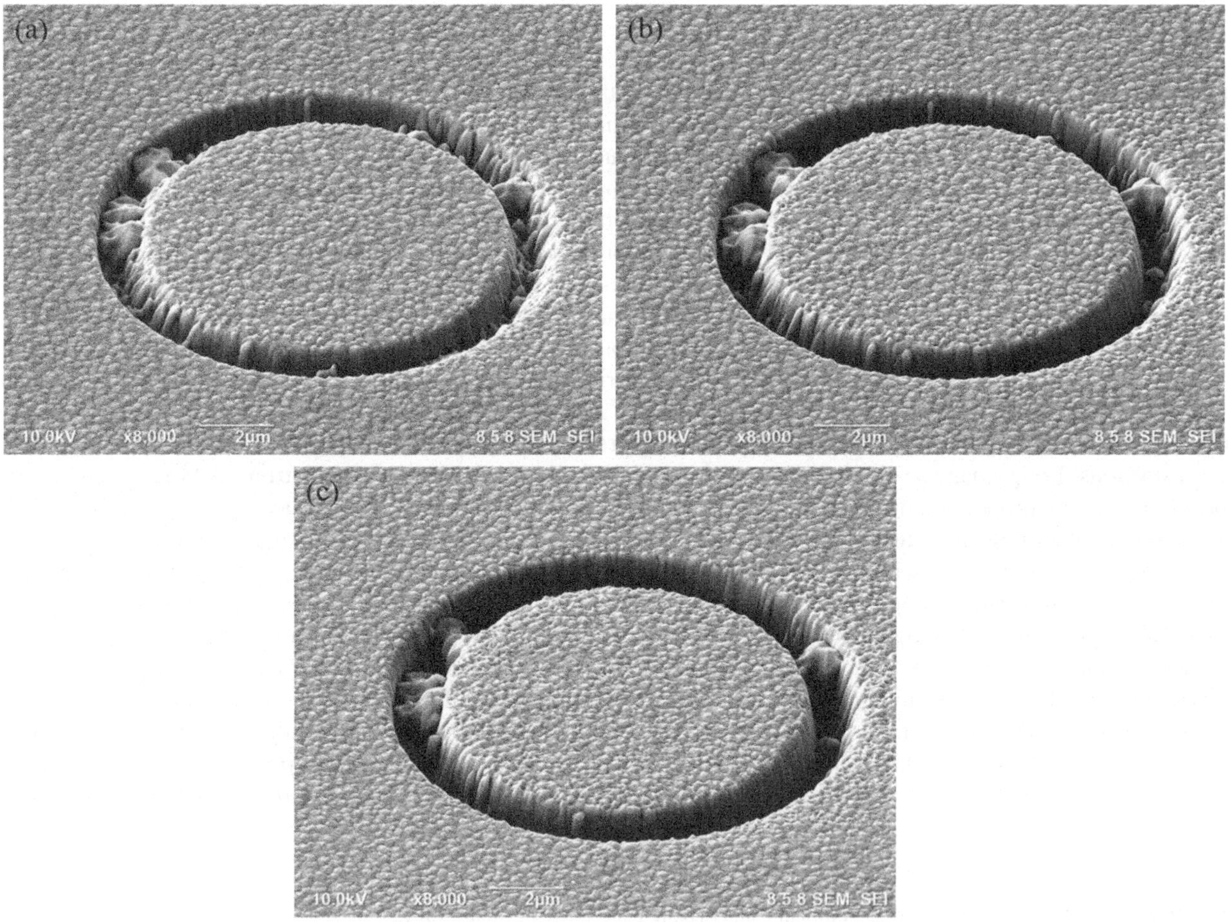

Fig. 15.2 SEM micrograph of the patterned ring core surface of the sample after milling depth of (**a**) 0.626 μm (**b**) 0.852 μm (**c**) 1.027 μm

Table 15.1 Summary of the residual stress values of Ag thin films determined from previous studies and the results tested here

	50/50 nm Ag/Ni bilayer	7/7 nm Ag/Ni multilayer	1200 nm Ag Film	5 μm Ag/Ni multilayer film	1.5 μm Ag film
FIB/DIC of this study (MPa)					+150 (0.5 μm) +450 (1 μm)
XRD texture analysis (MPa)	+130	+400		+310 (10 layers) +520 (4 layers)	
Curvature method (MPa)			310 ± 10		

images taken after milling depth of (a) 0.626 μm (b) 0.852 μm (c) 1.027 μm on it. These images were used for DIC and relaxation strain calculation to obtain the residual stress of each milling depth. The results were used to compare with the literatures values and summarized in Table 15.1. These stress values were consist to those obtain from XRD and curvature measurement results.

References

1. Bemporad, E., Brisotto, M., Depero, L.E., Gelfi, M., Korsunsky, A.M., Lunt, A.J.G., Sebastiani, M.: A critical comparison between XRD and FIB residual stress measurement techniques in thin films. Thin Solid Films. **572**, 224–231 (2014)
2. Massl, S., Keckes, J., Pippan, R.: A direct method of determining complex depth profiles of residual stresses in thin films on a nanoscale. Acta Mater. **55**, 4835–4844 (2007)
3. Kabiri, M.: Measurement of residual-stresses by the hole-drilling method-influences of transverse sensitivity of the gauges and relieved-strain coefficients. Exp. Mech. **24**, 252–256 (1984)
4. Schajer, G.S.: Application of finite element calculations to residual stress measurements. J. Eng. Mater. Technol. **103**, 157–163 (1981)

Chapter 16
Measurement of Local Strain Distribution and Its Variation Near Eyes During Blink Using Digital Image Correlation

Kasumi Sakai, Yuelin Zhang, Satoru Yoneyama, Yukihiro Miyazaki, Yuko Nagai, and Takanori Igarashi

Abstract The strain distribution of the facial skin surface at the local area near eyes at the instant of blinking are measured for evaluating the effect of cosmetics products. The motion of the skin at the local area as well as the whole face is observed using multiple digital high-speed cameras. Images of the typical blink motion are extracted from the mass of the local images by evaluating the blink motion in the whole face images. Two-dimensional digital image correlation is used for measuring strain. Irregularities and textures of the skin near the eyes are utilized as random patterns in digital image correlation. The strain distribution and its variation with time under various condition of skin such as bare skin, skin care, and foundation are obtained. The effect of cosmetic product on the skin behavior is elucidated by performing the series of the measurement for several subjects.

Keywords Facial skin · Strain · Digital image correlation · Skin condition · Blink

16.1 Introduction

The deformation of the facial skin surface is occurred due to the change of the facial expression, and it is known that the deformation is different under each condition such as moisture level or applied cosmetics [1]. Therefore it is considered that the deformation is potential indicator to know the skin conditions. In this study, the deformation on the facial skin surface of daily life is expressed by strain.

Human skin surface is sensitive, so a non-contact and non-invasive method is suitable to measure the strain. There are some methods to measure the strain such as digital image correlation (DIC) [1] and optical coherence topography [2]. The previous study [1] shows that DIC is effective for measuring the strain distribution on the whole face. However, the method can be regarded as a non-contact but not a non-invasive since random pattern is painted on the skin to measure the deformation of the face. To propose the non-invasive method by using DIC, authors attempt to substitute a pattern naturally present on the skin such as textures or irregularities for the paint pattern [3]. The results show that it is effective to measure the time histories of the strain in the local area around the eye (250 μm^2). The final goal of this study is to propose a non-contact and non-invasive method to measure the strain of the skin in-site by using DIC. In this paper, to show the effectiveness of the proposed method, the time histories and distributions of strain around the eyes during the blinking under various skin conditions are measured.

K. Sakai · Y. Zhang · S. Yoneyama (✉)
Department of Mechanical Engineering, Aoyama Gakuin University, Sagamihara, Japan
e-mail: yoneyama@me.aoyama.ac.jp

Y. Miyazaki · Y. Nagai
Makeup Products Research, Kao Corporation, Odawara, Japan

T. Igarashi
Skincare Products Research, Kao Corporation, Tokyo, Japan

L. Lamberti et al. (eds.), *Advancement of Optical Methods & Digital Image Correlation in Experimental Mechanics, Volume 3*,
Conference Proceedings of the Society for Experimental Mechanics Series, https://doi.org/10.1007/978-3-319-97481-1_16

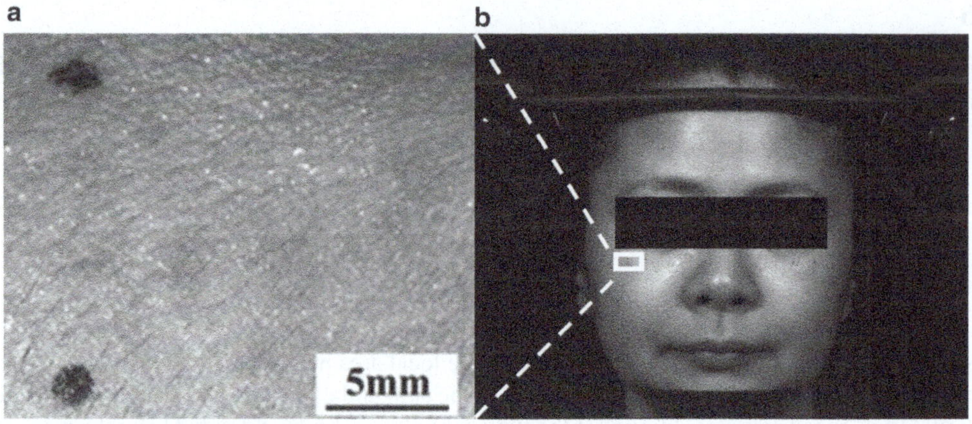

Fig. 16.1 The images of the local area near right eye and the whole face. (**a**) The local area; (**b**) whole face

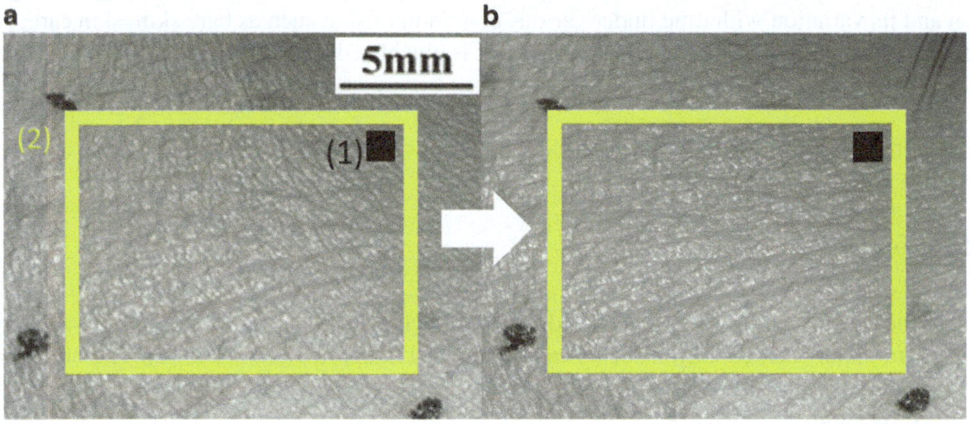

Fig. 16.2 Analyses areas of time history (1) and distribution (2) of strains. (**a**) Eye open (reference image); (**b**) eye close

16.2 Methods

16.2.1 Outline of Experiment

In this experiment, the deformation during blinks of whole face and the local area around eye (Fig. 16.1) of two subjects (female, 40s) are captured by two cameras with 200 fps (k7-USB, Kato Koken Corp.). The size of images is 1240 × 1048 pixs. The skin textures and irregularities is successfully observed at the local area (Fig. 16.1a), and they are used as the pattern to calculate the displacement of the skin. The measurements are performed under three skin conditions: after washing the face (bare skin), after applied the lotion (skin care) and further applying the foundation (foundation).

16.2.2 Analysis Condition

The reference image of the analyses is set to an image when the eye open (Fig. 16.2a), and 100 images including one blink are used to calculate the displacements. A subset of 21 × 21 pixs are used to calculate the displacements of the analyses. The time histories of the strain of the local area (101 × 101 pixs, approximately 1.2 × 1.2 mm) which is shown in the Fig. 16.2(1) are calculated, and the gauge length used is 20 × 20 pixs. The distributions of the principal strains and the principal direction of the wide area (781 × 581 pixs approximately 9.4 × 7.0 mm) which is shown in the Fig. 16.2a(2) are calculated, and the gauge length used is 100 × 100 pixs. ε_1, ε_2 and θ represent the max principal strain, the minimum principal strain and the principal direction, respectively.

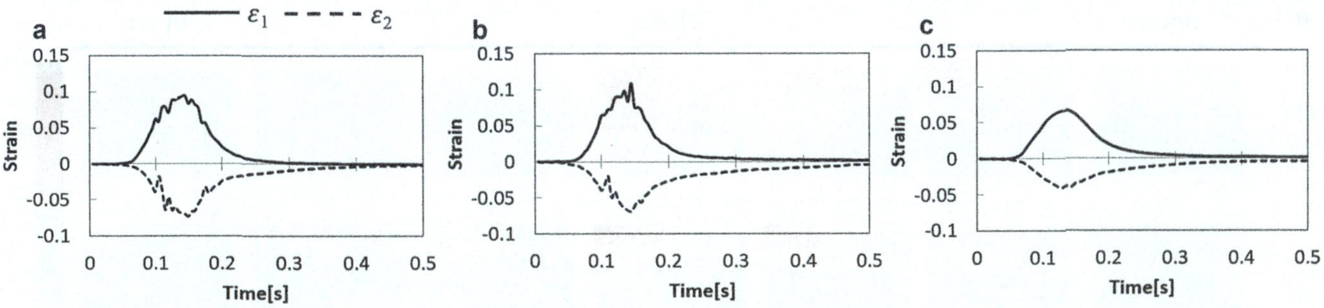

Fig. 16.3 Time histories of the strains of subject A. (**a**) Bare skin; (**b**) skin care; (**c**) foundation

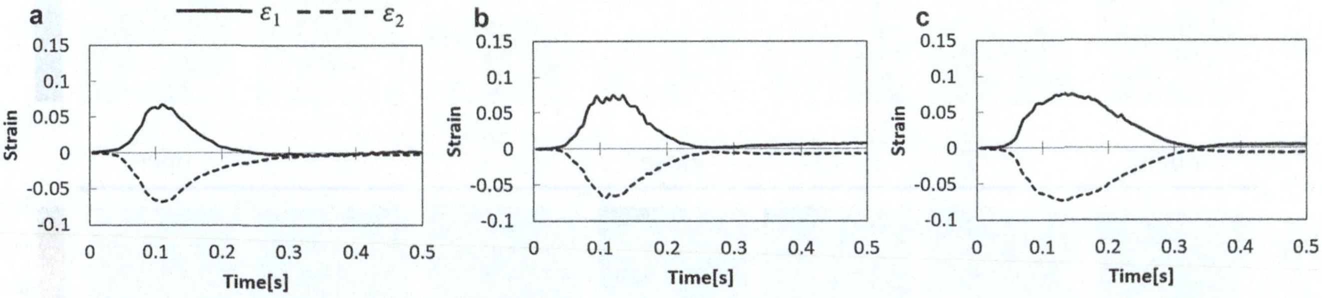

Fig. 16.4 Time histories of the strains of subject B. (**a**) Bare skin; (**b**) skin care; (**c**) foundation

16.3 Results and Discussion

16.3.1 Time Histories of the Strain Near Eye

The results of the strain time histories of subject A and B are shown in Figs. 16.3 and 16.4, respectively. In the figures, (a), (b) and (c) show the results of bare skin, skin care and foundation conditions, respectively. The strains decline after blinking under all of the skin conditions, so it can be considered that the method substituted skin textures and irregularities for painted pattern and effectively measures the time histories of strains.

The maximum value of the principal strain ε_1 of subject A is observed at 0.15 s after starting blinking, and the value under skin care condition is more than 0.1 (Fig. 16.3a) but under foundation condition is about 0.07 (Fig. 16.3c). The values of strain of the subject A are different under different skin conditions. The maximum value of the strains of subject B are about 0.075 under all the skin conditions (Fig. 16.4), so the difference of the strain between the each skin conditions is smaller than A.

16.3.2 Distribution of the Principal Strain and Direction

The results of ε_1, ε_2 and θ of subject A and B under all of the skin conditions are shown in Figs. 16.5 and 16.6, respectively. In the figures, 7 images are picked up from 100 pictures during a blink the forth images are the results when the eyes are completely closed and the displacements become maximum. The principal strains become maximum at the center image and decreases from the center. The both ε_1 and ε_2 of the subject A tend to increase at the local upper right region (Fig. 16.5a, b), whereas the both ε_1 and ε_2 of the subject B tend to increase in overall area (Fig. 16.6a, b). Moreover, there have not much difference of θ of each subject under the three conditions (Figs. 16.5c and 16.6c), so it can be said that the degree during blinks under the three conditions are similar.

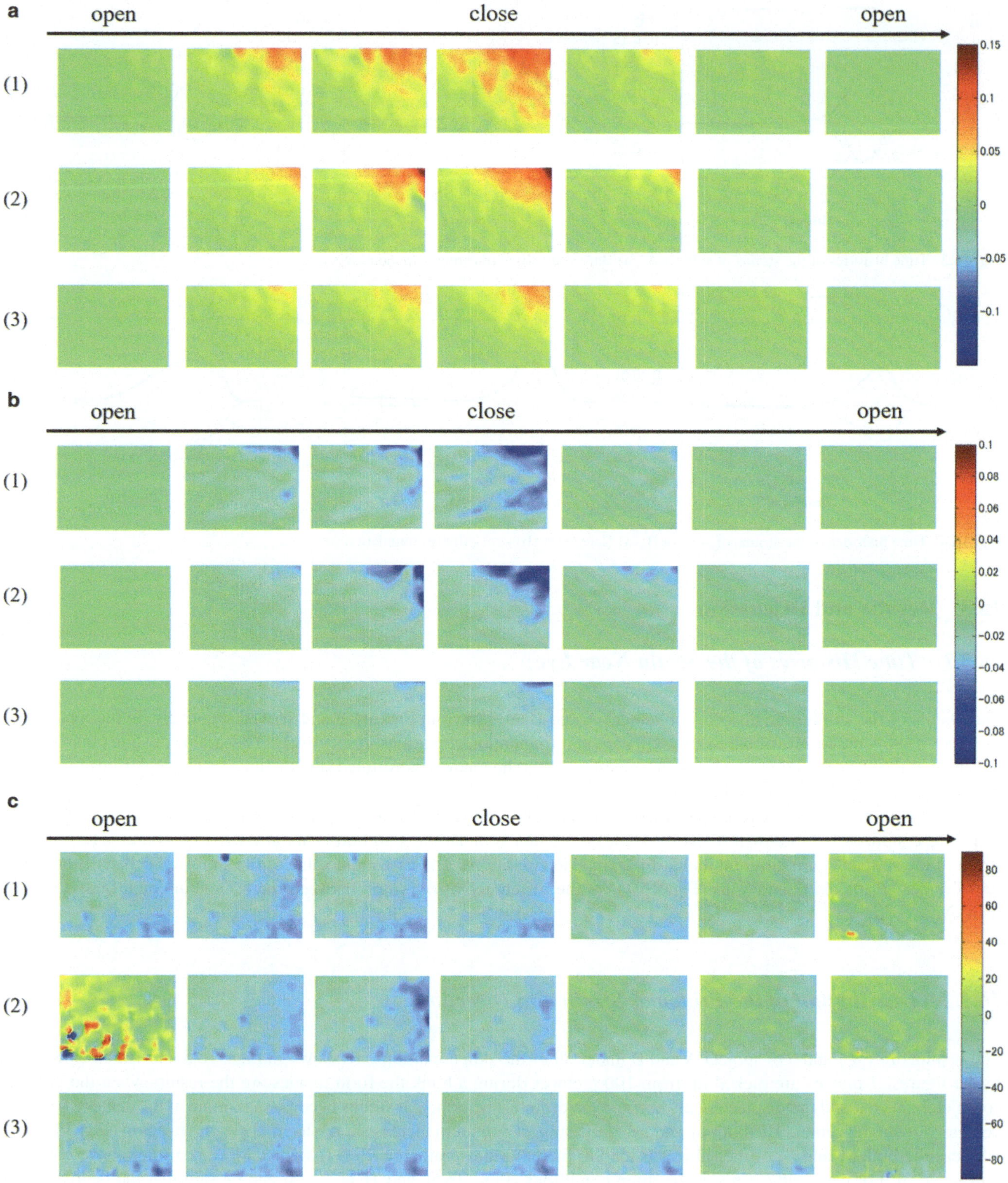

Fig. 16.5 Principal strains and direction distribution of the subject A (1), (2) and (3) indicate the bare skin, skin care and foundation condition, respectively. (**a**) Principal strain ε_1; (**b**) Principal strain ε_2; (**c**) Principal direction θ

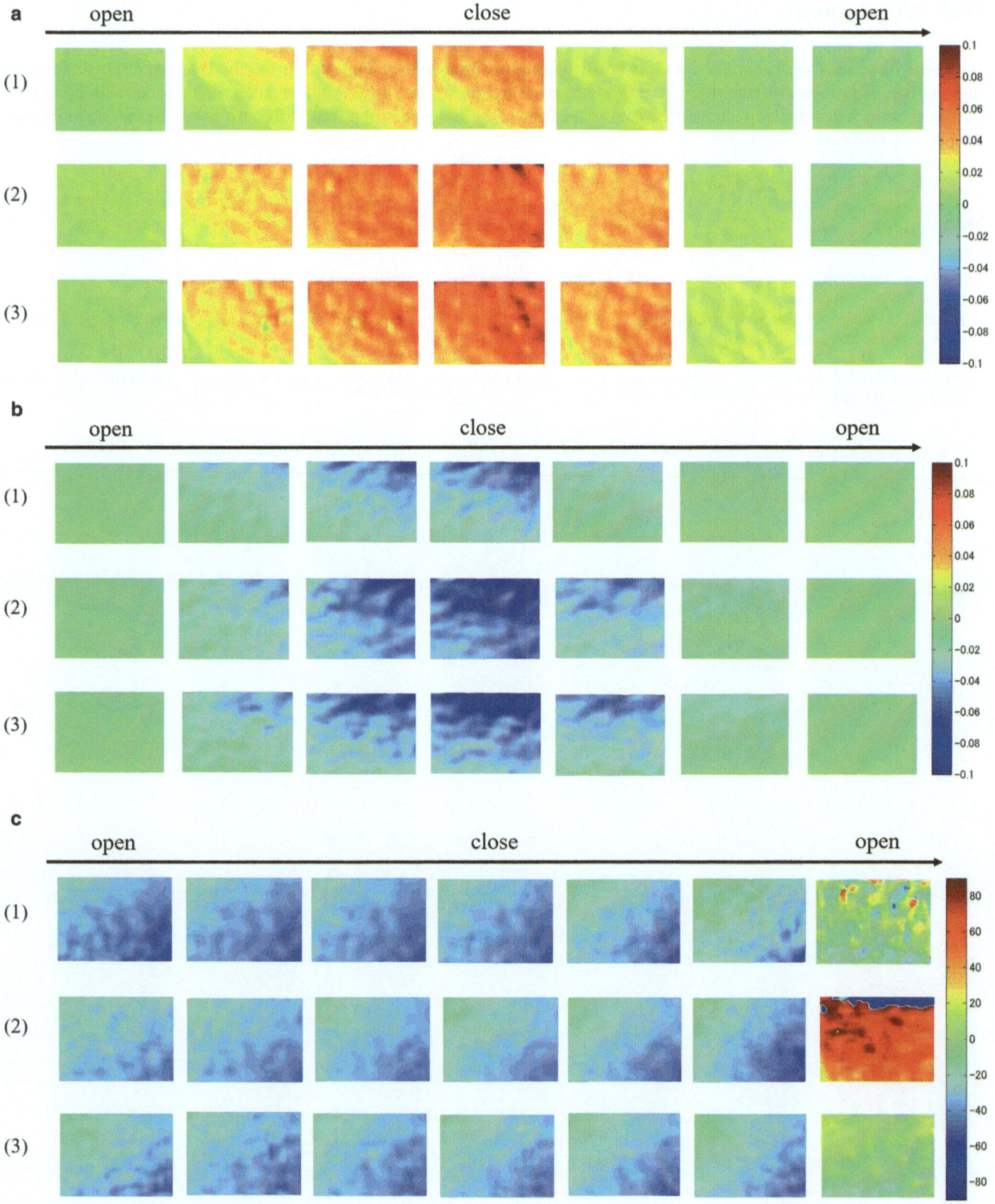

Fig. 16.6 Principal strains and direction distribution of the subject A (1), (2) and (3) indicate the bare skin, skin care and foundation condition, respectively. (**a**) Principal strain ε_1; (**b**) Principal strain ε_2; (**c**) Principal direction θ

16.4 Conclusions

A method for measuring the time histories and distributions of strain on the facial skin surface by DIC using skin textures and irregularities is proposed in this study. According to measure the strains of two subjects under the various skin conditions, the results shown that the individual differences in the time histories and distributions of the strains can be measured by using the proposed method.

References

1. Miura, N., Sakamoto, T., Aoyagi, Y., Yoneyama, S.: Visualizing surface strain distribution of facial skin using stereovision. Theor. Appl. Mech. Lett. **6**(2), 167–170 (2016)
2. Mahmud, J., Evans, S.L., Holt, C.A.: An innovative tool to measure human skin strain distribution in vivo using motion capture and delaunay mesh. J. Mech. **28**(2), 309–317 (2017)
3. Hiruma, U., Zhang, Y.L., Yoneyama, S., Miyazaki, H., Nagai, Y., Igarahsi, T.: Strain measurement during human wink by using digital correlation method (in Japanese). In: Proceedings of the 29th Bioengineering Conference, Annual Meeting of BED/JSME, vol. 29, Paper Number: 1D13, 2017

Chapter 17
Contribution to Fatigue Striation Phenomenon Analysis by Using Image Processing

Benoit Ruellan, Eric Robin, Jean-Benoit Le Cam, Isabelle Jeanneau, Frédéric Canévet, Gérard Mauvoisin, and Didier Loison

Abstract Since the use of energetic approaches for the prediction of the number at macro-crack initiation in elastomers, a special attention is paid on fatigue crack growth at the microscopic scale. In filled natural rubber, failure surfaces exhibit wrenchings and striations (Le Cam et al., Int J Fatigue 52:82–94, 2013). Both are assumed to be due to strain-induced crystallization (SIC). Only four studies address fatigue striations (Le Cam et al., Int J Fatigue 52:82–94, 2013; Le Cam and Toussaint, Macromolecules 43:4708–4714, 2010; Flamm et al., Int J Fatigue 33:1189–1198, 2011; Muñoz-Mejia, Dissertation, Université Claude Bernard, Lyon I, 2011), while they could provide information of importance to better understand how SIC enables natural rubber to resist the crack growth. As striations are similar to fringe patterns, this study aims at using a phase extraction algorithm from a single fringe pattern to analyse the striation morphology (Robin et al., Appl Opt 44:7261–7269, 2005; Takeda et al., J Opt Sot Am 72:156–160, 1982; Servin et al., Appl Opt 36(19):4540–4548, 1997; Robin et Valle, Appl Opt 43(22):4355–4361, 2004; Valle et al., Strain 46(2):175–183, 2008). This phase extraction methodology is split into three steps. The first one consists in extracting the wrapped phase without orientation. The second step is devoted to the determination of the fringe pattern orientation from a classic unwrapping algorithm. The third and last step consists in using an unwrapping algorithm (Zuo et al., Opt Lasers Eng 85:84–103, 2016; Menese et al., Appl Opt 44(7):1207–1215, 2005) and to compute the difference between the unwrapped phase processed and a plane in order to analyse the evolution of the striation morphology. This methodology has been applied to characterize the striation morphology observed at the failure surface of specimen tested under different fatigue loading conditions.

Keywords Fatigue striation · Phase demodulation · Single fringe pattern · Experimental mechanics · Rubber

17.1 Introduction

Fatigue striation patterns observed at the failure surface of samples submitted to fatigue loading conditions exhibit a similar morphology as a fringe pattern one, in the sense that they exhibit a periodic pattern [1–4]. This is illustrated in Fig. 17.1.

B. Ruellan (✉)
Univ Rennes, CNRS, IPR (Institute de Physique de Rennes)–UMR 6251, Rennes, France

LC-DRIME, Joint Research Laboratory, Cooper Standard—Institut de Physique UMR 6251, Rennes Cedex, France

Cooper Standard France, Rennes, France
e-mail: benoit.ruellan@cooperstandard.com

E. Robin · J.-B. Le Cam
Univ Rennes, CNRS, IPR (Institute de Physique de Rennes)–UMR 6251, Rennes, France

LC-DRIME, Joint Research Laboratory, Cooper Standard—Institut de Physique UMR 6251, Rennes Cedex, France

I. Jeanneau · F. Canévet
LC-DRIME, Joint Research Laboratory, Cooper Standard—Institut de Physique UMR 6251, Rennes Cedex, France

Cooper Standard France, Rennes, France

G. Mauvoisin
Laboratoire de Génie Civil et Génie Mécanique EA 3913, IUT-Université de Rennes 1, Rennes, France

D. Loison
Univ Rennes, CNRS, IPR (Institute de Physique de Rennes)–UMR 6251, Rennes, France

L. Lamberti et al. (eds.), *Advancement of Optical Methods & Digital Image Correlation in Experimental Mechanics, Volume 3*,
Conference Proceedings of the Society for Experimental Mechanics Series, https://doi.org/10.1007/978-3-319-97481-1_17

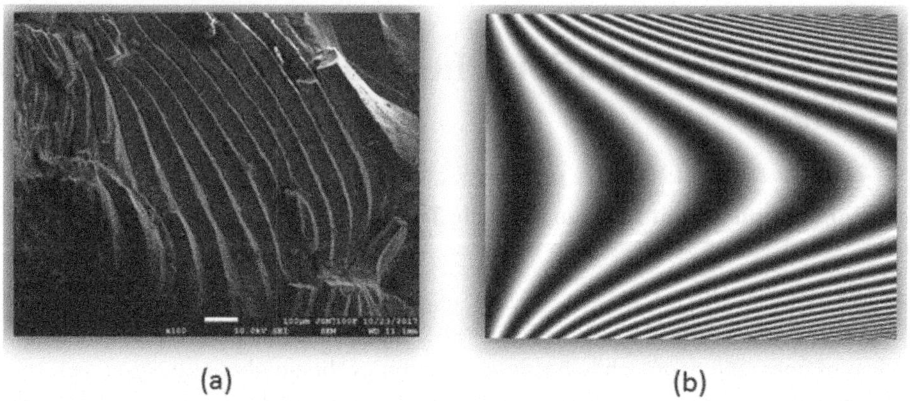

Fig. 17.1 (**a**) Striation pattern (**b**) fringes pattern

In an image (for instance that in Fig. 17.1a), the fringe pattern-like is issued from fatigue phenomenon and the corresponding phase cannot be shifted. This means that analysing fatigue striation as fringe pattern has to be carried out from a single image. Phase extraction from a single image was made possible by the development of analysis method such as the pMPC developed in [5–9]. The question addressed in the present paper is to know whether such analysis method can be directly applied to a SEM image of fatigue striation.

17.1.1 Fringe Pattern Method Used for Analyzing Fatigue Striation: Modulation Phase Correlation (pMPC)

The digital fringes pattern can be defined as follow:

$$I(i,j) = A(i,j)\cos(\varphi(i,j)) + B(i,j) \tag{17.1}$$

where $A(i,j)$, $B(i,j)$ and $\varphi(i,j)$ stand for the amplitude modulation, the background and the phase, respectively. The method is based on the measurement of the similarity degree between the fringe pattern studied and a virtual fringe pattern that is composed of parallel fringes. In other words, A, B and φ are determined by minimizing the cost function:

$$\Psi(A,B,\varphi) = \iint \left(Val_{exp} - (A\cos(\varphi) + B)\right)^2 \tag{17.2}$$

In practice, since the cost function describes in Eq. (17.2) is not unimodal, $\cos(\varphi) + B$ is replaced by its Maclaurin series expansion [5]. Indeed, we cannot ensure the unicity of A, B and φ obtained from the minimization process, but this is the case with a polynomial form. The polynomial form writes as follows:

$$\begin{aligned} &A\cos\left(\tfrac{2\pi}{p}X + \varphi\right) + B = \\ &A\cos(\varphi) + B - \tfrac{2A\sin(\varphi)\pi}{p}X - \tfrac{2A\cos(\varphi)\pi^2}{p^2}X^2 + \tfrac{4A\sin(\varphi)\pi^3}{3p^3}X^3 + O\left(X^4\right) \end{aligned} \tag{17.3}$$

where function $X = (x - \xi)\cos(\alpha) + (y - \gamma)\sin(\alpha)$ and α is the fringe inclination. In these conditions, variables optimizing the new cost function are the polynomial form's coefficients, in other words C_0, C_1, C_2, and C_3, and the cost function rewrites as follows:

$$\psi_{pMPC}(c_0, c_1, c_2, c_3, \alpha) = \int_{N_{xy}} \left(I(x-\xi, y-\gamma) - P\left(\xi, \gamma\right)\right)^2 d\xi d\gamma$$

with:

$$P(\xi,\gamma) = c_0 + c_1X + c_2X^2 + c_3X^3 \tag{17.4}$$

and

$$X = (x - \xi)\ \cos(\alpha) + (y - \gamma)\sin(\alpha)$$

In this formulation, A, B and φ are not directly obtained from the optimization process but only C_0, C_1, C_2, C_3 and α. φ is then obtained by the analytical resolution of the equation system below:

$$\begin{cases} C_0 = A\cos(\varphi) + B \\ C_1 = -\frac{2A\sin(\varphi)\pi}{p} \\ C_2 = -\frac{2A\cos(\varphi)\pi^2}{p^2} \\ C_3 = \frac{4A\sin(\varphi)\pi^3}{3p^3} \end{cases} \quad (17.5)$$

and φ chosen as:

$$\varphi = \arctan\left(\frac{\pi C_1}{pC_2}\right) \quad (17.6)$$

As the parameter α ranges between 0 and π during the optimization process the maps giving the phase field unwrapped is not correctly oriented [8]. The correctly orientated phase field is obtained with the help of unwrapping process applied to α field. During this unwrapping process [10, 11] of α field, φ field correctly orientated can be calculated by removing φ by (2π−φ) when π discontinuity is detected [8].

17.1.2 Applying pMPC to SEM Images

In a first approach, a single striation pattern SEM image has been processed with the pMPC algorithm. The SEM used for the purpose of the study is a JSM JEOL 7100F. Results are presented in Fig. 17.2.

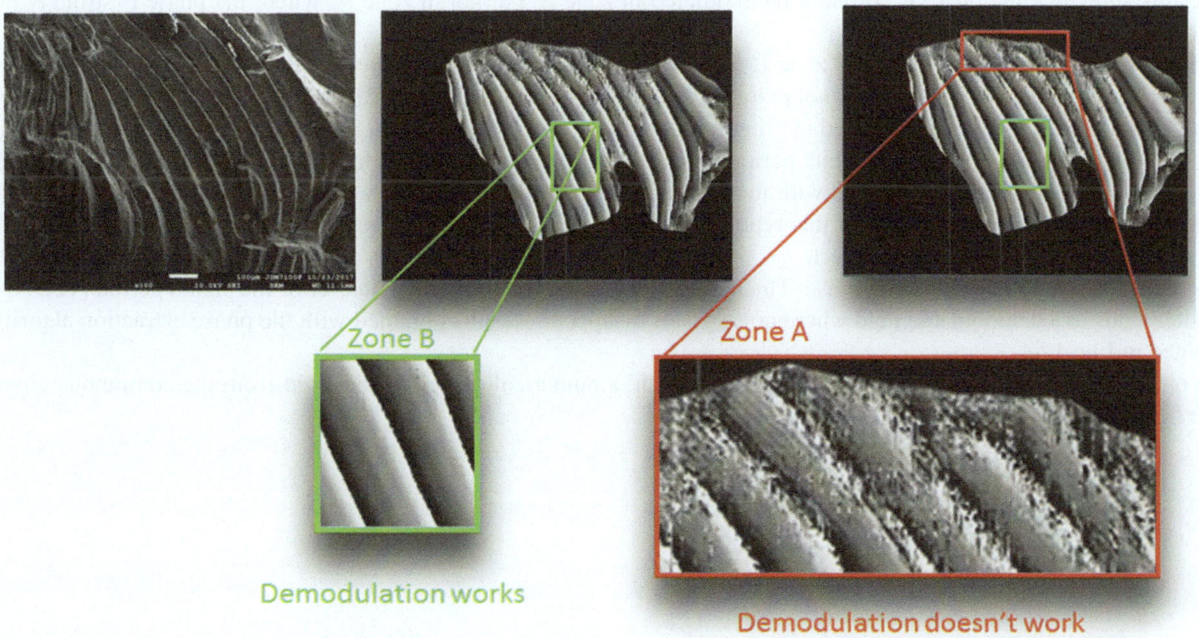

Fig. 17.2 Striation image directly analysed by pMPC

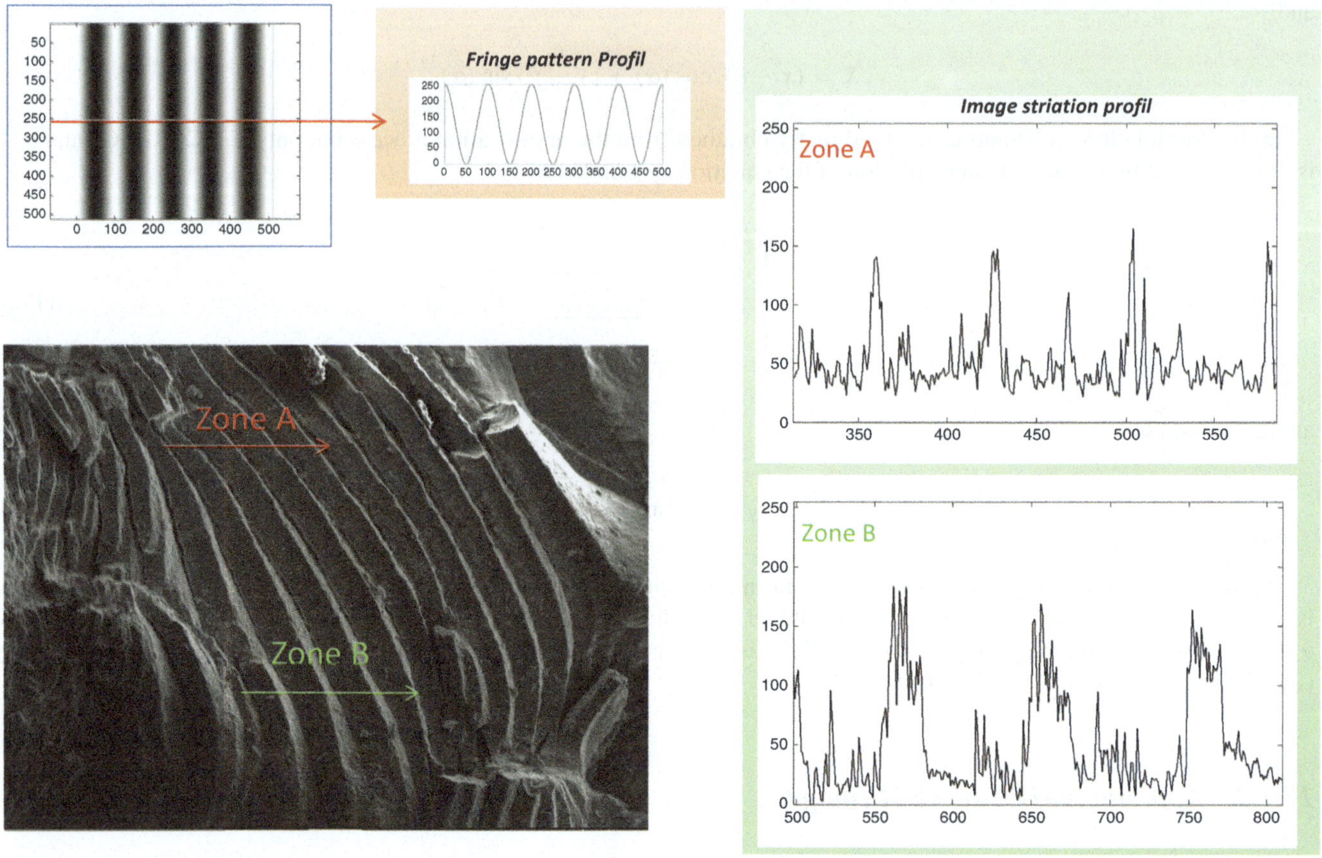

Fig. 17.3 Profiles of Zone A and B

Results show that the phase is correctly demodulated in zone B, but not in zone A, where the phase is strongly noisy or not correctly demodulated.

As shown by profiles in zone A and zone B in Fig. 17.3, the variation of the grey levels does not correspond to a fringe pattern. This explains why the phase is not extracted correctly in zone A (Fig. 17.2). Images, more especially the grey level dynamics, have therefore been processed in order to fit the grey level dynamics of a fringe pattern. Examining zone A profile shows that the signal is noisy but exhibits periodic peaks. These peaks should be larger and less noisy. The morphological modification of each peak is performed with the help of mathematical morphology in grey level. As morphology of profile becomes a square periodic signal with this type of morphological operation, we propose to use the edge effect of median filter in order to remove this square effect.

Note that such a filter reduces the noise. The kernel size has to be lower than a quarter of the signal period [12]. Applying this filter leads to a demodulated field where noisy zones disappear. Results obtained with the phase extraction algorithm are presented in Fig. 17.4.

In order to highlight the variation in the striation pitch, a numerical plane is subtracted from the continuous unwrapped phase field.

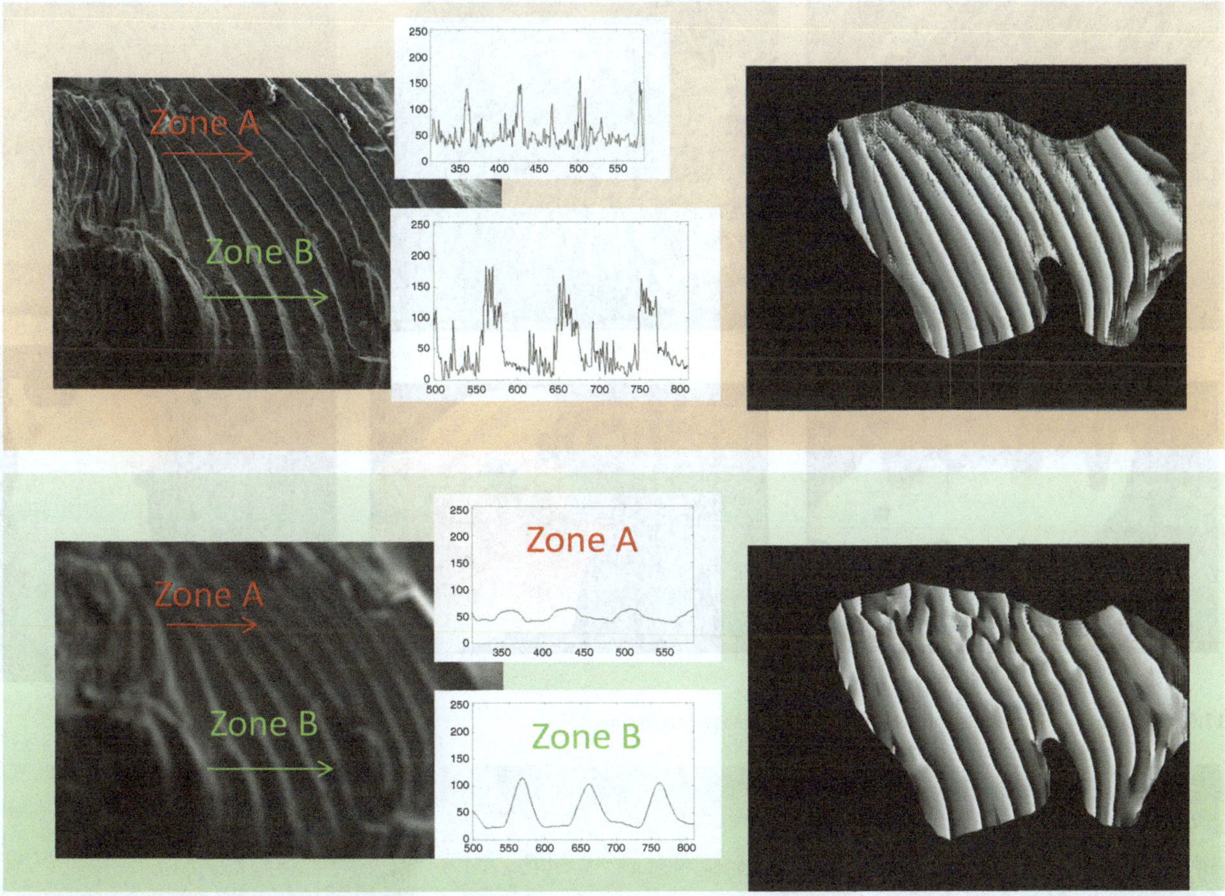

Fig. 17.4 Striation pattern processed for phase demodulation

17.2 Results

The methodology previously described has been applied on six different SEM images exhibiting fatigue striations. These striations have been obtained from different loading conditions applied to Diabolo samples (see [13]). Figures 17.5 and 17.6 present results obtained, which will be more precisely detailed and discussed during the presentation.

17.3 Conclusion and Perspectives

In this paper, a non-destructive technique is proposed to study fatigue striation based on a phase extraction algorithm. Results presented remain qualitative and not quantitative. Indeed, SEM images are affected by perspective effects (the different zones at the surface observed are not in the same plane), which induces the detection of apparent striation pitch variation by the algorithm. In order to access quantitative data, the striation topology is required. It can be characterized by different metrological devices, but the small striation height and the large out-of-plane zones makes difficult the measurement. This is why topology characterization of fatigue striation with SEM is currently under study in our lab.

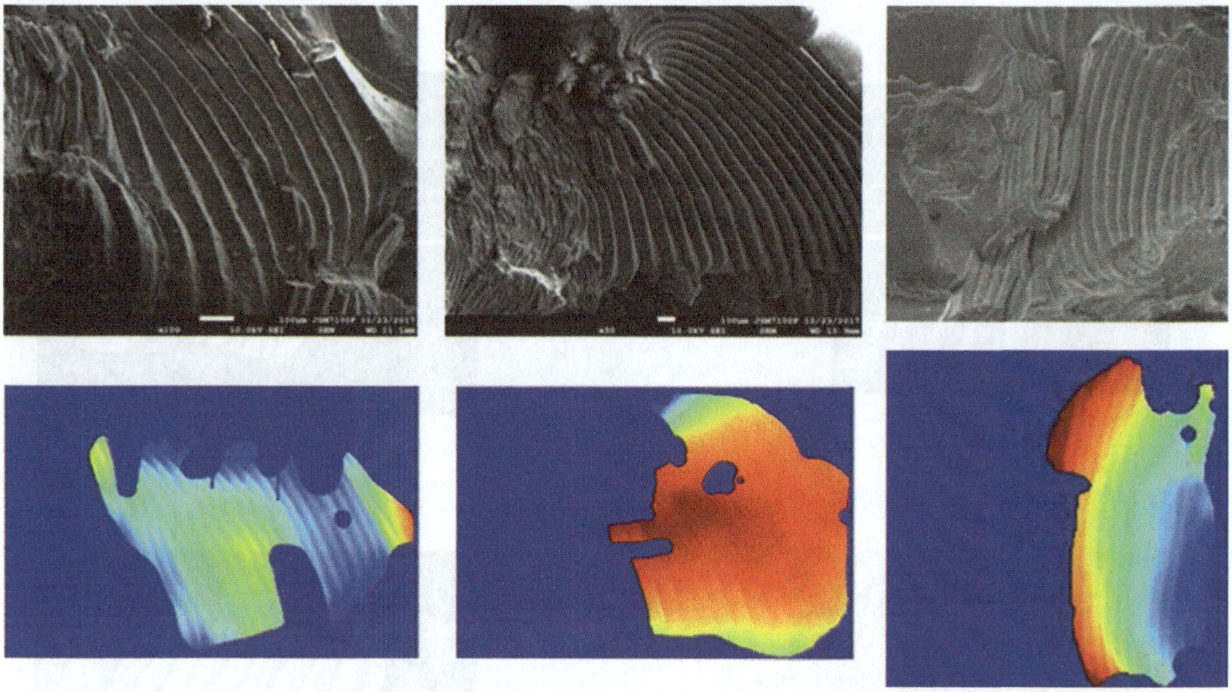

Fig. 17.5 SEM images analysis by phase demodulation image processing

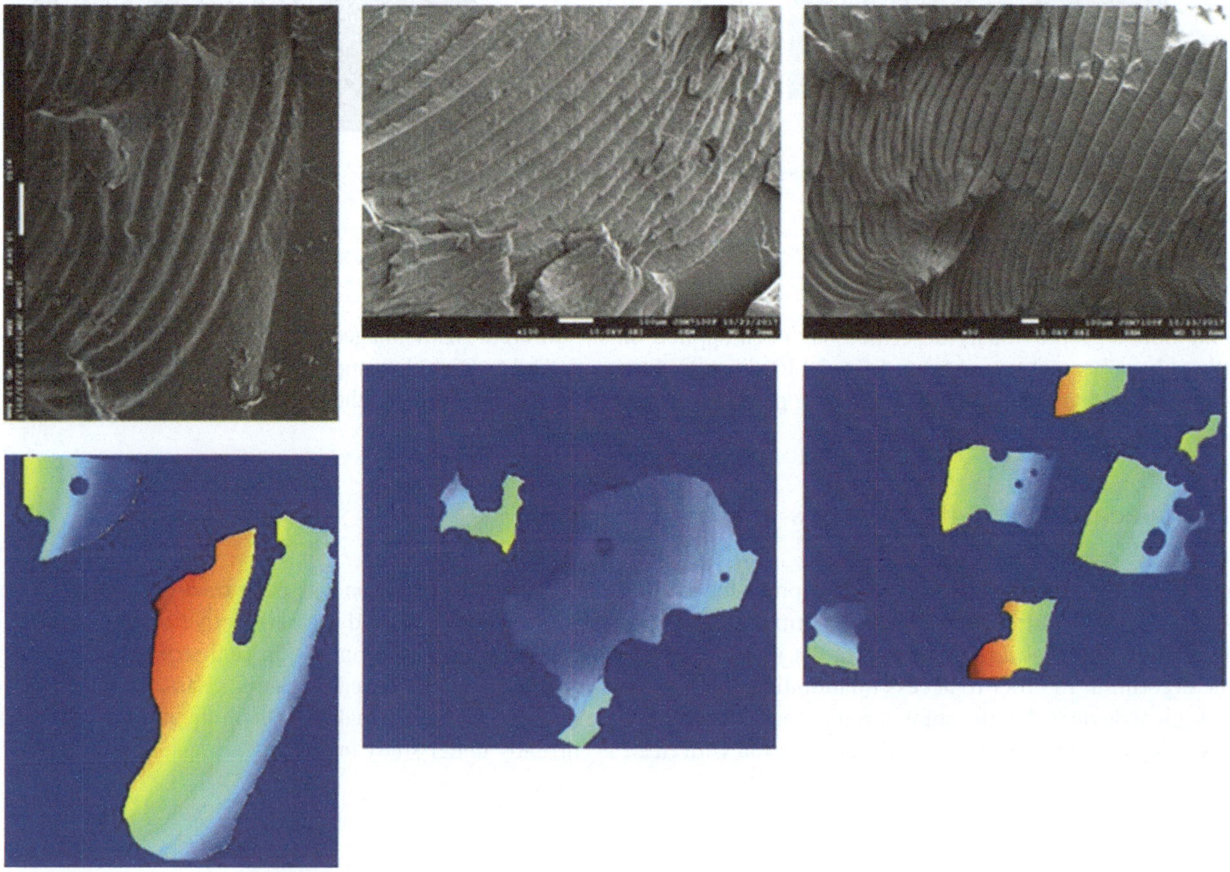

Fig. 17.6 SEM images analysis by phase demodulation image processing

References

1. Le Cam, J.-B., Huneau, B., Verron, E.: Fatigue damage in carbon black filled natural rubber under uni- and multiaxial loading conditions. Int. J. Fatigue. **52**, 82–94 (2013)
2. Le Cam, J.-B., Toussaint, E.: The mechanism of fatigue crack growth in rubbers under severe loading: the effect of stress-induced crystallization. Macromolecules. **43**, 4708–4714 (2010)
3. Flamm, M., Spreckels, J., Steinweger, T., Weltin, U.: Effects of very high loads on fatigue life of NR elastomer materials. Int. J. Fatigue. **33**, 1189–1198 (2011)
4. Muñoz-Mejia, L.: Étude expérimentale des mécanismes d'endommagement par fatigue dans les élastomères renforcés. Ph.D. dissertation, Université Claude Bernard, Lyon I (2011)
5. Robin, E., Valle, V., Brémand, F.: Phase demodulation method from a single fringe pattern based on correlation with a polynomial form. Appl. Opt. **44**, 7261–7269 (2005)
6. Takeda, M., Hideki, I., Kobayashi, S.: Fourier-transform method of fringe pattern analysis for computer based topography and interferometry. J. Opt. Sot. Am. **72**, 156–160 (1982)
7. Servin, M., Marroquin, J.-L., Cuevas, F.-J.: Demodulation of a single interferogram by use of a two-dimensional regularized phase-tracking technique. Appl. Opt. **36**(19), 4540–4548 (1997)
8. Robin, E., Valle, V.: Phase demodulation from a single fringe pattern based on a correlation technique. Appl. Opt. **43**(22), 4355–4361 (2004)
9. Valle, V., Robin, E., Bremand, F.: Analysis of digital speckle-pattern interferometry fringe patterns issued from transitory mechanical loading. Strain. **46**(2), 175–183 (2008)
10. Zuo, C., Huang, L., Zhang, M., Chen, Q., Asundi, A.: Temporal phase unwrapping algorithms for fringe projection profilometry: a comparative review. Opt. Lasers Eng. **85**, 84–103 (2016)
11. Menese, J., Gharbi, T., Humbert, P.: Phase-unwrapping algorithm for images with high noise content based on a local histogram. Appl. Opt. **44**(7), 1207–1215 (2005)
12. Corvec, G., Robin, E., Le Cam, J.-B., Sangleboeuf, J.-C., Lucas, P.: Improving spatio-temporal resolution of infrared images to detect thermal activity of defect at the surface of inorganic glass. Infrared Phys. Technol. **77**, 193–202 (2016)
13. Ruellan, B., Le Cam, J.-B., Robin, E., Jeanneau, I., Canévet, F., Mortier, F.: Influence of the temperature on lifetime reinforcement of a filled NR. In: Conference and Exposition on Experimental and Applied Mechanics SEM 2018, Greenville, 4–7 June 2018

Chapter 18
Ultra-High Speed Imaging for DIC Measurements in Kolsky Bar Experiments

Paul Moy and Timothy Walter

Abstract Full-field, stereo DIC were applied to measure the dynamic deformations for a series of compression Kolsky bar experiments at strain rates from 10^2/s to 10^3/s. A pair of Specialised Imaging's Kirana cameras were configured in a stereo configuration which helps to account for the large radial deformation. Key specifications for these CMOS-based cameras are the ability to record at a max frame rate of two million frames per second at full sensor resolution of 924 × 768 with a total of 180 frames. Common challenges that are associated with high speed imaging are adequate lighting throughout the duration of test and accurate triggering to ensure the event is properly captured in the allotted record time. In our setup, continuous, high-intensity LED lighting provided the needed light source. Combination of location of the strain gages and knowing the incident bar wave speed, the trigger is established by a delay time until the loading pulse reaches the test sample. In this work, the DIC methodologies and results for two class of materials, a set of formulated epoxy blends and steel alloy FeMnAl, are presented and discussed.

Keywords Digital image correlation · Kolsky bar · Strain-rate · Ultra-high speed imaging

18.1 Extended Abstract

Full-field, high speed stereo digital image correlation (DIC) were applied to measure the dynamic deformations for a series of compression Kolsky bar experiments at strain rates from 10^2/s to 10^3/s. Stereo DIC is ideally preferred over 2D since it will help account for large radial deformation (i.e. out-of-plane) which is quite evident for ductile polymers such as polycarbonate, [1]. Pankow et al. developed a high speed 3D technique from a single camera through the use of optical mirrors [2]. In our experiments, a pair of Specialised Imaging's Kirana cameras was used. Some of the key specifications for these CMOS-based cameras are the ability to record at a max frame rate of five million frames per second at full sensor resolution of 924 × 768 with a total of 180 frames. However, one concern that is typically associated with these Kirana cameras are the temporary burn-in of previous images overlaying subsequent frames in particularly when significant translation of objects are present. This effect is generally called "ghosting". The ghosting becomes more apparent when images contain shapes with high contrast. Figure 18.1a, b shows the first frame and the ghosted frame at 200 μs, respectively, for a rubber sample using a set of high intensity LED lighting. Other common challenges associated with high-speed imaging are having adequate lighting throughout the duration of test and precise triggering to ensure the event is properly captured in the allotted record time [3].

Contrary to the preferred high contrast patterns for DIC, grayer tone background and speckles are better suited for use with the Kirana cameras. Dark speckles are set to a minimal of about 100 counts and in the range of 500–700 counts for the background. To further alleviate ghosting effects, flash or pulsed laser lighting was used rather than continuous lighting from the start. It is understood that the camera performs ideally when the sensor is only exposed to lighting during image capture and not before being triggered. DIC results demonstrating the improved anti-ghosting techniques are presented in Fig. 18.2.

SEM Symposium on "The Role of Digital Image Correlation in Experimental Mechanics" in Honor of Prof. Sutton.

P. Moy (✉) · T. Walter
U.S. Army Research Laboratory, Weapons and Materials Research Directorate, Adelphi, MD, USA
e-mail: paul.moy.civ@mail.mil

L. Lamberti et al. (eds.), *Advancement of Optical Methods & Digital Image Correlation in Experimental Mechanics, Volume 3*, Conference Proceedings of the Society for Experimental Mechanics Series, https://doi.org/10.1007/978-3-319-97481-1_18

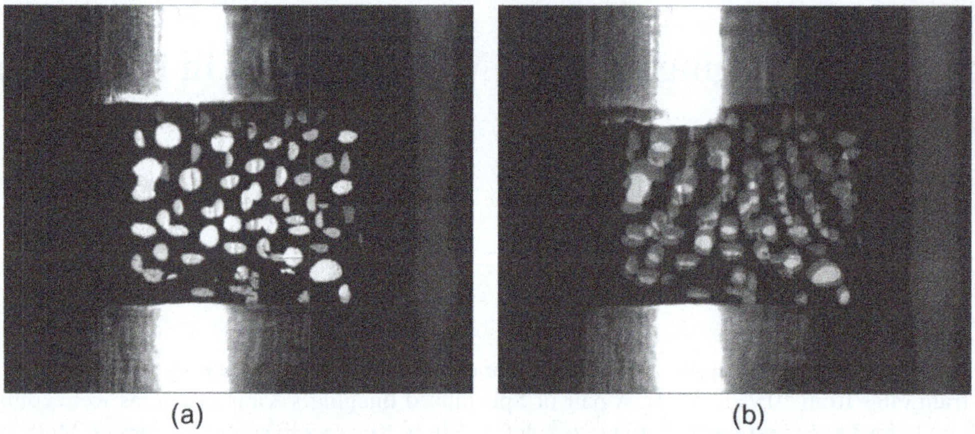

Fig. 18.1 (**a**) Initial frame at 0 μs and (**b**) ghosted image at 200 μs for epoxy captured at 100,000 fps

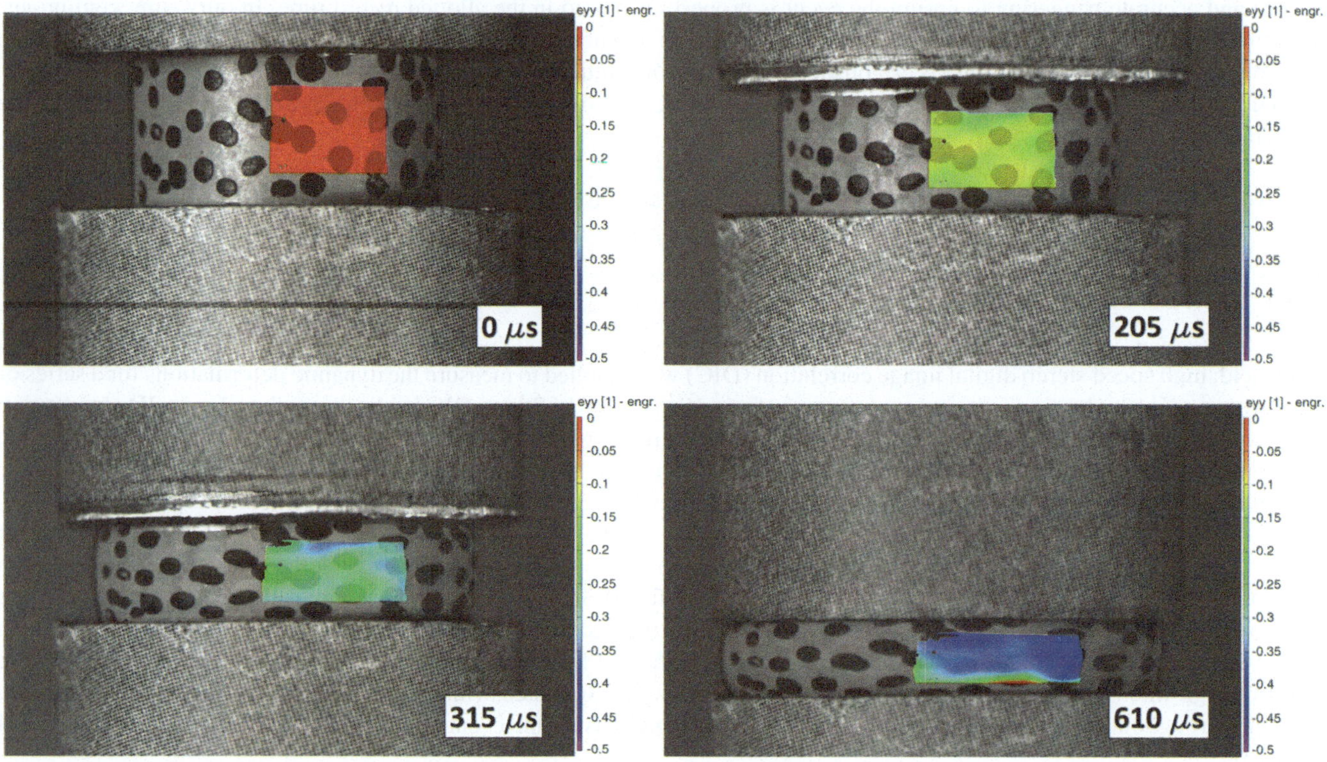

Fig. 18.2 High rate compression of PC with stereo DIC measurements

In this work, the DIC methodologies (e.g. lighting, pattern, exposure) and results for two classes of materials, a set of formulated epoxy blends and steel alloy FeMnAl, are presented and discussed.

References

1. Sutton, M.A., Yan, J.H., Tiwari, V., Schreier, H.W., Orteu, J.J.: The effect of out of plane motion on 2D and 3D digital image correlation measurements. Opt. Lasers Eng. **46**(10), 746–757 (2008). https://doi.org/10.1016/j.optlaseng.2008.05.005
2. Pankow, M., Justasson, B., Waas, A.M.: Three-dimensional digital image correlation technique using single high speed camera for measuring large out-of-plane displacements at high framing rates. Appl. Opt. **49**(17), 3418–3427 (2010). https://doi.org/10.1364/AO.49.003418
3. Reu, P.L., Miller, T.J.: The application of high speed digital image correlation. J. Strain Anal. Eng. Design. **43**(8), 673–688 (2008). https://doi.org/10.1243/03093247JSA414

Chapter 19
Application of Digital Image Correlation to Structures in Fire

Christopher M. Smith and Matthew S. Hoehler

Abstract The behavior of engineering structures in fire is commonly studied through large-scale experiments. Full-field, noncontact measurement techniques such as Digital Image Correlation (DIC) are potentially ideal for such experiments; however, the presence of light emitted by the flames, thermal radiation from the heated structure, and convective thermal gradients in the air make this a challenging application for DIC. A simple method has been developed to enable the use of DIC in large, low-soot, fires using narrow-spectrum blue light and spectrally-matched bandpass optical filters to increase signal-to-noise ratio and filter undesired radiant energy before it reaches the camera. The method is applied to full-scale experiments in which a 6-m long W16 × 26 steel beam is supported over a 700 kW fire from a natural gas diffusion burner. The resulting images are temporally and spatially averaged during post-processing to smooth out false distortions of the images caused by the thermal gradients in and around the flames before DIC techniques are applied to resolve strain.

Keywords Digital image correlation · DIC · Fire · Narrow-spectrum illumination · Blue light

19.1 Introduction

The behavior of engineering structures in fire is commonly studied through large-scale experiments. However, temperature-varying material properties, spatially-varying thermal loading and complicated structure geometries result in these structures deforming in complex and often unpredictable ways. This makes it difficult to fully characterize the response of the structure with traditional (point) sensors [1]. Full-field, noncontact measurement techniques such as Digital Image Correlation (DIC) are potentially ideal for such experiments; however, the presence of light emitted by the flames, thermal radiation from the heated structure, and convective thermal gradients in the air make this a challenging application for DIC.

19.2 Imaging Through Fire Using Narrow–Spectrum Illumination

A simple and robust method has been developed to enable the use of DIC in large, low-soot, fires to measure targets at temperatures up to 800 °C. This method, which builds on work by Grant [2] and Pan [3], uses narrow-spectrum blue light (450 ± 10 nm at one standard deviation) and spectrally-matched band-pass optical filters to increase signal-to-noise ratio and filter undesired radiant energy before it reaches the camera [4]. Figure 19.1a shows a benchtop-scale demonstration of this method. The distance from the camera to the target is about 1 m. Figure 19.1b contains three images of the same natural gas flame with different illumination and filtering methods applied. In the image on the left, no filters are used, and the target is illuminated only by the flame and ambient lighting in the room. In the middle image, a 10 W 450 nm light emitting diode (LED) source is added to illuminate the target; improving the visibility of the target. In the image on the right, a band-pass optical filter is placed in front of the camera, blocking the light from the flame and allowing the target behind the flame to be seen clearly. This technique can reduce the observed intensity of the flame by a factor of 10^4, which is sufficient to image full-scale structural experiments using less than 50 W of applied illumination [4].

C. M. Smith
Berkshire Hathaway Specialty Insurance, Boston, MA, USA

M. S. Hoehler (✉)
National Institute of Standards and Technology, Gaithersburg, MD, USA

L. Lamberti et al. (eds.), *Advancement of Optical Methods & Digital Image Correlation in Experimental Mechanics, Volume 3*,
Conference Proceedings of the Society for Experimental Mechanics Series, https://doi.org/10.1007/978-3-319-97481-1_19

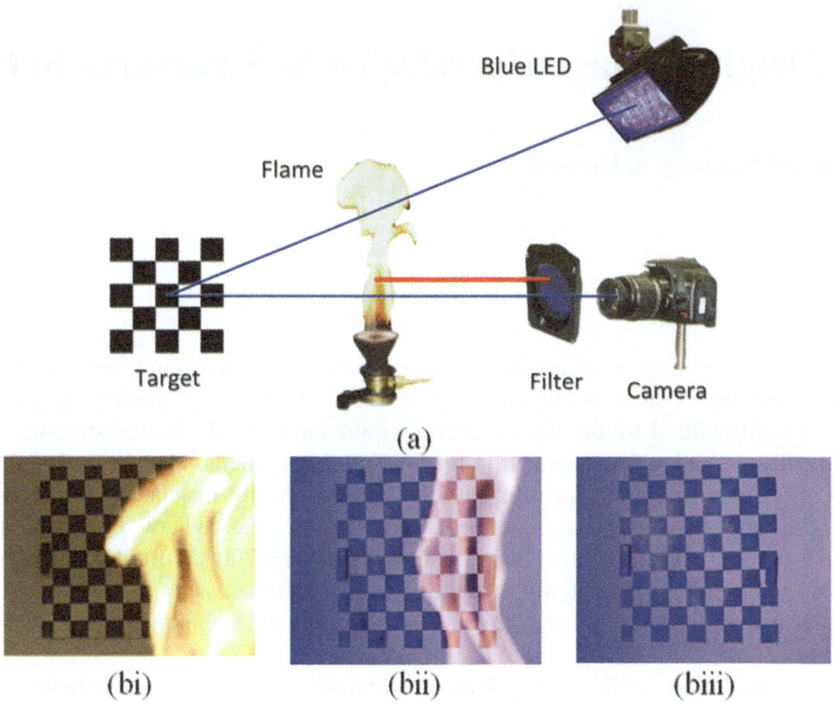

Fig. 19.1 Demonstration of imaging method: (**a**) schematic of setup and (**b**) target viewed through 5 kW natural gas flame when illuminated with (i) ambient light, (ii) 10 W of 450 nm light, and (iii) 10 W of 450 nm light and imaged through a band-pass filter

19.3 Applications to Optical Metrology

The proposed filtering method allows DIC measurements to be made through flames that have low soot content. Converting the images to quantitative measurements poses additional challenges because the gas temperature gradients optically distort the images causing fictitious (apparent) strains. While there is a steady-state aspect of this distortion caused by the overall shape of the fire, in a large diffusion flame, the turbulent flow (flicker) of the flame is responsible for most of the distortion. The characteristic frequency of the flame flicker is on the order of ~10 Hz at the small scale [5, 6], and lower frequency transient distortions can occur due to variations fire ventilation at the large (1 m) scale. Meanwhile the characteristic dimension of individual flame eddies is on the order of 5 cm. As the strain rates in structural testing are slow relative to the flame motion, temporal and spatial averaging can be used to improve the accuracy of measurements.

19.4 Application to Full-Scale Experiments

The method described above was applied to several repeated full-scale experiments in which a 6 m long W16 × 26 steel beam was supported over a 700 kW fire from a natural gas diffusion burner (Fig. 19.2a). Measured gas temperatures in the flame exceeded 1400 °C and the steel beam reached temperatures over 700 °C [7]. A commercial DIC system, with optical filters installed, was placed at 3 m stand-off and the specimen was illuminated using two 100 W blue LED arrays. The target area was approximately 0.4 m high by 1 m long. Prior to painting a pattern to enable DIC, the surface oxide layer ('mill scale') on the beam was removed using sandblasting, as it naturally flakes off when heated above 400 °C or when the steel yields. A base layer of white high-temperature spray paint was applied followed by a random pattern stenciled in black spray paint to provide contrast (Fig. 19.2a). The combination of a white base coat with a black pattern was found to be more temperature-resistant than the inverse. The optical distortion from the flame, and necessary time averaging, have the effect of blurring the image to be processed with DIC, so the pattern was designed to be coarser (10–20 pixels/feature) than necessary in an unobstructed test (3–5 pixels/feature).

During the experiment, the beam was heated by the flame, and the thermal strains were measured using DIC. Images were recorded at 1 Hz and time-averaged over 30 s to remove transient flame effects. The time-averaging was performed on the raw

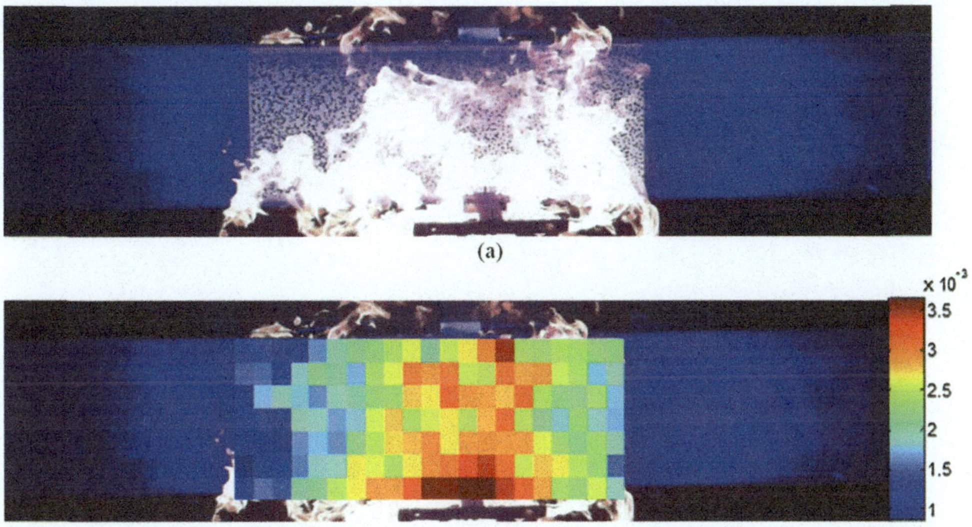

Fig. 19.2 Full-scale validation tests: (**a**) test setup and (**b**) example of measured longitudinal strain field (units in mm/mm)

images, which were subsequently reimported into the DIC system for processing. After performing the DIC, the resulting strains were spatially averaged over 4 cm regions (100 pixels). As illustrated in Fig. 19.2b, this method was sufficient to differentiate thermal strains in the steel of 3 microstrain from the flame-induced distortion. The tests also suggest that time averaging is preferable to spatial averaging, at least for structural experiments for which spatial resolution is preferable to temporal resolution. Further work is needed to better understand the post-processing options and methods for this application and to quantify the uncertainty associated with these processes.

19.5 Conclusions

Noncontact measurement techniques, such as Digital Image Correlation (DIC), are desirable for use in structural fire experiments because deformation and failure in large structures subject to fire is often unpredictable. The presence of light emitted by flames, thermal radiation from the heated structure, and convective thermal gradients in the air have challenged the use of DIC. A simple and inexpensive method to improve the signal-to-noise ratio (10^4 compared to ambient lighting) in images of objects taken through low soot-yield fires is described and has been successfully deployed on large scale fire tests. The resulting images were temporally and spatially averaged during post-processing to smooth out false distortions of the images caused by the thermal gradients in and around the flames before DIC techniques were applied to resolve strain. Additional work to quantify the uncertainty associated with these processes is needed.

References

1. McAllister, T., Luecke, W., Iadicola, M., Bundy, M.: Measurement of temperature, displacement, and strain in structural components subject to fire effects: concepts and candidate approaches. (2012). https://doi.org/10.6028/NIST.TN.1768
2. Grant, B.M.B., Stone, H.J., Withers, P.J., Preuss, M.: High-temperature strain field measurement using digital image correlation. J. Strain Anal. Eng. Design. **44**, 263–271 (2009). https://doi.org/10.1243/03093247JSA478
3. Pan, B., Wu, D., Wang, Z., Xia, Y.: High-temperature digital image correlation method for full-field deformation measurement at 1200 °C. Meas. Sci. Technol. **22**, 15701 (2011). https://doi.org/10.1088/0957-0233/22/1/015701
4. Smith C.M., Hoehler M.: Imaging through fire using narrow-spectrum illumination. Fire Technology. Posted online July 23, (2018). https://doi.org/10.1007/s10694-018-0756-5
5. Ballantyne, A., Bray, K.N.C.: Investigations into the structure of jet diffusion flames using time-resolved optical measuring techniques. Symp. Combust. **16**, 777–787 (1977). https://doi.org/10.1016/S0082-0784(77)80371-8
6. Buckmaster, J., Peters, N.: The infinite candle and its stability—a paradigm for flickering diffusion flames. Symp. Combust. **21**, 1829–1836 (1988). https://doi.org/10.1016/S0082-0784(88)80417-X
7. Choe, L., Ramesh, S., Hoehler, M., et al.: National fire research laboratory commissioning project: testing steel beams under localized fire exposure. (2018). https://doi.org/10.6028/NIST.TN.1983

Chapter 20
Full-Field Determination of the Taylor-Quinney Coefficient in Tension Tests of Ti-6Al-4V at Strain Rates up to 7000 s^{-1}

Jarrod L. Smith, Jeremy D. Seidt, and Amos Gilat

Abstract The percentage of plastic work converted to heat during plastic deformation, represented as β, and known as the Inelastic Heat Fraction (IHF) or the Taylor-Quinney coefficient is investigated in tension samples of Ti-6Al-4V tested to failure at strain rates ranging from 1 to 7000 s^{-1}. Quasi-static tests at strain rates at 1 s^{-1} were completed on a load frame while tests ranging from 500 to 7000 s^{-1} were completed on two separate split-Hopkinson bars. For each test, strain measurements were recorded on one side of the sample using Digital Image Correlation (DIC) while simultaneous temperature measurements were captured on the opposite side of the sample via infrared thermography. A method to determine β by combining the full-field temperature and strain measurements with the recorded force data is presented. Results demonstrating the dependence of β on plastic strain and strain rate are also described.

Keywords Infrared thermography · Digital image correlation · Plastic work · Thermomechanical · Stored energy

20.1 Introduction

The plastic deformation of metals at short time scales leads to a large increase in the temperature of the material due to a phenomenon known as thermomechanical coupling. As mechanical work is done on the material, a portion of the work is spent to deform the microstructure and increase the latent energy while a larger portion is dissipated as heat to the surroundings. The fraction of plastic work that is converted to heat during deformation is often denoted by the variable β and is known as the Inelastic Heat Function (IHF) or the Taylor Quinney coefficient, as it was first studied by Farren and Taylor [1] and Taylor and Quinney [2] during their experiments on the cold working of metals. In these experiments β was determined to be a constant 0.8–0.9 for all metals, however, these first tests were performed quasi-statically at low levels of strain and produced large levels of scatter. With the advent of high speed infrared detectors with response times as low as 1 μs many researchers began to investigate the temperature rise in materials during split-Hopkinson bar experiments in torsion and compression where adiabatic heating and homogeneous deformation can be assumed. Mason [3] investigated β for Ti-6Al-4V in compression at strain rates ~3000 s^{-1} using a compression split-hopkinson bar and an InSb infrared (IR) detector consisting of eight 0.16 × 0.16 mm elements in a linear array. The material was found to have inhomogeneous deformations which led to inhomogeneous temperature increases measured across the multiple detectors during deformation. The calculated beta ranged from 0.8 to 0.98 then decreased to 0.8 as the strain increased. McDougal and Harding [4] measured β in torsion tests of Ti-6Al-4V using a similar setup and found that the beta increased from 0.1 to 0.8 as plastic shear strain increased from 0 to 40%. It is important to point out that these calculations of β are defined as β_{diff} which differs from another calculation of β known as β_{int} This difference in Beta calculation was first noted by Rittel [5] and is described by the following equations

$$\beta_{int} = \frac{\rho c_p \Delta T}{\int_0^t W_p} \tag{20.1}$$

or

$$\beta_{diff} = \frac{\rho c_p \dot{T}}{d\dot{W}_p} \tag{20.2}$$

J. L. Smith (✉) · J. D. Seidt · A. Gilat
Department of Mechanical Engineering, The Ohio State University, Scott Laboratory, Columbus, OH, USA
e-mail: smith.6575@osu.edu

L. Lamberti et al. (eds.), *Advancement of Optical Methods & Digital Image Correlation in Experimental Mechanics, Volume 3*, Conference Proceedings of the Society for Experimental Mechanics Series, https://doi.org/10.1007/978-3-319-97481-1_20

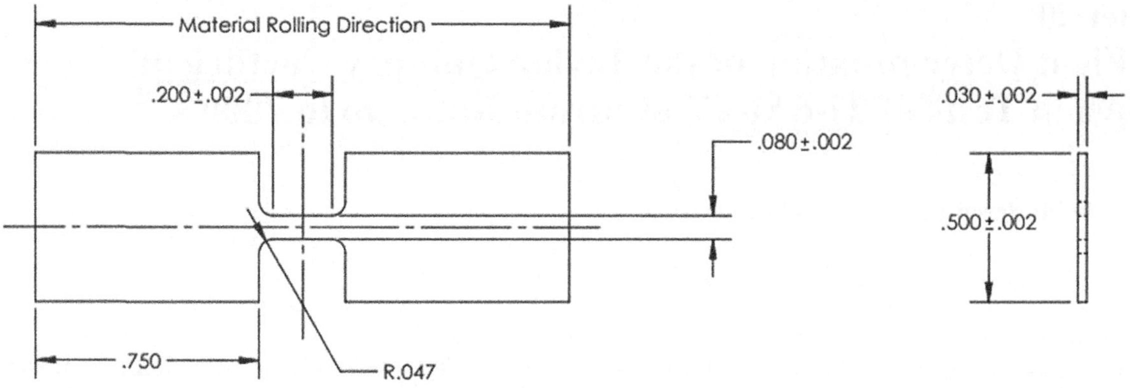

Fig. 20.1 Specimen geometry

where ρ is the density of the material, c_p is the specific heat of the material, ΔT is the change in temperature,$\dot{T}$ is the rate of the change in temperature, $\int_0^t W_p$ is the plastic work and $d\dot{W}_p$ is the mechanical power. The first is a relation of the thermal dissipation to mechanical work and requires simultaneous strain and temperature measurements while the second is a ratio of the thermal rate of dissipation to mechanical power and is found by taking the derivate of the rates. Rittel [6] investigated the β_{int} for Ti-6Al-4V in tension, compression, and shear at strain rates from 1500 to 3400, 2000, and 5000 to 7000 s^{-1} respectively. In these tests the β_{int} ranged from 0.5 to 0.4 which increasing plastic strain in compression and shear. The β_{int} for tension was calculated only during homogeneous deformation in the sample before necking and was found to be 0.38–0.4. In the current paper, the beta is calculated for four different tension tests at nominal strain rates of 1, 500, 2000, and 7000 s^{-1} using simultaneous DIC and IR measurements as in the test set up previously described in Seidt [7].

20.2 Experimental Procedures and Techniques

Tensile tests are conducted on thin flat Ti-6Al-4V specimens at strain rates from 1 to 7000 s^{-1}. During each test, deformation is recorded on one side of the specimen using 3D or 2D DIC while the temperature is recorded on the opposite side using a high speed Telops FAST-IR camera. The specimen is also illuminated using an LED light and covered with infrared blocking curtains to reduce noise in the infrared images. Since the specimen is thin it is assumed the temperatures and deformations on each side of the specimen are uniform. Tests at 1 s^{-1} are completed by attaching the specimen to a servo hydraulic load frame via two hydraulic wedge grips. 3D-DIC data is recorded using dual Photron MC2 cameras running synchronously with the IR camera at 2000 fps (Fig. 20.1).

Tests ranging from 500 to 2000 s^{-1} are completed using a direct tension SHB. A tensile wave is generated in this bar by applying a direct load to the incident bar while it is clamped in place. Once the clamp is broken, an elastic wave propagates through the bar to the specimen and the force is calculated by measuring the wave in the transmitted bar. Similarly, the tests at a strain rate of 7000 s^{-1} are completed using a split Hopkinson bar, however, the wave in this bar is generated by shooting a striker tube into a flange at the end of the incident bar. Due to the short time scales at which these split Hopkinson bar tests take place a high speed Shimadzu HPV-X2 is used to capture DIC data at frame rates from 200,000 to 1,000,000 Hz and the frame rate of the infrared camera is increased from 50,000 to 900,000 Hz while compromising the resolution and exposure time of the camera. An overview of the three different set ups is shown in Fig. 20.2 while the parameters for the various optics used are shown in Tables 20.1 and 20.2.

20.3 Infrared Camera Calibration

The infrared camera used in these experiments records data in either Radiometric temperatures or in-band radiance values (W-sr) which assume that the target of the camera is a black body and its emissivity is 1. However, the book values for Ti-6Al-4V can range from 0.19 to 0.6 depending on the level of polish or oxidation on the surface. In previous works several researchers have adding a coating such as paint or soot to the surface to increase the emissivity and insure a uniform finish.

Fig. 20.2 (Top) testing configuration on MTS frame. (Middle) testing configuration for 7000 s^{-1} tests. (Bottom) testing configuration for 500 and 2000 s^{-1} tests

Table 20.1 Infrared camera specs for the various tests

Strain rate (1 s−¹)	Pixel resolution	Frame rate (Hz)	Exposure time (μs)
1	256 × 256	2000	10
500	64 × 20	50,000	5
2000	64 × 16	62,500	5
7000	64 × 4	90,000	5

Table 20.2 Visible camera specs for the various tests

Strain rate (1 s−¹)	Camera	Pixel resolution	Frame rate (Hz)
1	Photron MC2	512 × 512	2000
500	Shimadzu HPV-X2	400 × 250	200,000
2000			625,000
7000			1,000,000

Fig. 20.3 Image of the calibration setup

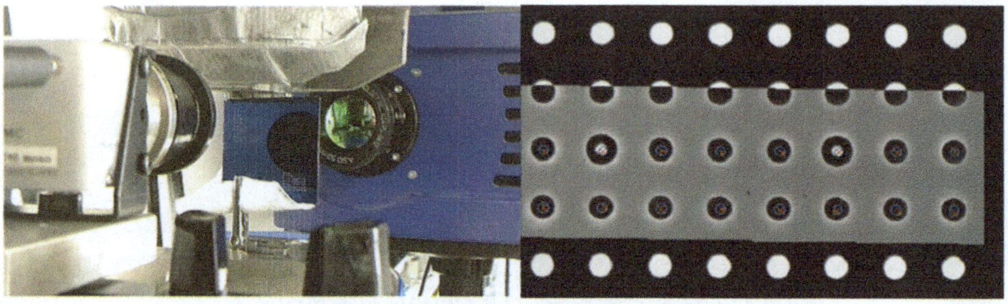

Fig. 20.4 (Left) alignment procedure for each test. (Right) result of the alignment procedure

However, we found that the surface acts as an inhibitor to heat transfer during the higher strain rate tests and gives artificially low temperature readings. To calibrate the samples used in these experiments a thermocouple was attached to the surface of each sample while the temperature of the sample was increased using a hot plate as shown in Fig. 20.3. Meanwhile, the infrared camera was used to capture average in-band radiance values of the surface and was operated using the same parameters as the tension tests. Using these measurements, a calibration curve was created to convert in-band radiance values into surface temperatures for each test.

The temperature and deformation fields calculated from the IR and DIC measurements were aligned by taking both visible and infrared images of a glass DIC alignment grid as shown in Fig. 20.4.

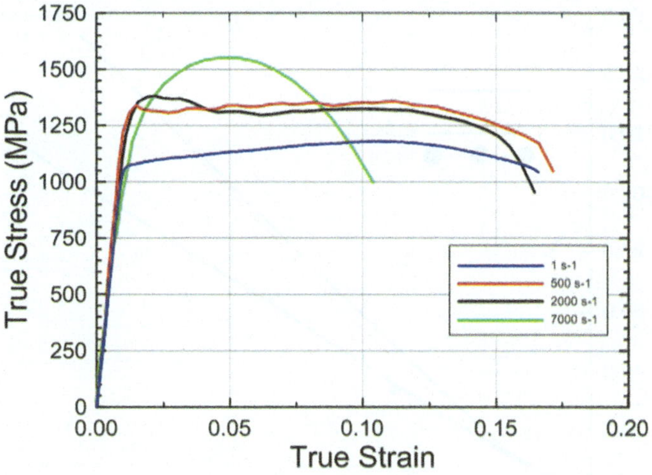

Fig. 20.5 True stress vs. true strain for strain rates from 1 to 7000 $s{-}^{1}$

By matching the centers of each circle in both images a 2D affine transform was created and applied to each IR image. The temperature at each DIC data point was then interpolated from the transformed infrared image resulting in temperature and strain measurements at each point on the surface.

20.4 Results and Discussion

The measured true stress and true strain of a 4 mm extensometer for each strain rate test is displayed in Fig. 20.5. The material exhibits strain rate effects as shown by the increase in ultimate stress with increasing strain rate. The transition point from elastic to plastic strain is at approximately 1.75% true strain for each test.

The local hencky strain versus the change in temperature for a DIC data point located at the point of failure for each strain rate test is shown in Fig. 20.6. The 1 $s{-}^{1}$ test shows an increase of 120 °C with a 0.4 increase in strain while the 500 and 2000 $s{-}^{1}$ tests show increases of 125 and 155 °C with 0.3–0.375 increases in strain. The temperature in the 500 and 2000 $s{-}^{1}$ tests also increases at a great rate than that of the 1 or 7000 $s{-}^{1}$ tests as shown by their larger slope. Due to the short duration of the 7000 $s{-}^{1}$ test and the limitations of the infrared camera, only one temperature point was measured during the test at a temperature increase of 30 °C and a local strain increase of 0.15.

The β_{int} was calculated at along several lines along the gage section within ±1.15 mm of the localization as shown by Fig. 20.7. Along the lines closest to the localization the β_{int} increases from 0.4 to 0.5 until a plastic strain of 0.1 is reached and then remains relatively constant until failure. The artificial increase in the β_{int} for the lines furthest from the localization is due to heat transfer from the localization section which increases the temperature of these sections while plastic deformation has stopped.

Similar lines are used on the 500, 2000, and 7000 $s{-}^{1}$ tests as shown in Fig. 20.8. In the 500 $s{-}^{1}$ test the β_{int} increases from 0.35 to 0.6 as the plastic strain increases from 0 to 0.1 then remains almost constant until failure. In these tests the β_{int} does not artificially increase the temperature in sections away from the localization as the duration of the test is much shorter and heat transfer is negligible leading to nearly adiabatic conditions.

A comparison of the β_{int} for the center point of the line closest to localization for the four different tests is shown in Fig. 20.9. The β_{int} for 500 and 2000 $s{-}^{1}$ shows some rate sensitivity as the β_{int} increases from 0.5 to 0.6 as the strain energy and temperature both increased. The β_{int} for the 7000 $s{-}^{1}$ test is 0.48 at a plastic strain of 0.125 and does not follow trend of the other tests. At strain rates above 5000 $s{-}^{1}$ the material exhibits abnormal behavior in tension as shown in the stress-strain curve in Fig. 20.5. Further study will need to build on these results to determine if a phase change in the material may be the cause of this behavior.

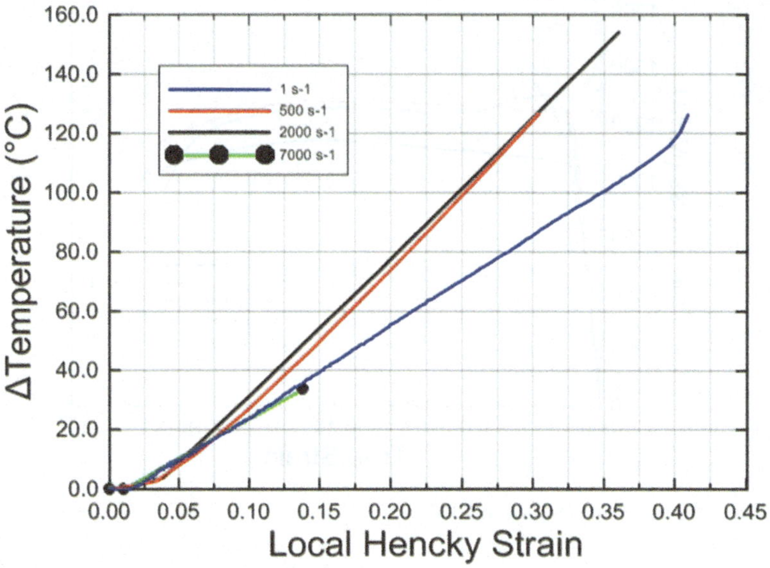

Fig. 20.6 True stress vs. true strain for strain rates from 1 to 7000 $s-^1$

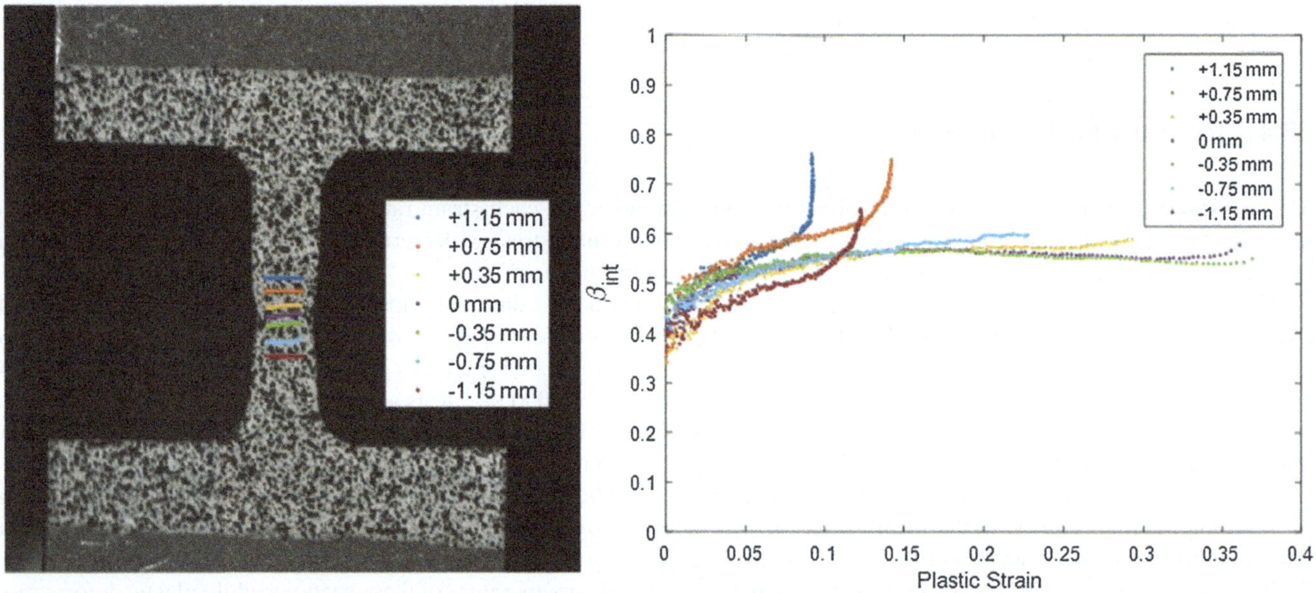

Fig. 20.7 (Left) location of beta calculations for strain rate of 1 $s-^1$ test. (Right) beta versus plastic strain for strain rate of 1 $s-^1$ test at various locations

20.5 Summary and Conclusions

Full-field deformation measurements with corresponding temperature measurements were completed on tension tests of Ti-6Al-4V at strain rates ranging from 1 to 7000 $s-^1$. Utilizing these measurements, the β_{int} at several points along the gage section of the specimen was calculated. The β_{int} increased with increasing plastic strain and increasing strain rate. Tension tests at 1 $s-^1$ cannot assume adiabatic behavior as there is an increase in temperature in partially deformed regions due to heat transfer from the localization region. Simulations should be completed to validate the temperatures and β_{int} measured for the different strain rate tests.

Acknowledgements The research was supported by the U.S.A. Federal Aviation Administration, Grant No. 11-G-004. The authors are grateful to Mr. William Emmerling, and Dr. Chip Queitzsch for their support and involvement.

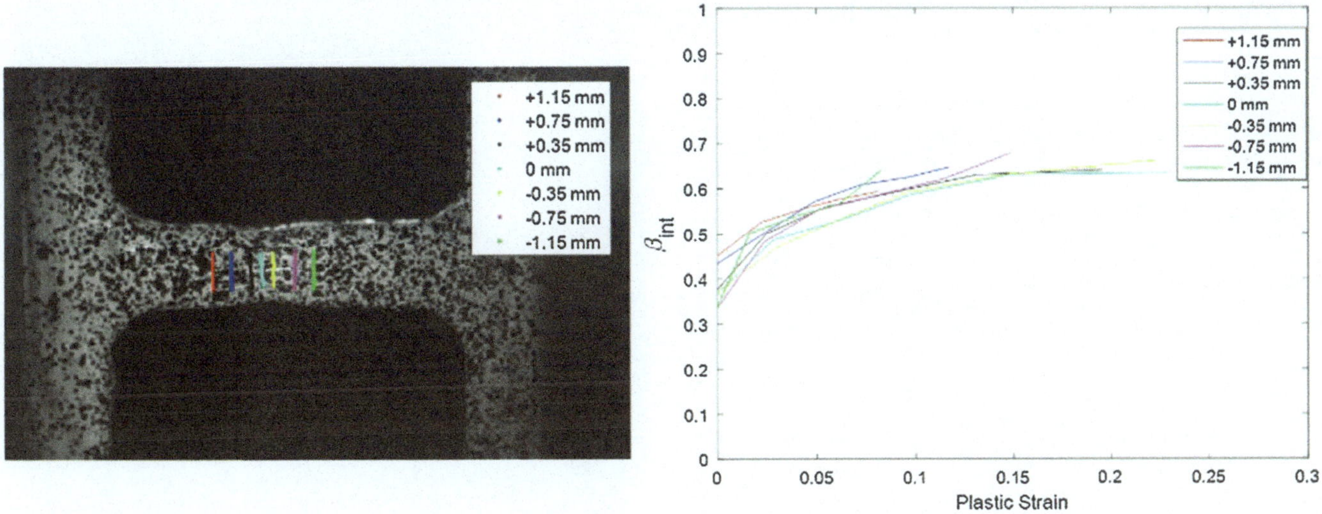

Fig. 20.8 (Left) location of beta calculations for strain rate of 500 s−^{1} test. (Right) beta versus plastic strain for strain rate of 500 s−^{1} test at various locations

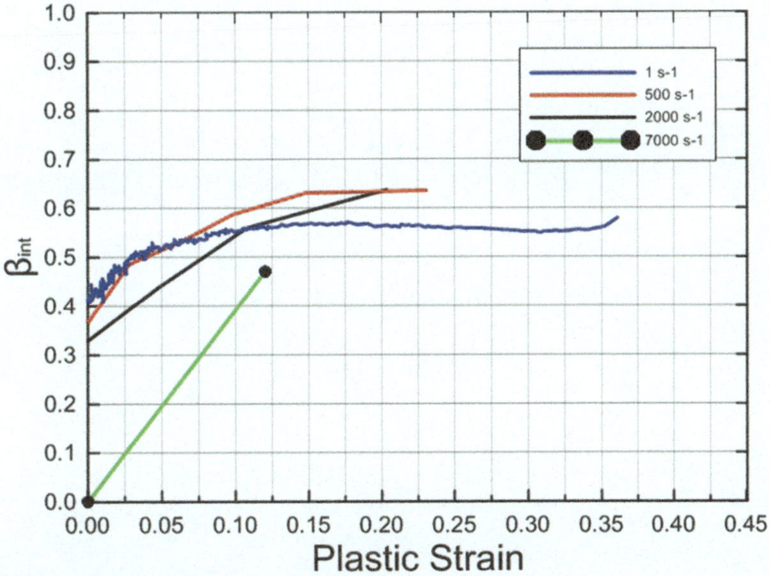

Fig. 20.9 Comparison of beta vs. plastic strain

References

1. Farren, W.S., Taylor, G.I.: The heat developed during plastic extension of metals. Proc. R. Soc. **A107**, 422–451 (1925)
2. Taylor, G.I., Quinney, H.: The latent energy remaining in a metal after cold working. Proc. R. Soc. London. **143**, 307–326 (1934)
3. Mason, J.J., Rosakis, A.J., Ravichandran, G.: On the strain and strain rate dependence of the fraction of plastic work converted to heat: an experimental study using high speed infrared detectors and the Kolsky bar. Mech. Mater. **17**, 135–145 (1994)
4. Macdougall, D.A.S., Harding, J.: The measurement of specimen surface temperature in high-speed tension and torsion tests. Int. J. Impact Eng. **21**, 473–488 (1998)
5. Rittel, D.: On the conversion of plastic work to heat during high strain rate deformation of glassy polymers. Mech. Mater. **31**, 131–139 (1999)
6. Rittel, D., Zhang, L., Osovski, S.: The dependence of the Taylor–Quinney coefficient on the dynamic loading mode. J. Mech. Phys. Solids. **107**, 96–114 (2017)
7. Seidt, J.D., Kuokkala, V.-T., Smith, J.L., Gilat, A.: Synchronous full-field strain and temperature measurement in tensile tests at low, intermediate and high strain rates. Exp. Mech. **57**, 219–229 (2017)

Chapter 21
Laser and White-Light Speckle Techniques: A Tutorial Review

Gary L. Cloud

Abstract This instructional review begins with descriptions of simple experiments that illustrate the phenomenon of coherent-light speckle and its properties that cause it to be useful for measurement of displacements and strains, namely the sensitivity of speckle brightness to object motion and the relationship between speckle size and imaging aperture. The concepts of laser speckle interferometry, speckle shearography, speckle photography, and white-light speckle photography are presented, and basic apparatus setups for implementation are illustrated. Exceptional data density and flexible displacement sensitivities cause these techniques to be favored for whole-field non-contacting measurements of displacement, strain, and vibrations, as well as for nondestructive inspection. The concept of using phase shifting to increase the precision of speckle interferometry far beyond that offered by analyzing fringe patterns is presented along with details of one method of execution. A simple MATLAB® code for phase-shifting interferometry is included. Illustrative and suggestive applications are drawn from studies in geomechanics, biomechanics, fluid mechanics, NDI, strain analysis, residual stresses, and vibration characterization.

Keywords Speckle interferometry · Speckle photography · White light speckle photography · Speckle shearography · Phase shifting

21.1 Introduction and Purpose

Beginning shortly before 1970 and continuing for more than three decades, laser speckle interferometry; speckle shearography; speckle photography; and their kindred method, white-light speckle photography, developed into go-to techniques for whole-field non-contacting measurements of displacement, strain, and vibrations, as well as for nondestructive inspection. These approaches still have much to offer and can be considered to be forerunners and augmenting companions of digital image correlation. They are characterized by exceptional data density and flexible displacement sensitivities, reaching from a small fraction of the wavelength of light to meters. Thus, they can be adapted to a broad spectrum of measurement problems. Some variants can be easily implemented in the field or factory, being tolerant of environmental effects.

A brief history of this class of techniques can be traced beginning with the landmark elucidation of the physics underlying speckle phenomena by E. Archbold, J.M. Burch, A.E. Ennos, and J.W. Goodman; its advancement as the basis of interferometric measurement techniques by E. Archbold , J.A. Leendertz, J.N. Butters, and Y.Y. Hung, plus several others; and its subsequent blossoming into a class of powerful electronic methods that incorporate digital capture and processing plus phase shifting, all easily accomplished now using standard software and devices.

Short demonstrations illustrate the properties of laser speckle that cause it to be useful to experimental mechanicians, namely the sensitivity of speckle brightness to object motion and the relationship between speckle size and imaging aperture. The basic concepts and theories of laser and white-light measurement techniques are presented, with emphasis on common fundamentals beginning with the phenomena of interference and diffraction. Some laboratory and vendor-supplied setups for digital speckle interferometry and shearography are illustrated, and a simple MATLAB® code for phase-shifting interferometry is presented. Relationships with moiré techniques are mentioned in passing. Some illustrative and suggestive applications drawn from studies in geomechanics, biomechanics, fluid mechanics, NDI, strain analysis, residual stresses, and vibration characterization are described in brief.

G. L. Cloud (✉)
Michigan State University, Composite Vehicle Research Center, Lansing, MI, USA
e-mail: cloud@egr.msu.edu

L. Lamberti et al. (eds.), *Advancement of Optical Methods & Digital Image Correlation in Experimental Mechanics, Volume 3*,
Conference Proceedings of the Society for Experimental Mechanics Series, https://doi.org/10.1007/978-3-319-97481-1_21

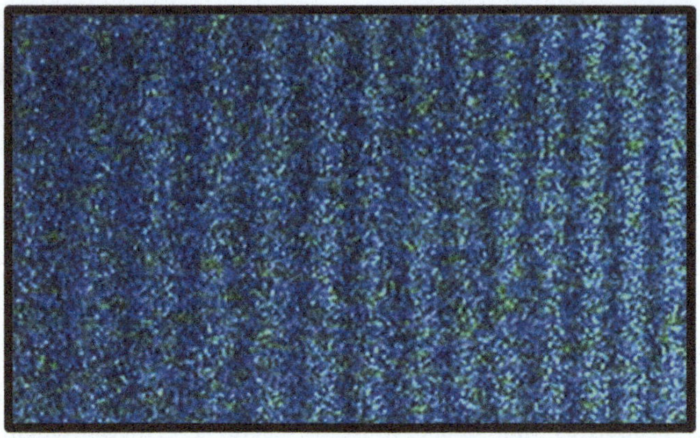

Fig. 21.1 Coherent-light speckle in an interference fringe pattern created using an argon laser and Lloyd's mirror

Given the tutorial nature of this presentation, its length, and the fact that most of it is drawn from the experience of the author and his colleagues, the usual listing of sources is not included. The most significant contributors are named in the text. We hope that none are offended by this practice.

21.2 What Is Laser or Coherent-Light Speckle?

When the laser was invented, scientist and photographers were ecstatic, because it seemed that, finally, we had a source of light that was intense, collimated, narrow, monochromatic, and coherent. Disappointment soon set in. Images of objects illuminated by a laser were degraded by a grainy structure that severely limited the effective spatial resolution. Figure 21.1 shows one such example. Even our amazing 3-dimensional hologram images and laser-based photos were contaminated. Considerable time and effort was expended to understand this vexing phenomenon, which came to be called the "speckle effect," with the aim of eliminating it from images taken with laser light.

Happily, the early investigators mentioned above taught us that these speckle patterns carry much information about the illuminated object as well as other components of the optical system. Applications were soon invented, and this formerly problematical phenomenon rapidly evolved into the basis of a whole family of measurement technologies. These new techniques, which are loosely grouped under the name "speckle methods," utilize speckle fields to generate data about motion, deformation, and strain.

21.3 Demonstrations of Important Properties of Laser Speckle

Laser speckle is one manifestation of the interference that occurs when two beams of coherent light are brought together. Interference is one of the two fundamental pillars of optical physics, the second being diffraction, and we assume here that you have at least a basic knowledge of these phenomena.

21.3.1 Observation of Laser Speckle

Before discussing how coherent-light speckle is used, consider some simple experiments that illustrate fundamental speckle behaviors that cause it to be useful. The only apparatus needed for these demonstrations is a small laser, a beam expander, and a stable screen. A low-power HeNe laser suffices, and some laser pointers might work. The beam expander is easily implemented with a decent microscope objective lens. The goal is to produce on a screen a nicely illuminated patch that is large enough for you to see comfortably. The screen can be any matte surface such as a painted wall, a sheet of poster board, or even a solid projection screen. A most important requirement is that both laser and screen be fixed so that they do not wobble or vibrate. Darken the room for ease of viewing.

Sit very still and just stare at the illuminated patch on the screen. You should see that it is filled with tiny bright and dark specks. The bright ones are sometimes called sparkles, but the whole thing is called a speckle pattern. If you do not see the speckle, then check the stability of the laser and screen and yourself. If you do not possess a definite master eye, then you should try closing one of them.

21.3.2 Speckle Motion with Lateral Movement

Now, move your head slightly sideways. You will see that the speckles move sideways also. A conclusion is that the speckle moves with the viewer. An obvious suggestion is that if the eye is held stationary and the illuminated specimen moves, then the speckles will move with it. Such is indeed the case, and this phenomenon forms the basis of a class of methods called "speckle photography." Lateral motion of individual speckles is related to local specimen motion. That is, the speckles act as a very fine array of surface markers. If the motions of the markers can be recorded and ascertained, then the lateral specimen displacement field can be mapped.

21.3.3 Dependence of Speckle Size on Aperture

The objective of the next experiment is to demonstrate that the aperture of the imaging system affects the apparent size of the speckles. There are three ways of accomplishing this, including:

1. While staring at the illuminated patch, bring your eyelids close together so that you are looking through a narrow slit.
2. Form a small aperture by curling your index finger against the base of your thumb or by laying the index and middle finger of one hand crosswise over the same combination of the other hand so as to form a small square aperture.
3. Punch several small holes in a card, such as by pushing a pencil through it, so as to form small apertures of different sizes.

You will notice that the speckles in the pattern appear on average to become larger as the aperture is made smaller, although the contrast between light and dark speckles might seem diminished. Figure 21.2 illustrates such behavior for a two aperture settings (f-stops). Close examination shows that, even in this reduction, the smallest details are much larger for the smaller $f/16$ setting than they are for the larger aperture. This apparent inverse relationship between speckle size and aperture is useful for two reasons, namely: (1) it allows control of speckle size to match pixel size in digital imaging systems; and (2) we can adjust the speckles to the small sizes that are required for speckle photography techniques where the speckles are used merely as markers, or we can make the speckles comparatively large to facilitate speckle interferometric methods.

21.3.4 Brightness Change with Longitudinal Movement

Most people find this demonstration difficult to execute; others can do it only on good days, while some cannot do it at all. Sit and stare at the speckle pattern as outlined above. Use a small aperture to make the speckles as large and as stationary as you can. Focus on only one large dark speckle blob and gently move your head a very small amount away from the screen. With luck and practice, you might notice that the brightness of that speckle changes from dark to light and probably back to light again.

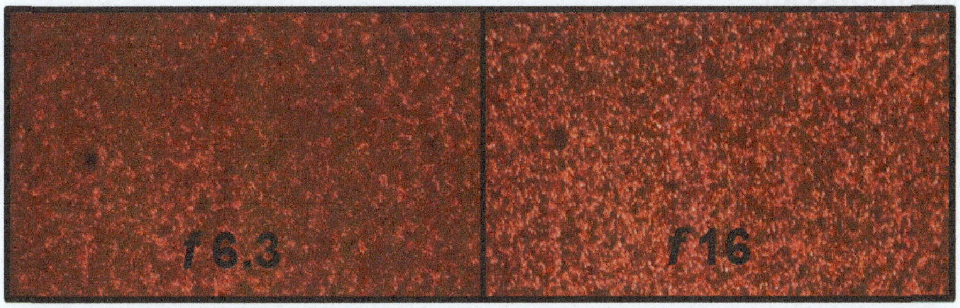

Fig. 21.2 Two photographs taken with different apertures of a speckle pattern created by projecting Helium-Neon laser light onto a screen. The effect of camera aperture on apparent speckle size is evident

That speckle brightness depends on distance from object (screen) to viewing system is the basis of a broad family of speckle measurement techniques that includes digital and electronic speckle pattern interferometry, laser shearography, and related phase shifting methods. The speckle is conditioned by adding a reference beam. Then, one merely records the beginning and final brightnesses of each and every speckle in the field and subtracts final intensity values from initial values to determine the displacement map for the object.

21.4 Classes of Speckle Measurement Techniques and the Naming Problem

One complication which arises upon examining the history of speckle measurement is inconsistency in naming the method. There have been divergences of viewpoint regarding the particular interferometric processes involved. It seems, for example, that several early researchers did not realize that double aperture speckle "interferometry" can be seen as basically a moiré method. The major cause of misunderstanding results from the use of the terms "speckle interferometry" to refer to a variety of related techniques. A simple solution to the terminology issue was offered by K. Stetson in 1975. If the technique involves incoherent recording of two speckle fields and if no reference beam is employed, then it is called "speckle photography". If a reference beam is present, then the speckle fields are still recorded incoherently, but the fields are being modified by the reference beam. In this case, the term "speckle interferometry" is used. This definition must be expanded to account for the valuable cases where two object illumination beams are employed. Still, the two approaches are somewhat unified by a common body of theory; and optical interference is important in both, as it is fundamental to the formation of speckle and in the recovery of data from recorded speckle patterns. Some variants, in particular the "dual aperture speckle" approach actually fall within the realm of moiré techniques, so they can be called "speckle-moiré" methods, and are not studied here. For some reason, they seem to have been largely forgotten, but they are powerful and easy to implement. Table 21.1 is an attempt to distinguish between variants of speckle photography and speckle interferometry, but we beg indulgence because the list is not complete and it leaves room for confusion because of overlaps in the nomenclature.

21.5 The Displacement–Strain Problem

At the beginning of any discussion of optical measurement techniques within the context of experimental mechanics, especially those involving interferometry, it seems wise to remind ourselves of some fundamental facts, which may be summarized as follows.

1. Except when buckling is an issue, engineers usually design in terms of stress and strain.
2. Validation of numerical or analytical models should not be done on the basis of displacement if strength is an issue.
3. Experimental mechanicians must, to a large degree, measure either stress or strain.
4. Photoelasticity is the only whole-field technique that offers the possibility of measuring stress or strain directly.
5. All other optical methods, including the various interferometric techniques, yield only displacement or, in the case of shearography, slope.
6. The displacement data must be differentiated to obtain a map of strain.
7. Differentiation of experimental data is always problematical because it is a "roughening" process.
8. Displacement data must be highly dense and reasonably smooth in order to extract from it reliable strain measurements.

21.6 The Physics of Speckle

As suggested above, the experimental mechanician has at his disposal several different classes of speckle measurement techniques, each of which utilizes one or two of the special properties of the speckle. Understanding of the fundamental physics of speckle phenomena helps us avoid confusion and allows us to choose and adapt a method that is appropriate to the problem at hand. So, before proceeding with descriptions of the various speckle techniques, we undertake a review of how coherent-light speckles are created and their fundamental properties.

Table 21.1 Classification and naming of speckle techniques

Name(s) of technique	How speckles are created	How speckles are used	What is measured
1. Laser speckle photography 2. Coherent-light speckle photography	Coherent light interference	As surface markers	Displacement map: in-plane and out-of-plane
1. White-light speckle photography 2. Incoherent-light speckle photography	1. Painting or appliques 2. Oblique lighting to create shadows of surface structure	As surface markers	Displacement map: in-plane and out-of-plane
Speckle interferometry (using photographic film)	Coherent light interference	Changes of brightness caused by deformation-induced phase change used to form correlation fringes	Displacement map: in-plane and out-of-plane
1. Electronic speckle interferometry 2. ESPI	Coherent light interference	Changes of brightness caused by deformation induced phase change used to form correlation fringes	Displacement map: usually out-of-plane
1. Electronic speckle pattern interferometry 2. ESPI 3. Digital speckle pattern interferometry 4. DSPI 5. Phase shifting speckle interferometry 6. Phase stepping speckle interferometry	Coherent light interference	Successive changes of brightness caused by deformation-induced phase change recorded with the addition of known phase changes at each specimen state	Displacement map: in-plane and out-of-plane
1. Laser speckle shearography 2. Speckle shearing interferometry	Coherent light interference	Changes of brightness caused by deformation induced phase change used to form correlation fringes	Map of slope of surface
1. Electronic speckle shearography 2. Digital speckle shearography 3. Phase shifting speckle shearography	Coherent light interference	Successive changes of brightness caused by deformation-induced phase change recorded with the addition of known phase changes at each specimen state	Map of slope of surface
1. Dual-aperture speckle interferometry 2. Speckle-moiré	Coherent light interference	Speckles overlap to create moiré gratings	Map of in-plane displacement

21.6.1 Formation of Coherent-Light Speckle

Two types of speckle patterns exist. The first is called *objective speckle* because it is formed on any surface that catches the light that is scattered from an object that is illuminated by a coherent source. No imaging system is involved. This type can be seen only indirectly in ways that do not incorporate a lens, including the lens of our eye. It is of limited usefulness, so we will not discuss it further here. The second type, which is the one that is important for us, is called *subjective speckle* because it is created when forming an image of the illuminated object using a lens.

Figure 21.3 shows a simple model of how subjective speckles are formed by interference of light waves passing through a lens.

An expanded laser beam is directed onto an object that has a matte surface so as to scatter the incoming waves. Some of the scattered waves are collected by a lens and redirected onto a screen, a photographic film, an analog television sensor, or a digital detector array. Assume for temporary convenience that the distances between object, lens, and screen are proportioned so that a focused image of the object appears on the screen.

The important fact in this case is that the waves scattered from any single point of the object are focused to a corresponding point of the image, as suggested by the sketch. A cone of waves originating at point A of the object and entering the lens (here shown as dashed lines) are all redirected to converge at image point A′. These converging waves have traveled various path lengths, and they arrive at the image point with a variety of phases. They interfere with one another to produce a particular brightness at the convergence point. Perhaps those arriving at point A′ experience predominately destructive interference

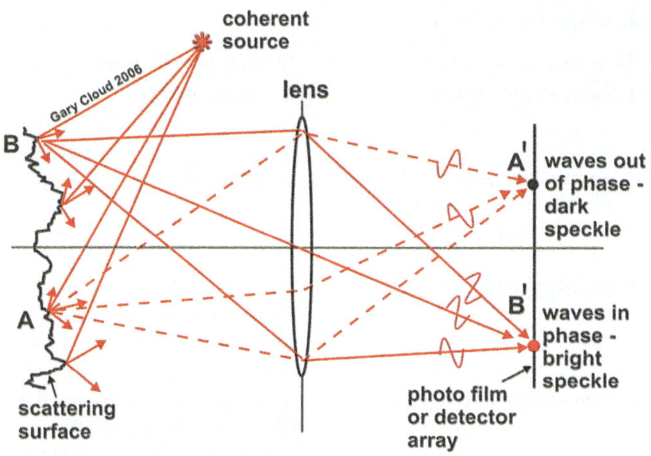

Fig. 21.3 Formation of subjective speckles by interference of light waves scattered from a surface that is imaged by a lens

and so image point A′ will be dark. At screen point B′, which is the image of object point B, the waves (solid lines) might interfere constructively to produce a bright speckle. Other points of the image will experience other mixtures of phases, and the brightnesses will lie between totally black and maximum intensity. The brightness distribution will be random, and the brightnesses of adjacent points are not related to one another.

Missing from this simplified model is the fact that a lens cannot map an object *point* to an image *point*. Rather, it creates an image of a small *patch* of the object that is defined by the diffraction limit or, more precisely, the resolution limit that includes the effects of aberrations. What actually happens, then, is that a collection of waves is mixing together in each image patch. Thus, within each image patch, we have entering a chaotic assembly of phases and incidence angles.

21.6.2 Controlling Speckle Size

Having some knowledge and control of the size range of a speckles that are formed by coherent light is of considerable importance in the design of optical measurement systems, as mentioned above. This determination can be executed in several different ways, all of which give essentially the same result. None of them are entirely satisfactory from a physics standpoint, but they give results that, as the saying goes, "are close enough for an engineer." We deal here only with the simplest model, and we limit consideration to only subjective speckle.

21.6.2.1 Defining Speckle Size

The first problem is to adequately define the metric to be used in gauging speckle size. We inherit an understanding of size that serves well when shoes or pizzas are being purchased. We naturally tend to extend this concept of size to speckle patterns with the result that we attempt to gage the dimensions of, say, the black spots. This common approach is both difficult and prone to error, partly because the pattern is intricate and partly because the perceived size of a speckle is affected greatly by the recording medium, whether the eye, a photo emulsion, or an electronic sensor. Let us solve this vexing problem by defining speckle size as the center-to-center spacing of adjacent dark spots or of adjacent light spots. At the minimum, this uncommon approach greatly reduces the effects of sensor response on apparent speckle size.

21.6.2.2 A Fundamental Assumption

Recall the formula for the distance d_v between the adjacent dark fringes that are produced when two collimated beams of light of wavelength λ meet at angle Ψ,

$$d_v = \frac{\lambda}{2\sin\frac{\Psi}{2}} \tag{21.1}$$

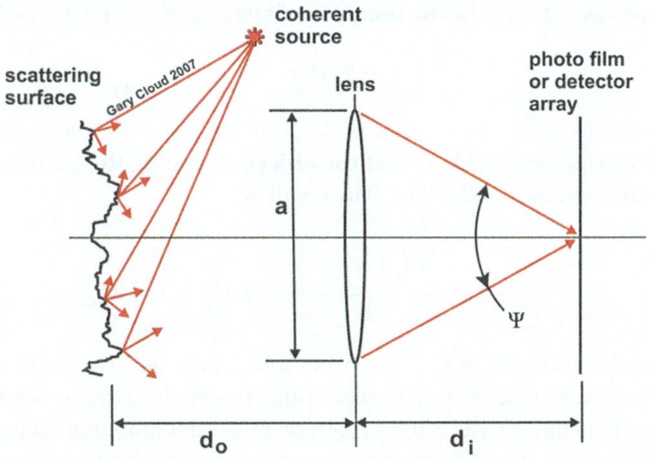

Fig. 21.4 Estimation of speckle size using oblique interference of waves meeting at the extreme angle of incidence

A reasonable assumption is that the minimum fringe spacing produced by oblique interference will be the dominant size, defined as above, of the speckles produced by the complex interference systems extant in the speckle situation. We need to calculate the fringe spacing produced by the waves that approach a screen point from the extremes of incidence angles, because only the ones with the greatest included angle are important as far as speckle size is concerned. Larger speckles that are produced by waves converging at smaller included angles will probably be modulated or broken up by the smaller speckles that are created by the waves converging at the extreme included angle.

21.6.2.3 Size of Subjective Speckle

Figure 21.4 is similar to Fig. 21.3 except that only the waves passing through the outer edge of the lens are shown to converge at the center of the image, these being the waves that meet at the greatest included angle. The added dimensions show that a is the lens diameter, d_o is the object-to-lens distance, and d_i is the distance from the lens to the sensing screen. The dimensions of the illuminated area on the object are not relevant for subjective speckle. The lens captures light from every scattering point that is illuminated.

The idea that the smallest speckle size approximates the smallest fringe spacing in oblique incidence applies, so Equation 21.1 can be used. First, assume that the object is quite far from the lens so d_o is much larger than d_i. By definition, d_i becomes the focal length F of the lens. If the lens diameter a is significantly smaller than F (paraxial approximation), then the sine of $\Psi/2$ approaches $a/2F$. Recall the definition of the "f-stop" or "f-number" of a lens as the ratio of focal length to diameter. The size of the smallest subjective speckles in the image plane is calculated to be,

$$S_{subj} = \lambda\,\frac{F}{a} = \lambda\, f \tag{21.2}$$

In the laboratory, of course, the object-lens distance is likely not significantly larger than the lens-image distance, and the calculation must be extended for finite conjugate distances. In this case, at the image plane, $\sin(\Psi/2)$ is approximated by $a/2\,d_i$, where again the lens diameter is taken to be much smaller than the image distance. Equation 21.1 yields,

$$S_{subj} = \lambda\,\frac{d_i}{a} \tag{21.3}$$

This result is unwieldy, so bring the lens equation into the picture,

$$\frac{1}{d_i} + \frac{1}{d_o} = \frac{1}{F} \tag{21.4}$$

and the definition of magnification,

$$M = \frac{d_i}{d_o} \tag{21.5}$$

to obtain an estimate of subjective speckle size **in the image** for finite object and image conjugates.

$$S_{subj} = \lambda \frac{F\,(1+M\,)}{a} = \lambda\, f\ \ (1+M) \tag{21.6}$$

Sometimes, estimates of the corresponding speckle size **at the object** are useful. Repeat the calculation for the cone of waves from the center of the object to the extremes of the lens. The result is,

$$S_{subj} = \lambda \frac{F\left(1+\frac{1}{M}\right)}{a} = \lambda \frac{f}{M}\ \ (1+M) \tag{21.7}$$

If the paraxial approximation does not obtain in a given lab setup, then the size of the speckles will vary over the field because of the variation of convergence angles for the waves coming from the extremes of the field. Further, manufacturing constraints, lens aberrations, and lens flaws reduce resolving power, so the minimum subjective speckle size in real life is larger, perhaps much larger, than that calculated from any of our models. That said, speckle size estimation is useful as well as interesting. For one thing, it helps us to approximately match speckle size to sensor pixel size in order to obtain a properly modulated image for speckle interferometry and similar procedures. Final tweaking of the lab setup is usually needed to obtain the best modulation possible.

21.6.3 Estimating Speckle Brightness

Recall that each speckle is formed from a large number of waves that arrive at the speckle point with a random distribution of phase and amplitude. There is uniform likelihood that any particular wave will arrive with phase between zero and π radians. It is convenient to assume that the amplitudes of the waves are equal, but that is not necessary. We do apply the restriction that the waves are of identical polarization so that they are able to interfere. The problem is to predict the resultant brightness (irradiance, intensity) of any individual speckle. Given the randomness of the process, the resultant for one speckle in the field is not related to that of any other.

This problem is one of a class called the "random walk" or, by some folks who know of such things, it is described as the "drunkards walk." Considerable thought by generations of wise scholars has been devoted to solving problems of this type, which are ubiquitous in many areas of endeavor including meteorology, finance, physics, gambling, biology, geology, and so on. Some of these problems remain unsolved, or have been solved only numerically.

Starting from the origin, the walker takes an arbitrary step in a random direction. His step has magnitude and direction, and it seems reasonable to think of these parameters as analogous to amplitude and phase angle in the physics problem at hand. He then takes another random step in a random direction. And so on. The problem is to figure out what the resultant of the accumulation of many such step vectors is. That is, "Where does the walker end up?"

Space limitations preclude a full discussion of the solution, so we pass quickly to the most basic result for a speckle within a pattern that is created by a single illuminating beam. Figure 21.5 shows in a semi-quantitative way the probability that the normalized intensity of a single speckle will have a brightness (intensity, irradiance) lying in the range between I and $I + dI$.

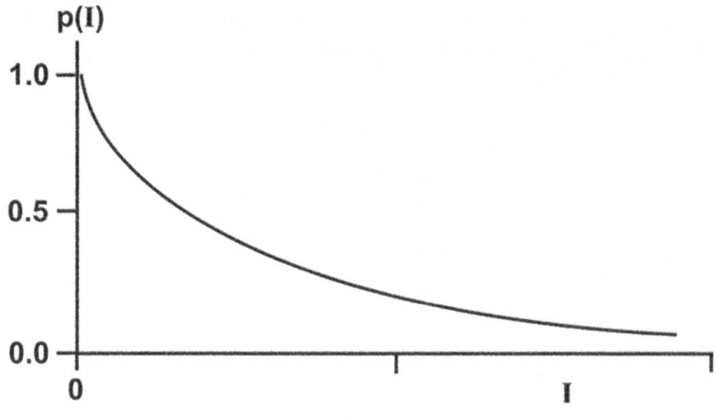

Fig. 21.5 Probability that a coherent-light speckle will have a given intensity

It seems startling that the most probable speckle intensity is zero, meaning dark. In other words the random walker will eventually get back home. The probability of very bright speckles approaches zero. This fact carries some implications for experimental mechanists. It is a reason why double-exposure speckle photography works very well for determining displacement fields.

If more than one illuminating beam is used, as in the various forms of speckle interferometry where an object beam and a reference beam are mixed together, then the statistics of speckle brightness change. If the two beams are coherent, then both the speckle size and brightness change, but dark speckles still predominate by far. If, however, the two speckle fields are not coherent, then dark speckles do not appear at all. This latter case, especially, places bounds on what can be done with speckle in the domain of measurement.

Close examination of a recorded speckle pattern raises a serious question as to the correctness of the prediction of speckle brightness. Bright spots seem to be at least as common as dark spots. There are several reasons for this apparent disagreement, some being a result of the way we create and process the patterns and others being a manifestation of the nonlinear response of our visual system.

21.7 Speckle Measurement Methods Made Easy

Before attacking the details of speckle measurement techniques, it seems worth mentioning that they can be carried out with very simple equipment, at least for demonstration or pilot study purposes. Do not be overwhelmed by the fancy setups described in the usual research papers. For example, the scatter plate interferometry technique that is described later in this article can be executed with a laser or even a good laser pointer, a lens, a piece of glass, a web or phone camera, and a computer with standard photo processing software that offers picture subtraction or similar logical functions. The addition of a beam splitter and some mirrors facilitate more sophisticated work. Even phase shifting interferometry can be executed using manual mirror adjustments—tricky but possible.

21.8 Laser Speckle Photography

Speckle photography is probably the easiest and most direct exploitation of the speckle phenomenon for whole-field non-contacting measurement of displacement and deformation, because it uses the speckles simply as microscopic markers of points on the surface of the object being studied.

Because we are detecting the movement of the speckles, the sensitivity of the technique is limited by the speckle size. If the speckles overlap in the images recorded for the two specimen states, then we cannot discern the displacement that the speckle has undergone. Hence, to obtain maximum sensitivity in speckle photography, the smallest practicable speckles are required. The imaging lens must have a wide aperture.

A single speckle is a unique signature derived from the local characteristics of a small area of the object surface as well as the geometry of the optical system and the numerical aperture of the recording system. If a speckled image is created, then the speckle near a point in the image is uniquely identified with the corresponding point on the object. Then, if the point on the object moves within certain limits, and if the optical system is not changed, the speckle moves with the point, and the motion is apparent in the image. The speckle is neither lost nor reformed. If the speckle is recorded for two states of the specimen, then the displacement of the speckle corresponds to the local displacement of the surface. These ideas are easily extended to map the displacements of all points of the image through the use of ordinary photography. We consider here only the single-aperture double-exposure implementation using analog recording in photographic emulsion media.

21.8.1 Data Recording

An optical system for recording speckle data on a photographic plate is shown in Fig. 21.6. There is nothing special about this setup, which is an advantage of the method. Simply illuminate the object with an expanded laser beam and take a picture. As mentioned, the aperture of the recording lens governs the speckle size, and therefore the basic sensitivity of the method. The lens numerical aperture must be chosen with an eye to the magnitudes of the displacements to be measured and the resolution capabilities of the film. For most purposes, a numerical aperture on the order of $f/4$ to $f/11$ seems appropriate.

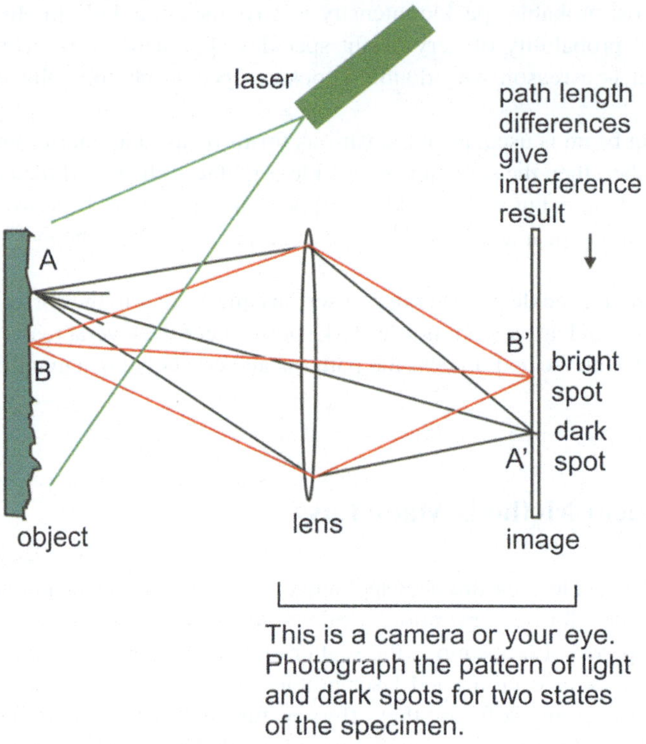

Fig. 21.6 Recording of laser speckle pattern for speckle photography

A typical double-exposure sequence is utilized, with the specimen being in a different state for each of the two exposures. The processed photograph, sometimes called a "specklegram" contains a speckled image of the specimen with no evident fringes.

21.8.2 Data Analysis Using Young's Fringes

Now consider from a physical viewpoint the structure that is recorded on the film. A single speckle pattern is a very fine random interference pattern of small patches which have the highest probability of being dark. The exposed film is a negative which will exhibit light spots (holes, actually) in the areas of the predominant dark speckles, so there will be a random array of numerous nearly transparent spots in the negative image. The second exposure will add a similar array of light spots which are shifted with respect to the first set by the displacements and deformations of the object. If the shifts are not too large, but also if they are large enough so that speckles in the two systems do not overlap appreciably, then the resulting pattern will consist of a multitude of light speckle pairs. The spacing between the two elements of any pair is indicative of the magnitude of the local deformation as seen at the film plane. The orientation of the speckles in a pair with respect to one another indicates the orientation of the local displacement as seen at the film plane. Figure 21.7 is an illustrative cartoon that shows relationships between a few pairs for a cantilever beam that is being bent. The problem is to determine the local displacement vector over the extent of the specimen. Figure 21.8 shows a photomicrograph of just such a double-exposure speckle pattern. The pairing of the speckles can be seen, and, in fact, the dense pattern of speckle pairs seem to form a grating-like structure.

That the plate is a complex diffraction grating is readily ascertained by illuminating the plate in reflection or transmission with a point light source such as a pen light. Several orders of Young's fringes with spacing and direction dependent upon the displacement of each area of the specimen between exposures will be seen. This diffraction pattern forms the basis of one method of displacement measurement.

Suppose that a slender beam of coherent light is passed through the double-exposure photographic negative. A screen is placed some distance away from the negative for observation of the diffraction pattern. Figure 21.9 is a schematic of such an arrangement showing a laser beam impinging on only one speckle pair, something unlikely in practice. But, it makes no difference, because the spacings of the speckle pairs will be slowly varying within the area subtended by the laser beam. The pattern seen downstream of the speckle photo is the diffraction pattern or the optical Fourier transform of the illuminated

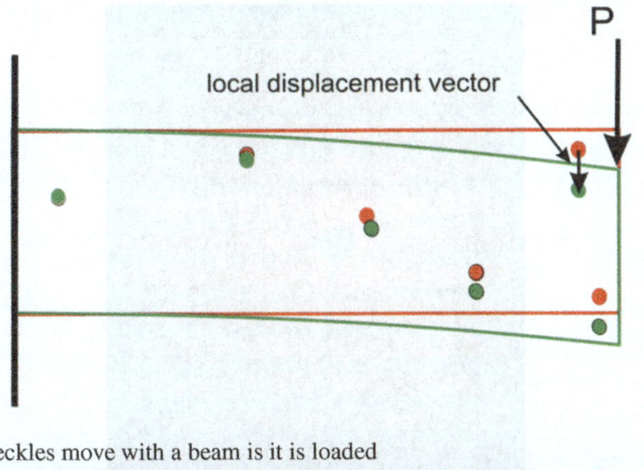

Fig. 21.7 Speckle pair formation as speckles move with a beam is it is loaded

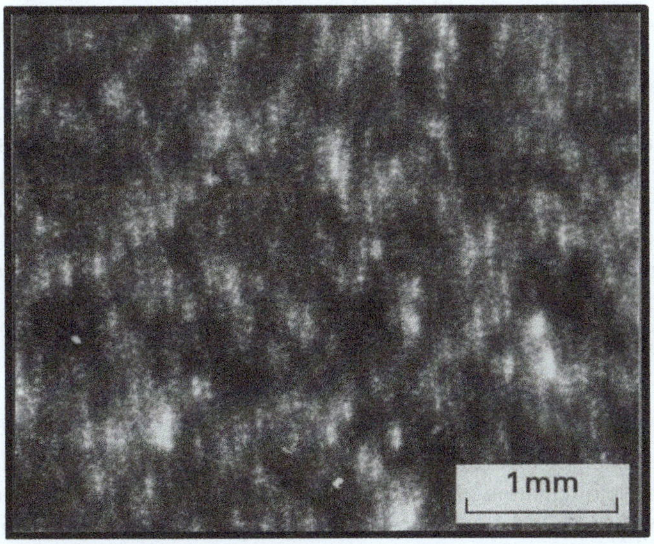

Fig. 21.8 Photomicrograph of a double-exposure laser speckle pattern

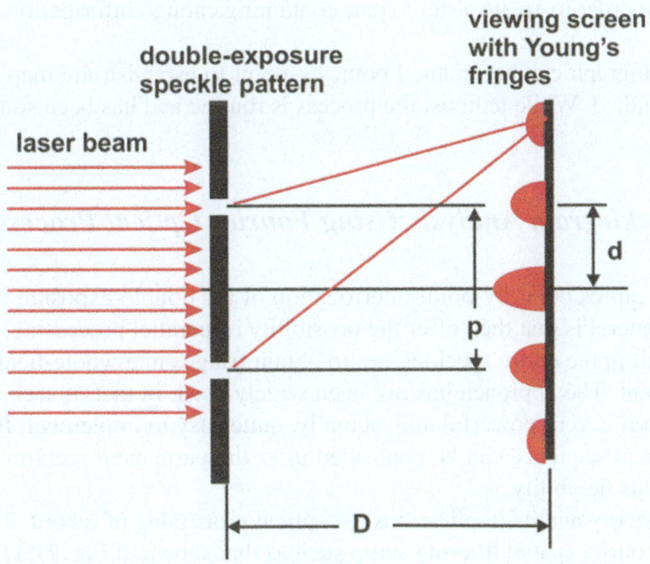

Fig. 21.9 Setup for observing Young's fringes from a speckle photograph

Fig. 21.10 Young's fringes from a specklegram showing displacement magnitude and direction at a point in an aluminum plate

speckle pattern, but it is named "Young's fringes" after the genius who first used it to explain the wave nature of light. Figure 21.10 is a photograph of such a pattern that was obtained from a speckle photograph of a sandblasted aluminum plate that was obliquely illuminated. The dark spot in the center is caused by a mask that blocks the strong central beam.

Now, if the distance *D* from the speckle pattern to the screen is significantly larger than the diameter of the laser beam having wavelength λ, then the distance *p* between the speckles of the pair, which is the magnitude of the local displacement of the specimen *in the image plane*, is directly related to the spacing *d* of the Young's fringes by the following equation, which may be derived directly from Huygens's principle or from Fourier analysis,

$$p = \lambda D/d \tag{21.8}$$

The *specimen* displacement at the point being interrogated can then be determined by using the image magnification factor. Furthermore, the orientation of the local displacement vector is normal to the inclination of the Young's fringes. Thus, it is important to know the relative orientation of the speckle photograph relative to the specimen coordinate system. This approach has been shown to be insensitive to moderate out-of-plane motions or specimen tilt. However, if the specimen surface is severely inclined or if large out-of-plane displacements occur, then the geometry of the setup is used to map the measured displacement vector back to specimen space. For 3-dimensional measurements, two speckle patterns are taken from different viewing angles in order to create a stereo pair containing enough information to allow mapping displacements back to specimen space.

In practice, the speckle photograph can be scanned point-by-point to establish the map of displacement vector over the extent of the specimen being studied. While tedious, the process is routine and has been somewhat automated.

21.8.3 Whole-Field Specklegram Analysis Using Fourier Optical Processing

The method described above requires point-by-point interrogation of the double-exposure speckle photograph. A favorable feature of optical methods in general is that they offer the possibility of parallel processing. In this instance, it is possible to spatially filter the data contained in the entire specklegram to obtain fringes in a whole-field image that are loci of points of constant displacement component. The approach has not been widely used; in fact, it seems to have been forgotten in both its analog and digital forms, but it is very powerful and, actually, quite easy to implement. It's outstanding advantage is that the sensitivity of the measurement sequence can be controlled *after* the permanent recording of the primary data. Very few experimental techniques offer this flexibility.

A full review of diffraction theory and its implications for optical processing of recorded data cannot be undertaken here, so, we summarize. An optical Fourier spatial filtering setup such as that shown in Fig. 21.11 is utilized.

The basic ideas, specific to the context of speckle photography, are as follows. The speckle photograph is illuminated with an expanded beam of coherent light. A lens is placed behind the illuminated photo to create in its focal plane the optical Fourier transform of the photo. Distance off-axis in the transform plane corresponds to spatial frequency, i.e. the number of speckles per unit distance, in the photograph. A blocking filter with a small aperture is placed off-axis in the transform plane to select only the data that corresponds to a certain spatial frequency. The inverse transform is then created from only

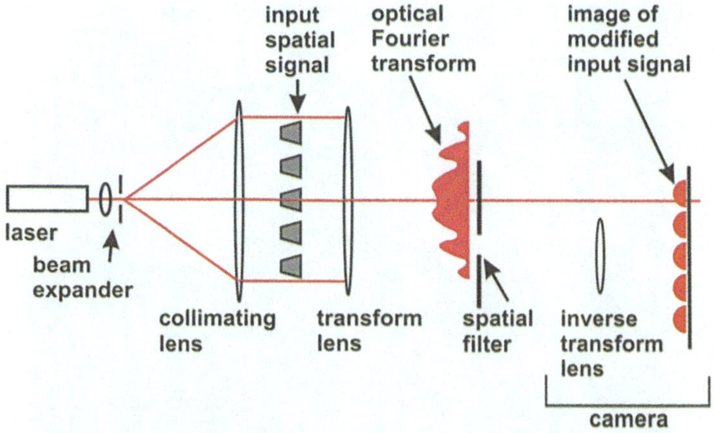

Fig. 21.11 Schematic of an optical Fourier spatial filtering setup for obtaining whole-field displacement maps from speckle photographs

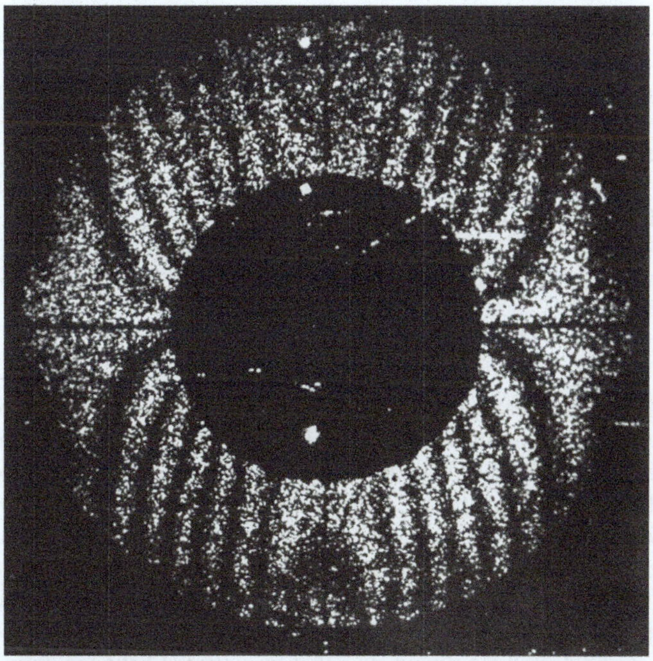

Fig. 21.12 Whole-field vertical displacement fringes from optical spatial filtering of a speckle photograph of an aluminum ring in diametral compression

that data using another lens. There will be constructed on the screen an image of the specimen with fringes showing loci of points having constant magnitude and direction of the local displacement vector throughout the specimen. The displacement amplitude sensitivity is established by the distance that the spatial filter is offset from the optical axis. The displacement direction sensitivity is set by the direction that the filter is offset. So, we obtain in a single step a complete map of a given magnitude and direction of the displacement vector distribution over the extent of the specimen. Figure 21.12 shows one such fringe pattern for a ring in diametral compression.

21.9 White-Light Speckle Photography

The speckle methods discussed so far have used the speckles formed in an image merely as fiducial markers that track motions. A pressing question is whether it is necessary to create the speckles by coherent light. Could not a fine set of random dots on the surface of the object serve just as well? The answer is yes! The question was answered in the affirmative by several investigators almost simultaneously, particularly by P. Boone and L. DeBacker. who applied the speckle photography

Fig. 21.13 Photomicrograph of double-exposure white-light speckle from a sandblasted aluminum disc

technique using a retroreflective surface which was illuminated with noncoherent light. They also showed that placing special dots or other materials on the surface was not necessary if the surface contained appropriate roughness features. When the surface is illuminated at a low angle, the peaks and valleys in the surface create shadows which can be used as a speckle pattern. At about the same time, J. Burch and C. Forno used similar procedures with slotted "tuned" apertures in a camera to perform moiré analysis on large objects. Other aspects and applications of noncoherent light speckle were soon reported by several others including F.P. Chiang, A. Asundi, and G.L. Cloud.

21.9.1 Date Recording in White-Light Speckle

The recording of double-exposure noncoherent-light speckle patterns in the laboratory is very simple. The speckles can be created in several different ways. Splattering the surface with fine blobs of ink or paint works, as does the idea of applying markers to the surface using decals or appliques that carry a random pattern of small dots. Most promising is that low-angle lighting of the test surface will create a useful speckle pattern provided the surface has appropriate fine roughness features. Since they are uniquely related to local surface topography, the speckles move with the surface as do laser speckles. One exposure is taken with the specimen in its beginning state, and the second is recorded for a subsequent state. If the specimen is deformed or moved between exposures, then the result will be a specklegram that is similar to those obtained by double-exposure laser speckle photography. Figure 21.13 shows a photomicrograph of a double-exposure speckle pattern that was obtained by this method from a sandblasted aluminum disc that was rotated between exposures. Note the similarity of this patterns to the one of Fig. 21.8, which was obtained in coherent illumination.

21.9.2 Data Analysis in White-Light Speckle Photography

Extraction of displacement data from a white-light double-exposure speckle pattern parallels that for coherent speckle photography with one exception. Because the speckles are usually relatively larger than coherent light speckles, and because there are fewer of them, the whole-field optical spatial filtering technique does not, usually, produce very good results.

21.9.3 Some Applications of White-Light Speckle Photography

A most useful aspect of white light speckle is that it can be used to measure displacements of objects which, by virtue of size, location, or fragility cannot be illuminated with laser light. The technique of creating the speckle patterns from naturally occurring surface features causes it to be almost totally noninvasive. Additionally, very large structures can be analyzed in the field using either natural or artificial light.

The following list offers three examples of unusual applications of white-light speckle photography that have been explored by the author and colleagues:

1. Mapping the velocity field in glaciers. The motion of the surface of a portion of the Nisqually Glacier, Mt. Rainier National Park, WA, USA was measured using white light speckle. The specimen is about 6 km long and 800 m wide, it exists in a hostile environment, and it is not readily accessible for conventional measurements, especially in the steep icefall area which is of interest. Cameras especially constructed for the purpose were anchored on the valley walls to command a view of the upper elevations. From these locations, double-exposure images were acquired on holographic plate over intervals of 24 h. Patterns were recorded when the sun was low so as to emphasize the undulations, sun cups, crevasses, and other surface features in the study area. The pictures were taken in pairs from the opposite valley walls so as to have a stereo-pair for data reduction. Displacements in the film plane were gathered in the laboratory from these specklegrams by interrogation with a laser beam so as to produce Young's fringes. The displacements measured at the film plane were corrected for magnification, then projected onto the glacier using topographical map data and a computer algorithm similar to those used in photogrammetry.
2. Measuring displacements in human skin *in vivo*. This pilot study was of interest to the medical people who deal with scarring and other maladies that affect the compliance of skin. The back of the subject's hand was lightly wiped with ink so as to highlight the pores and creases that occur naturally. As the hand was flexed, double exposure speckle photographs were recorded using sequenced flashlamps for illumination. The resulting specklegrams were then interrogated with a laser to establish the displacement profiles. The sensitivity was not great because of the coarse nature of the speckles. It was, however, sufficient because the displacements of skin are large even during small movements.
3. Mapping the velocity field within turbulence cells in thin sheets of rapidly flowing fluid. As fluid, in this case water, flows over obstacles, moving eddy currents and turbulence cells are created. Determining velocities profiles throughout the extent of these structures was, at the time, a problem of long standing. For this pilot study, the water was seeded with fine aluminum powder. A rigid camera was aligned normal to the sheet of flowing water. Illumination was by two flashlamps placed adjacent to the camera. A special circuit was designed and constructed so that the lamps could be triggered in a close controllable sequence. Double-exposure speckle photographs were recorded. A small segment of one such specklegram can be seen in Fig. 21.14. Flashlamp timing was measured so that the water displacements between exposures could be converted to velocities. Motion pictures were also recorded in order to identify areas of most interest within the field. Young's fringe patterns showing displacement magnitudes and directions at key points were obtained through pointwise interrogation of the specklegrams. One sample of the Young's fringes obtained in this manner appears in Fig. 21.15. Notice that the fringes are of lower quality than those of Fig. 21.10, but they are sufficiently good for the purpose.

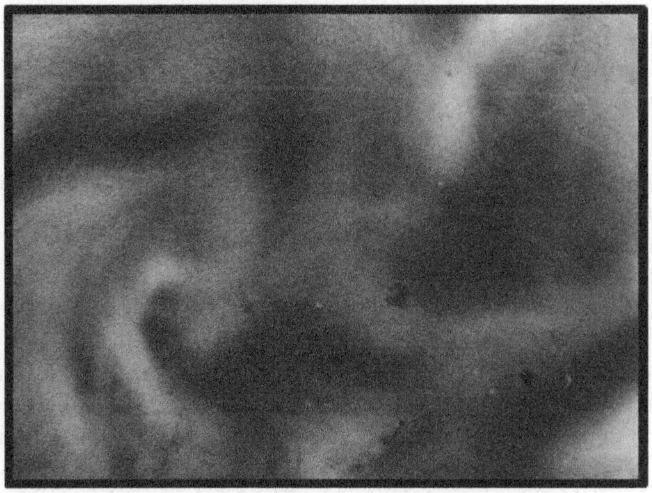

Fig. 21.14 Segment of a double exposure specklegram showing turbulent flow of water

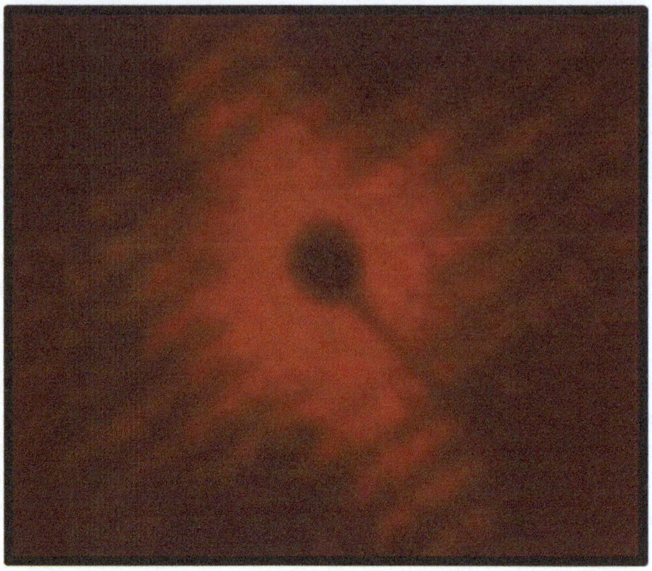

Fig. 21.15 Young's fringes showing displacement magnitude and direction at a point in turbulent flow of fluid obtained from a double exposure speckle photograph

21.10 Speckle Correlation Interferometry

So far, we have used laser speckles only as fiducial markers that move with the surface being studied. The simple experiments described at the beginning of this paper suggest that speckle brightness might also vary with certain types of specimen motion. This behavior suggests that variations of brightness could be used to map displacements and motions, perhaps with very high sensitivity. Our experience-based intuition might also lead us to think that all we need do in order to get a map of displacement is to record the initial brightness distribution of the speckle pattern, move or deform the specimen, record the second speckle brightness pattern, then subtract the first record from the second. In the early days of speckle-based measurement, many of us used up a lot of high-resolution photography film in such experiments. We were usually adding the images in double-exposure mode instead of subtracting them, but this approach worked for hologram interferometry, so why did it not work with speckle? Actually, the product of these experiments were doubly-exposed speckle pattern photographs of the type described above. They gave us, indeed, displacement maps, but the sensitivities were not even close to the what is expected of true interferometry. Where did we go wrong?

The answer was obvious, although, in retrospect, it took rather a long time for some of us to find it. We were trying to perform interferometric measurements with a one-legged interferometer. Three facts are relevant here, as follows:

1. All light detectors, including the eye, photographic film, and photo cells are not sensitive to phase, and phase changes caused by path length changes are what we are trying to measure. One reason for this fact is that the frequencies of light waves are far beyond the bandwidth of any detectors. Our detectors respond only to wave intensity (irradiance, brightness), which depends only on wave amplitude.
2. Absolute phase and phase changes in light waves can be preserved and measured only by converting the phase data to intensity values through the phenomenon of interference with other waves, perhaps some sort of reference beam.
3. The brightness of an individual speckle does not change rapidly with either lateral or line-of-sight motions of the illuminated surface. Our calculations of speckle size tell us that this is so. Imagine a small movement of the illuminated specimen in, say, the longitudinal direction. The phases of all the light waves that are gathered to form a given speckle will change, but by nearly the same amount. So, the interferences within a speckle will not change appreciably, meaning that the resultant brightness of that speckle will not change. Actually, this is the main reason why speckles seem to move with the surface and so make speckle photography possible. The brightness changes that we can see are only those that are caused by motions large enough to change the phase relationships within a speckle.

These facts lead to the conclusion that the waves forming a speckle pattern must be mixed with another beam, commonly called a reference beam, in order to observe and preserve the phase data. Doing so allows us to perform measurements with sensitivities on the order of the wavelength of light.

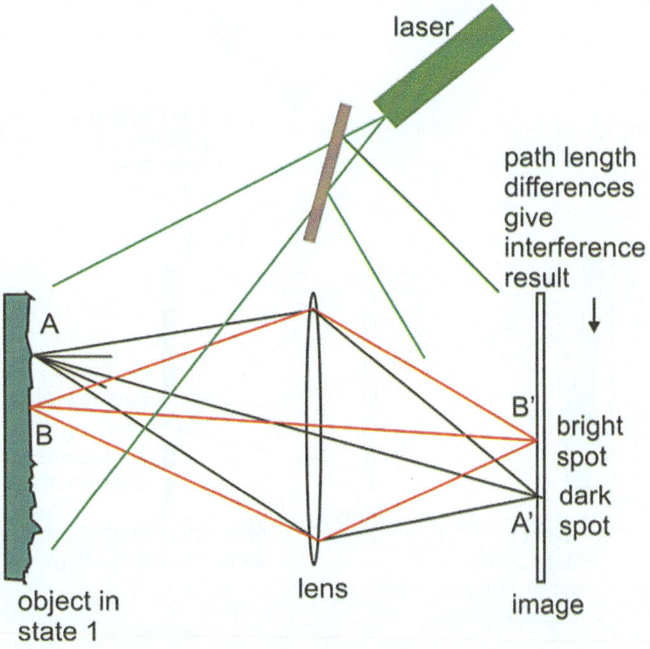

Fig. 21.16 Setup and recording of speckle interferometry pattern for initial state of specimen

21.10.1 Speckle Correlation Interferometry

Guided by these facts, we are able to devise procedures to perform basic real-time interferometric measurements of object displacement using coherent-light speckle. All that is required is to simply add a reference beam to the setup shown on Fig. 21.6 for implementing speckle photography. Controlling the geometry of the setup allows us to select whether it is sensitive to in-plane or out-of-plane motion and also facilitates adapting the measurement sensitivity to the problem at hand. For momentary convenience in gaining understanding, we will assume that the data recording is done using photographic emulsions. However, as subsequently discussed, speckle correlation interferometry is now implemented using electronic recording and processing of data.

21.10.1.1 Out-of-Plane Speckle Correlation Interferometry

Out-of-plane speckle correlation interferometry using photographic media for data recording was the earliest implementation of the concept, and improved versions are still very useful in mapping deformations and for NDI. We study it in some detail here because of its instructional value. If this method is understood thoroughly, then the learning makes it easy to fathom related techniques and even invent useful variants.

Consider the setup shown in Fig. 21.16, which is the same as that shown in Fig. 21.6 but with a reference beam that is created by splitting off some of the expanded laser beam using a partial mirror and causing it to impinge onto the photographic film along with the waves that are scattered by the specimen and collected by the lens.

The photographic plate recording of the speckle interference pattern is processed, and put back exactly into its original position. It is a negative of the speckle, so speckles that are bright convert to dark on the plate. If the plate is brought exactly into registration, then all the bright spots line up with dark spots on the negative, and the whole image will be dark. The specklegram is acting as a filter mask.

Suppose now that the specimen is shifted and distorted as illustrated in Fig. 21.17.

The speckle at position A′, which was formerly dark, is now bright and it passes through the hole in the negative filter mask. The speckle at B′, formerly bright, is now maybe changed to gray, but, whatever it is, it is blocked by the exposed negative, so it now appears black. We are, in effect, subtracting the original brightness distribution from the new one. It happens that the coherent addition of a the smooth reference beam with a random object beam produces a field in which the speckles will be predominately dark, so the negative image in the mask will contain mostly transparent holes, but that fact is of minor importance.

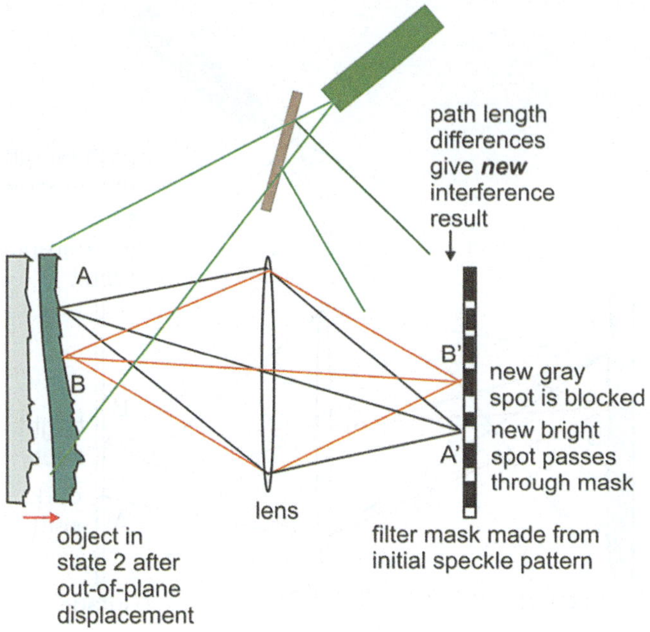

Fig. 21.17 How the speckle image from final state of specimen is subtracted from that of initial state by use of the filter mask

Now, extend this thinking to the entire field, which contains many thousands of speckles that have random brightnesses in both the initial and final states. Neither the initial nor the final speckle patterns make any sense. But, we are interested only in the *changes of brightness*. The key idea is that rigid body motions and typical deformations of the specimen will cause *phase changes*, and therefore speckle *brightness changes*, that are not random but that vary only relatively slowly over the field. The map of brightness change will be well-ordered, and the areas of uniform brightness will be assembled into smooth loci of constant brightness change, or, in other words, loci of constant phase change. Only in the vicinity of cracks or other discontinuities will these loci be not smoothly varying.

The loci of constant phase change look like the patterns that are typical of, say, photoelasticity, so they are usually called "fringes." Actually, they are not interference fringes at all, but we are stuck with the name.

Further, a little thought suggests that this technique will not work if the speckles shift very much laterally relative to the mask. We saw that separation of the initial and final speckles is key in speckle photography, so small speckles are used. Such separation in speckle interferometry is fatal. One ends up comparing the initial brightness in one speckle to the final brightness in some other speckle, which is meaningless. Such behavior is called "speckle decorrelation." In order to maintain correlation throughout an experiment, relatively large speckles are used, meaning that the imaging aperture must be rather small. Given the importance of keeping the patterns superimposed within certain limits, this technique and its variations are called "speckle correlation interferometry."

The fringes can be quite easily ordered in most cases, and the relationship between fringe order and specimen motion/deformation can be derived by simply calculating the path length change between the two states of the specimen. Assume that the lens aperture is small relative to the distance from object to image (paraxial approximation), take the viewing angle to be 0°, the specimen illumination angle to be θ_i, the wavelength of the laser as λ, the local fringe order as n, and let Δz be the displacement of the specimen along the line of sight. The phase change is calculated and converted to path length change. The result is that there will be a dark fringe wherever,

$$\Delta z = \frac{n\lambda}{1 + \cos\theta_i} \tag{21.9}$$

Surprisingly, the reference beam angle is not relevant. If a partial mirror is inserted into the system so that the illumination angle is zero, then this result becomes identical to that discovered by Newton to explain the fringes that bear his name, and the fringe pattern is a map of the change of contour of the specimen. In-plane motions do not affect the result as long as they not so large as to cause loss of speckle correlation.

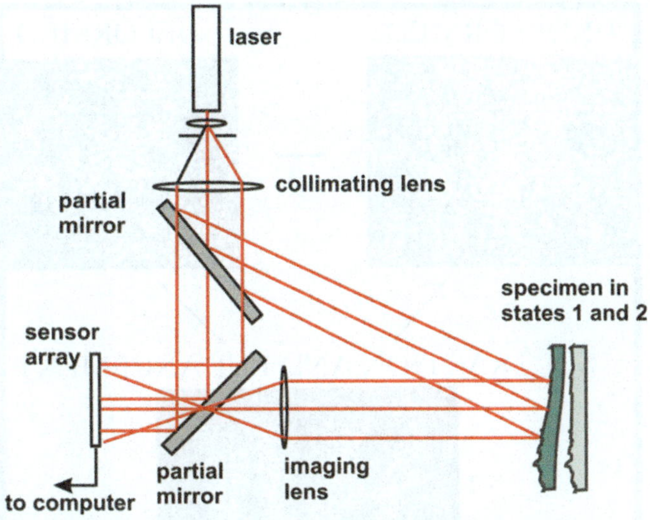

Fig. 21.18 Schematic of a basic electronic speckle correlation system for measuring out-of-plane displacements

Film-based speckle interferometry using a subtraction mask operates in real-time or live-fringe mode. Double-exposure or frozen-fringe mode, wherein the initial and final speckle brightness patterns are added together, can be made to work, but fringe visibility is very low. The same is true of time-averaged film-based speckle interferometry that is used to study objects that are vibrating.

21.10.1.2 Electronic Speckle Pattern Correlation Interferometry (ESPI)

Three problems arise when executing film-based speckle correlation interferometry of the type described above. One is that, because of the nature of the speckle brightness profile, the visibility of the correlation fringes is limited to about 14% unless nonlinear film processing is used. Even then, the visibility increases only marginally. The second problem is that, as with real-time holographic interferometry, various schemes must be used in order to get the developed mask back into exactly its original position. Third, the required photographic media are obtained with difficulty these days.

A striking solution to these problems was furnished by J. N. Butters and J.A. Leendertz plus their colleagues in about 1970. They recorded the speckle patterns on video tape and then manipulated them electronically to obtain the fringe patterns. Subsequently, digital sensor arrays and cameras arrived on the scene. These made it possible to record the speckle brightness patterns as arrays of numbers that could be stored in computer memory and later subtracted from one another to obtain a new array giving the required phase changes. When displayed on a computer monitor, the resulting digital photograph contains the correlation fringes. Visibility of the fringes is enhanced by additional processing. This digital acquisition and processing of speckle patterns quickly replaced film-based techniques and, further, facilitated great improvements in speckle interferometry. Figure 21.18 is a schematic of such a system that uses collimated light and two partial mirrors.

Figure 21.19 illustrates how the initial random speckle pattern is subtracted from the deformed final pattern to form an ordered array of "fringes."

In general, ESPI and its evolutionary variants can be made to operate in either real-time or frozen-fringe mode. Unlike the film-based procedure, the initial and final speckle brightness patterns can be stored and then the one subtracted from the other, thus offering frozen-fringe capability. These techniques can also be used to study mode shapes and the like in vibrating objects when operated in time-average mode.

21.10.1.3 In-Plane Speckle Correlation Interferometry

The ideas described above can be extended to measure in-plane displacements in such a way that the data are not contaminated by out-of-plane motion. J.A Leendertz in 1970 devised a setup that dispensed with the reference beam and instead used two beams to illuminate the object in a symmetrical fashion. The out-of-plane displacements cause equal optical path length changes in the two beams, and these are canceled when the initial speckle brightness pattern is subtracted from the final pattern. Such an arrangement is illustrated in Fig. 21.20.

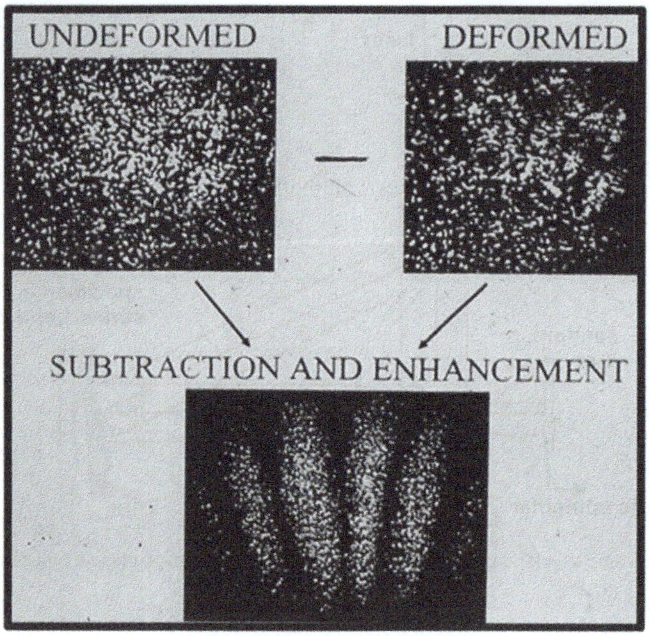

Fig. 21.19 How an ordered speckle correlation fringe pattern is formed by subtraction of one random speckle pattern from another

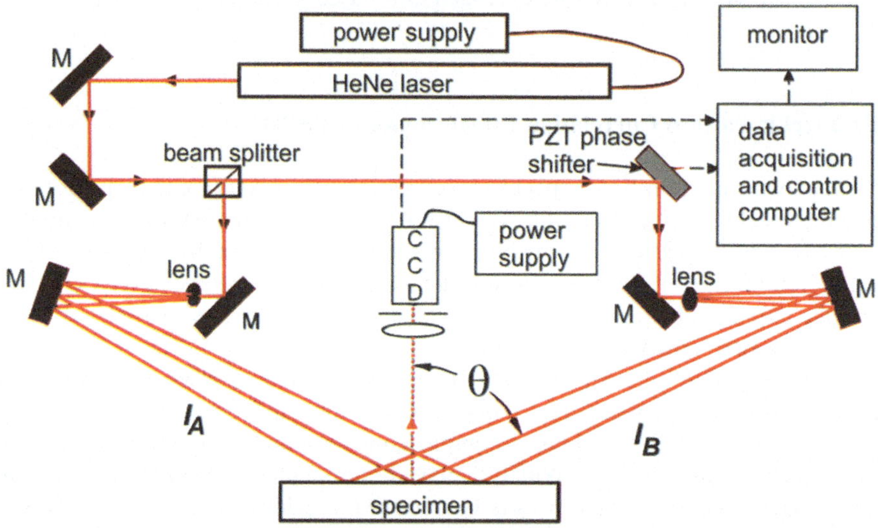

Fig. 21.20 Schematic of a setup for measuring in-plane displacements using electronic speckle pattern interferometry

Notice that the arrangement of Fig. 21.20 includes a phase shifting device, which will be discussed presently. If the phase shifter is not activated, it acts as a fixed mirror and only correlation fringes will be observed. Again, if the dark fringes can be rank-ordered and numbered, and if viewing is normal to the specimen, then path length difference calculations lead to the following result for the in-plane displacement, Δx,

$$\Delta x = \frac{n\lambda}{2\sin\theta} \tag{21.10}$$

where n is the fringe order, λ is the wavelength, and θ is the illumination angle.

Apparatus for performing electronic speckle pattern interferometry soon began to be marketed by vendors, notably Ealing, Ettemeyer, and Karl Stetson Assoc. Figure 21.21 shows a segment of a pattern of raw real-time correlation fringes that was acquired using one such device to detect impact damage in a composite natural gas fuel tank. The tank was rated to operate at 3000 psi, but a pressure change of only 30 psi was enough to quantify damage with the sensitivity desired.

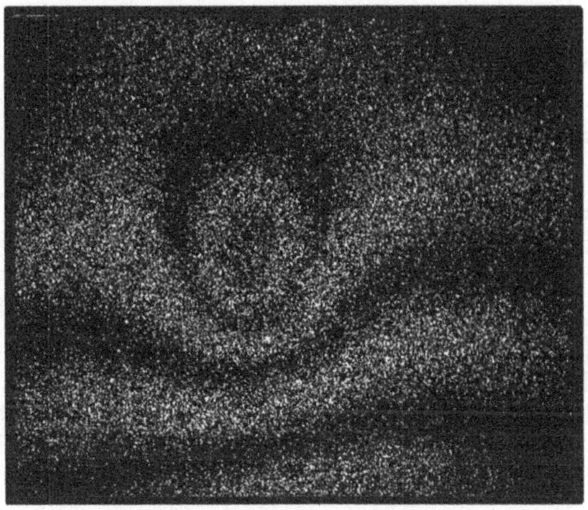

Fig. 21.21 Raw speckle correlation fringes showing damage in a composite NGV fuel tank

21.10.2 Speckle Interferometry in Noisy Environments

A characteristic of the speckle correlation interferometry and speckle phase-shifting interferometry implementations described so far is that they tend to be sensitive to disturbances such as vibrations and air currents. Most experiments with speckle interferometry have, indeed, been conducted on vibration-isolated platforms. The techniques have tended to be rejected for field and factory use because of the belief that good results cannot be realized in such environments.

The reason for this sensitivity is that the speckle interferometers discussed so far are like many others, including hologram interferometers, in that they are not "common path" systems. The object and reference beams, for example, follow separate paths. Disturbances in either beam will cause instabilities in the speckle brightness pattern that confuse the results. Photoelasticity is the only truly common path interferometric method we have for strain measurement. Laser speckle shearography, in its most usual form, is also a common path interferometer, which is why it is widely used for NDI in the field. T. Siebert, R. Wagner, and A. Ettemeyer introduced in 2001 a compact ESPI sensor that can be attached to the test object, thus eliminating vibration effects, and these proprietary systems were widely sold.

The author and colleagues have worked along three different lines to devise and developed alternative approaches that, singly and together, largely eliminate environmental effects on laser speckle interferometry. These approaches utilize the following concepts:

1. Converting the optical setup to a "nearly common path" design,
2. Utilizing the vibrations constructively and/or eliminating their effects during image processing,
3. Developing a phase shifting method that does not require known phase shifts.

The resulting implementations of laser speckle interferometry gave rise to exceedingly simple optical setups that are inexpensive and easy to set up. They have proven useful for measurement and inspection in field environments.

21.10.2.1 Scatter Plate Out-of-Plane Laser Speckle Interferometer

Figure 21.22 is a schematic of an out-of-plane speckle interferometer that utilizes a "scattering plate," which can be as simple as a piece of non-reflective picture frame glass, to create the reference beam. This plate is set close to the specimen, meaning that the only part of the optical path that is not "common" is the bit between the plate and the specimen. In fact, in some realizations, the plate can be attached directly to the specimen. Several other varieties of this concept have been described.

Notice that the scatter plate device includes a PZT driver that causes the plate to wobble or vibrate in order to induce phase shifts in the reference beam. If the driver is quiescent, then the system yields correlation fringe patterns as described above and shown in the example of Fig. 21.23, which is from an NDI study to find thinning defects on the back side of an aluminum plate.

If precise quantitative measurements are sought, then the PZT driver causes the plate to vibrate. Because the resulting phase shifts are not uniform over the field, and might not even be known, the data processing must be different from that usually used in phase-shifting interferometry, as will be discussed presently. Dr. X. Ding developed in 2001 an Improved

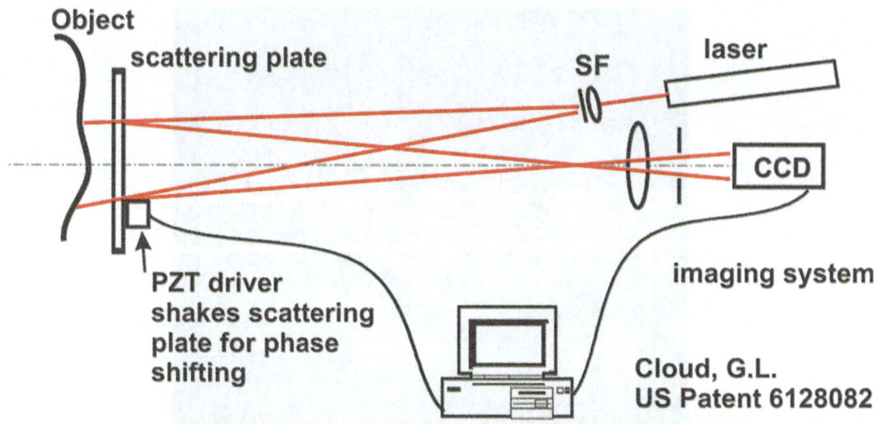

Fig. 21.22 Schematic of a scatter plate "nearly common-path" laser speckle interferometer for measuring out-of-plane displacements

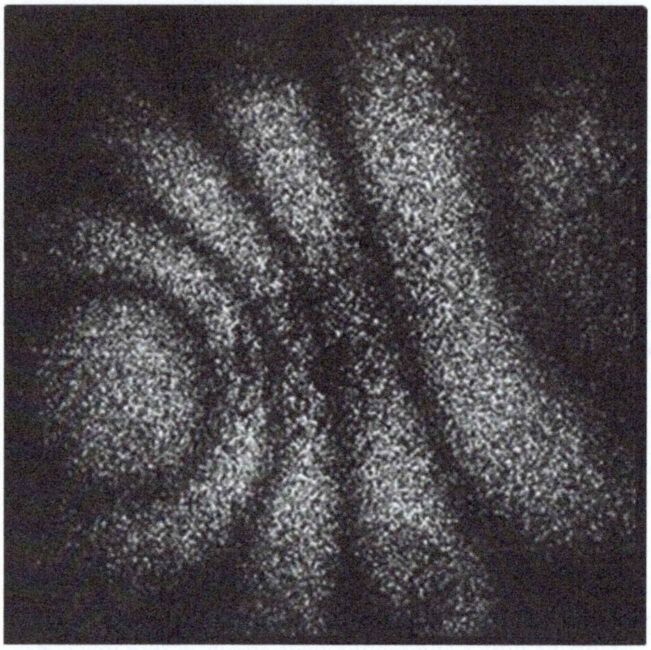

Fig. 21.23 Raw speckle correlation fringe pattern from NDI of an aluminum plate obtained using the scatter plate laser speckle interferometer

Max-Min Scanning (IMMS) phase shifting algorithm that serves the purpose nicely because it has excellent immunity to vibration noise as well as non-linearity and non-uniformity of the phase shifts.

21.10.2.2 Mirror Technique for In-Plane Speckle Interferometry

Figure 21.24 illustrates the use of a mirror to create the second beam required for in-plane displacement mapping. This arrangement was apparently introduced by C. Wykes and R. Jones in the early 1980s, but it was not used much owing to the difficulties in controlling the mirror for phase shifting. Curiously, a similar setup has long been used in moiré interferometry. If the mirror is placed close to, or, better yet, attached to the specimen, then the system becomes essentially a common-path speckle interferometer. Note that, in the version shown, the sensitivity will vary over the field because of the use of a diverging illumination beam. More precise results can be obtained by including a collimating lens in the setup downstream of the diverging lens. Functions of the instrument to obtain speckle correlation fringes or to perform phase-shifting speckle interferometry are similar to those described above for the scatter plate device.

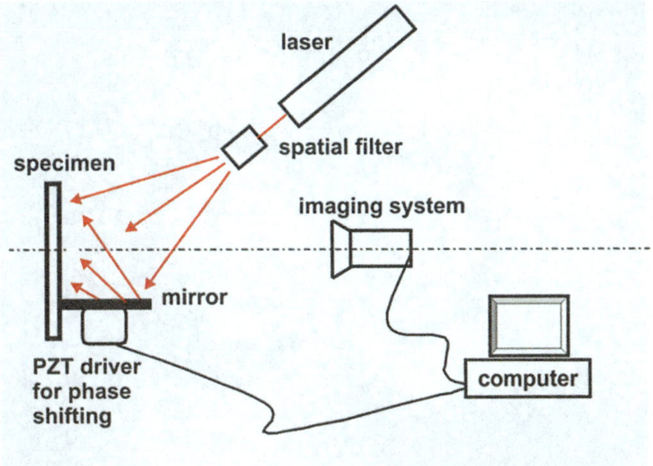

Fig. 21.24 Schematic of a setup using a mirror to create a common-path laser speckle interferometer for measuring in-plane displacements.

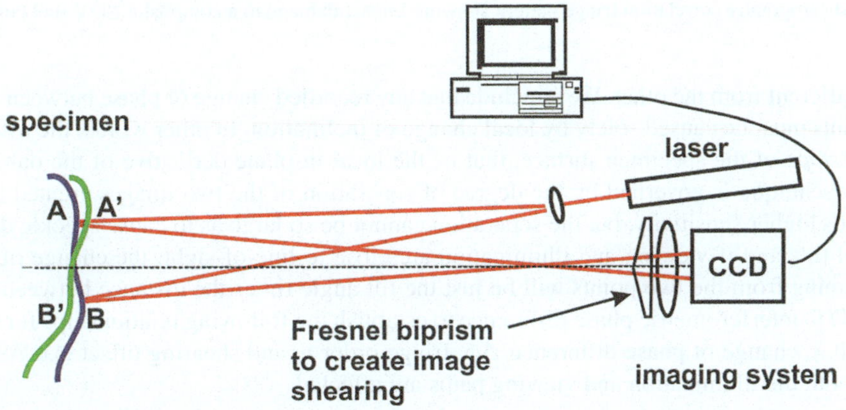

Fig. 21.25 Basic apparatus for laser speckle shearography

21.11 Laser Speckle Shearography

This powerful technique falls into a broad class known as shearing interferometry. Two separate speckle images are created simultaneously by a device such as a Fresnel biprism, a split lens, or a small Michelson interferometer and caused to overlap in the image plane with one image slightly displaced from the other. Thus, at each image point, waves coming from two adjacent object points are caused to overlap and combine interferometrically, and no separate reference beam is required. A result is that the setup for speckle shearography is remarkable in its simplicity, as suggested by the schematic of Fig. 21.25.

The intensity of each resultant speckle in the overlapped images is recorded. The specimen is deformed and the new intensity pattern recorded. The intensity maps are subtracted to give data related to the specimen deformation. Different setups can be used depending on the information desired. Described here is only the most common out-of-plane detection model in which the fringe orders are related to the in-plane derivative of the out-of-plane displacement (i.e. slope).

The relationship between fringe order, change of phase difference, and specimen motion or deformation can easily be derived by calculating the changes of path length for waves arriving at an image point. It is instructive, however, to reason out what happens from the physical viewpoint. As mentioned above, the waves arriving at a single image point originated from two object points, shown in exaggerated form as points A and B in the figure. The points are separated by some distance that is smaller than the speckle size, this separation being dictated by the shearing device in the imaging system. Consider what happens when the specimen is translated along the line of sight. The two waves will undergo identical phase shifts, so their initial phase difference will be maintained and no change of phase shift will be detected, i.e. a bright spot will stay bright. A similar argument applies if the specimen is shifted laterally.

Now, imagine that the specimen is tilted around an axis perpendicular to the shearing plane, as shown in an exaggerated way in figure above. One of the object points, such as A, will advance towards the imaging system to position A′. On the other hand point B might shift further away to position B′. The point is that tilting will cause one of the points to create a path

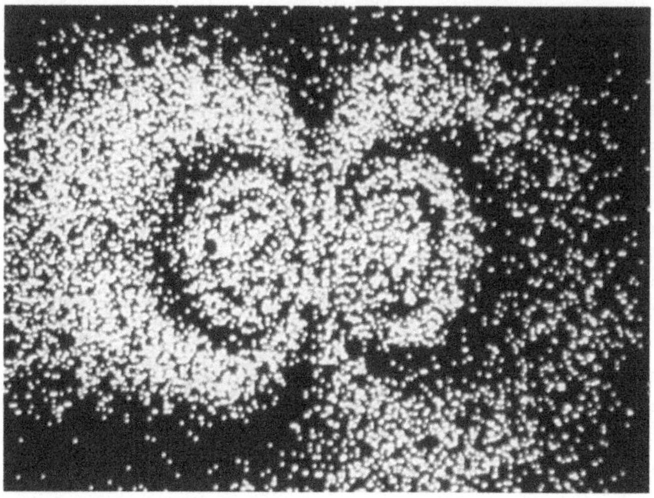

Fig. 21.26 Raw speckle shearography correlation fringe pattern showing impact damage to a composite NGV fuel tank

length change that is different from the other. We conclude that any recorded change of phase between waves originating from these two adjacent points must be caused solely by local change of inclination. In other words, the resulting change-of-phase map yields *change of slope* of the specimen surface, that is, the local in-plane derivative of the out-of-plane displacement. The sensitivity of the technique is governed by the degree of separation of the two images created by the shearing device. Larger separation yields higher sensitivity, but the separation cannot be so large as to cause speckle decorrelation.

For relatively small tilts and if viewing and illumination are close to line-of-sight, the change of path length difference between the waves coming from the two points will be just the tilt angle times the distance between the points. Pursue this idea and convert ΔPLD to interferometric phase difference to establish the following relationships for specimen slope change in terms of wavelength λ, change of phase difference $\Delta\phi$, fringe order n, and shearing offset Δx. As usual, the extra factor of 2 appears because both the illumination and viewing paths are affected.

$$\Delta\left(\frac{dz}{dx}\right) = \frac{\Delta\phi\lambda}{4\pi\Delta x} = \frac{n\lambda}{2\Delta x} \tag{21.11}$$

Unlike speckle interferometry, shearography is, in its out-of-plane measurement form, a common path interferometer. It is tolerant of noise and vibrations, and both stand-alone and hand-held speckle shearography devices are commercially available and very successfully used in field and factory for nondestructive inspection. A sample laser speckle shearography pattern showing impact damage to a composite natural gas vehicle fuel tank is presented in Fig. 21.26. Emphasis must be given the fact that this is a raw pattern that has not been optimized, filtered, or smoothed. In practice, phase-shifting, as described below for speckle interferometry, can be routinely incorporated into speckle shearography apparatus to enhance sensitivity and yield quantitative data.

A thought-provoking feature of shearography is that oblique incidence interference fringes are created within each of the combined speckles. If large speckles are used and if the pattern recording is done with sufficient resolution, superposition of the before- and after-deformation patterns would create moiré fringes, so, again, the relationships between speckle methods, geometric moiré, and moiré interferometry become apparent.

21.12 Phase-Shifting Speckle Interferometry

Before getting into the description of phase-shifting interferometry, a forbidding subject to some, attention is directed to the compact computer routine that appears in the Appendix to this article. This software package utilizes MATLAB® subroutines and incorporates the phase shifting driver, smoothing of data, calculation of change of phase, and phase change map unwrapping.

As is true of all interferometric measurement techniques, including speckle correlation, the quantity sought is either (a) the *absolute* phase difference or path length difference (PLD) between two waves or (b) the *change* of phase difference or PLD between two waves. When extended to the collection of waves that cover a field, these techniques yield whole-field maps of either or both these quantities. These maps are usually called fringe patterns. Lacking more sophisticated methods, determination of phase difference or change of phase difference at a point in a fringe pattern is implemented by counting whole fringe orders, itself not a trivial task, and estimating or somehow measuring the fraction of a whole order. These techniques have served well for a long time, for instance in photoelasticity; but they do not, usually, offer the sensitivity and accuracy of which interferometry is capable, that being some small fraction of the wavelength of light. Is there a better way?

The answer is that both precision and accuracy can be enhanced to a degree approaching the ideal through the use of phase-shifting or phase-stepping interferometry. We will use the phase-shifting terminology to cover both these approaches, which differ only in details. A complete study of phase-shifting interferometry would require more space than is available here, but let us try to fathom the fundamentals and end with a recipe for implementation and an example.

21.12.1 The Fundamental Problem

Consider the interference of two coherent beams that are brought together at some included angle as happens in all the speckle correlation interferometers pictured above. The mixed beams fall on, say, a screen or an array of photodetectors. If the beams are reasonably ordered, we would find a pattern of interference fringes on the screen, and we might be able to figure out the fringe order at any one point by the fringe counting procedure mentioned above. However, such is not the case with holography and speckle, because one or both the beams is scattered from the illuminated specimen. The result, as we know, is a complex fringe pattern in which the brightness varies wildly from one point to the next. How can we possibly obtain the phase difference between the waves that arrive at any single point-sensor?

To better understand the problem, calculate the intensity at a single point in the field that is created by two waves of the same wavelength and polarization, but that have different amplitudes, A and B. One of the waves lags behind the other by some relative retardation r, which translates to a phase difference of $\delta\phi = 2\pi\, r/\lambda$. The two wave equations are added together and the scalar amplitude is calculated. Detectors respond to intensity (irradiance), which is the square of the amplitude. Some trigonometric identities are called upon, and we can write the intensity I_s at the sensing point S in several different ways, the two most useful being,

$$I_s = I_A + I_B + 2\sqrt{I_A\, I_B}\,\cos\delta\phi \tag{21.12}$$

$$I_s = \frac{I_{\max} + I_{\min}}{2} + \frac{I_{\max} - I_{\min}}{2}\cos\delta\phi \tag{21.13}$$

Figure 21.27 is a plot of the intensity at point s as a function of both relative retardation and phase difference. It oscillates predictably, and its minimum value is not zero unless the two beams have exactly the same amplitudes.

Now, here is the problem. Suppose in a given interferometry setup, we measure an intensity value of I_o at some field point. What is the phase difference, $\delta\phi$, between the two interfering beams? It might be the value at point O (ca. 6.7π) as suggested in Fig. 21.27, but it might be any of the other points where the horizontal line through I_o intersects the cosine intensity curve. How can we establish which point is the correct one?

The question cannot be answered with the information so far in hand. In graphical terms, we cannot ascertain which cycle of the intensity curve the measured intensity lies on, that being the nearest whole fringe order. Further, we cannot establish the so-called fractional order because we do not know for sure the maximum and minimum values of the intensity curve. There are an infinity of cosine curves having the same period that would encompass intensity I_o, and we need to find the correct one.

In mathematical terms, there are three unknowns in each of Equations 21.12 and 21.13. So far, we have only one piece of data, that being I_o. No matter our viewpoint, the experiment cannot be completed unless we can gain not two but **three** more pieces of information, as follows:

1. The first two bits of data allow us to solve Equations 21.12 and 21.13, defining exactly the parameters of the intensity—phase difference relationship and therefore yielding the fractional fringe order. These data are obtained by *phase shifting*.
2. The final step is to determine which cycle of the intensity—phase angle plot we are dealing with. This information is obtained by *phase unwrapping*.

Let us attend to these demands in order.

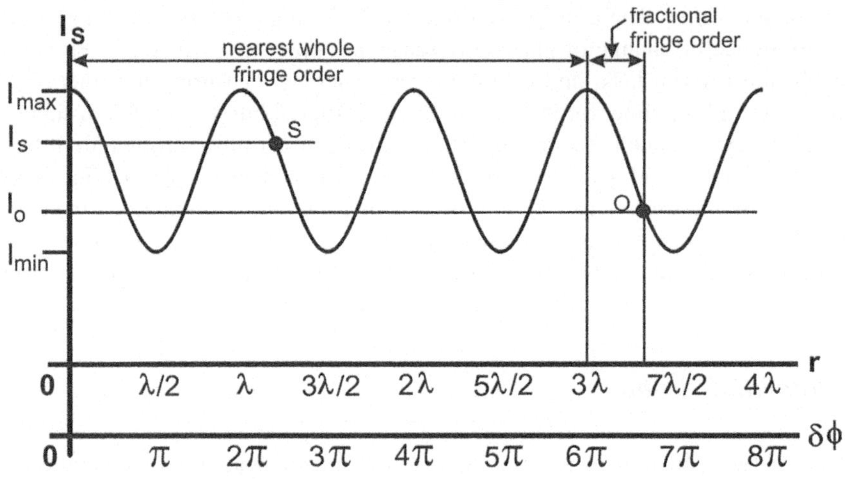

Fig. 21.27 Intensity as a function of PLD and phase difference between two interfering beams

21.12.2 Phase Shifting and Recording

The needed data to solve the first problem are obtained by introducing into one of the interfering beams at least two but usually more known phase shifts or steps and recording the new intensity at each step. At the hardware level, this can be accomplished by incorporating a mirror or other optical element that physically changes the optical path length. Such devices are already included in the systems shown schematically in Figs. 21.22, 21.24, and 21.25, so we do not need to modify them for converting from speckle correlation work to phase-shifting interferometry. Nowadays, the phase shifter is usually a mirror that is attached to a piezoelectric crystal driver (PZT). The PZT produces, on demand from the control circuitry, movements of the mirror that amount to only fractions of the wavelength of light. Usually, the PZT is controlled by the same computer that is used to capture the speckle images. Depending on the algorithm used in data reduction, it may or may not be necessary to calibrate the device.

It is convenient and conventional at this point to simplify notation by dropping the δ and writing the phase difference between the two beams arriving at a detector as just ϕ rather than $\delta\phi$. The result of the phase shifting procedure, when extended to all sensors in the detector array, will be a *phase difference map*, but it is usually called just a *phase map*. Later, the symbol $\Delta\phi$ will be used to identify the *change of phase difference*, which is the quantity usually needed to calculate the change of specimen state (e.g. displacement) that occurs during an experiment.

Several different phase shifting procedures and data reduction algorithms have been invented, and there is extant a huge body of literature on the subject. Noteworthy for insight and lucidity is the work performed in the early 1980s by K. Creath. We outline here only one such process, that being the so-called four-step technique, which provides some redundancy of data for checking and is easy to execute.

Write Equation 21.13 in the following form, which gives the intensity measured at the sensor located at field coordinates (x,y) in the array and where the definitions of I_{av} and I_v are obvious,

$$I\left(x,y\right)=I_{av}\left(x,y\right)\left[1+I_v\left(x,y\right)\cos\phi\left(x,y\right)\right] \tag{21.14}$$

The coordinate specifiers *(x,y)* are usually dropped for brevity, but please do not forget that the calculations must be performed for each and every detector in the array in order to map the phase difference over the field.

To execute the four-step technique, the intensity at zero phase step is used along with three others at $\pi/2$ increments, meaning intensity maps are recorded at phase steps of 0, $\pi/2$, π, and $3\pi/2$. Substitute these values into Equation 21.14 to establish the starting point,

$$\begin{aligned}
I_1 &= I_{av}\left[1+I_v\cos\phi\right]\\
I_2 &= I_{av}\left[1+I_v\cos\left(\phi+\tfrac{\pi}{2}\right)\right]\\
I_3 &= I_{av}\left[1+I_v\cos\left(\phi+\pi\right)\right]\\
I_4 &= I_{av}\left[1+I_v\cos\left(\phi+\tfrac{3\pi}{2}\right)\right]
\end{aligned} \tag{21.15}$$

Rather routine manipulation leads to the following expression for the desired phase difference at each detector,

$$\phi = \arctan\left[\frac{\sin\phi}{\cos\phi}\right] = \arctan\left[\frac{I_4 - I_2}{I_1 - I_3}\right] \qquad (21.16)$$

The trouble now is that the phase difference is known only to modulo π, which is not sufficient information. To belabor the point, suppose that $\tan\phi = 1$. Recall the shape of the tangent graph. We do not know whether the phase angle is 45° (first quadrant) or 180° + 45° = 225°, (third quadrant), and we must figure out which it is. Stated another way, we must convert the calculated phase from modulo π (ϕ lies between 0° and 180°) to modulo 2π (ϕ ranges from 0° to 360°). Two approaches are routinely used to assign the correct quadrant to the calculated phase difference. The first uses a look-up table that examines the signs of the numerator and denominator in Equation 21.16. A speedier approach for clearing the modulo π ambiguity is to utilize the "atan2(y,x)" function that is available in commercially available software such as MATLAB®, FORTRAN® and C®, to name only three. Use of these software packages means that a routine to capture data, control the phase shifter, determine phase difference, and plot the results can be set up with only a few lines of code.

The result of the phase-shifting procedure is a *wrapped phase map* that, if plotted, shows the phase difference for each sensor in the imaging array. Within the context of speckle interferometry, the results will seem nonsensical. The reason is that the phase differences in a speckle pattern are distributed randomly, and no sensible pattern can be perceived.

What we seek when performing speckle interferometry to measure displacements is a map of the *change of phase difference* for each point in the specimen. So, the map of phase difference for the initial state of the object is stored in computer memory. The specimen is then loaded or disturbed as befits the experiment. The procedure described above is repeated for this second specimen state, and the result stored. Finally, the initial phase map is subtracted from the second. Although each of the maps shows a random phase difference distribution, the change of phase difference will be ordered because the displacements within the specimen are smoothly ordered according to the laws of physics, except where there are discontinuities such as cracks. Because each of the phase difference maps was determined only to modulo 2π, the change of phase difference map will also be determined to modulo 2π, and it is called a *wrapped phase map*, but it actually is a *wrapped change of phase difference map*,

At this point we realize that, while the phase change maps and the subsequent displacement and strain maps are easy to plot and kindly to our analog eyes and brains, all the quantitative data are stored in computer memory to a high degree of precision. We no longer need to examine fringe patterns, do fringe multiplication, or interpolate to determine local displacements and strains.

Figure 21.28 shows an example of a raw wrapped phase change map obtained by this method, except that fiber optics were used in the setup. The pattern has not been filtered or smoothed, and the residual speckle effects are clearly evident. The "stair step" nature of the pattern shows that the "fringes" are derived from phase shifting. Note that the fringes are discontinuous where they cross the crack. The goal of the experiment was to develop a speckle processing algorithm for measuring crack length, an endeavor that proved successful with additional data processing. Figure 21.29 is a reproduction in false color of a smoothed wrapped phase change map for a composite specimen with a hole that is loaded via a pin. Notice the cyclic repetition of the colors, which correspond to the modulo 2π restriction.

21.12.3 Phase Unwrapping

The final step in a speckle interferometry experiment to map displacements is to get rid of the cyclic 2π boundaries. In principle, this step seems rather trivial. In practice, it can be troublesome, especially at specimen boundaries or in the vicinity of cracks or other discontinuities.

The idea is to begin at some sensor location (pixel) in the array and evaluate successively the difference between the phase change values of neighboring pixels across a transect of the wrapped phase change map until a sharp break greater than π radians in phase difference is found. Decide whether to add or subtract 2π to reduce the break to less than π, and apply the chosen correction to all the remaining values in the array. Continue this process to the end of the transect, then drop down a row and work back across the next transect. Repeat until the entire phase change array has been treated. Plot the results and examine them to see if they make sense to an engineer. Some repetitions and corrections will likely be necessary.

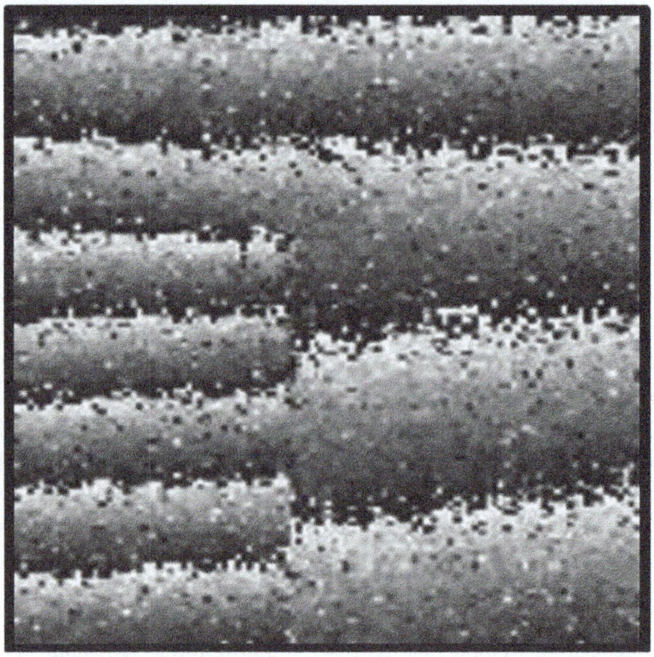

Fig. 21.28 Raw wrapped speckle interferometry phase change fringe pattern showing in-plane displacements in a specimen containing a crack: Photo by Dr. X. Ding

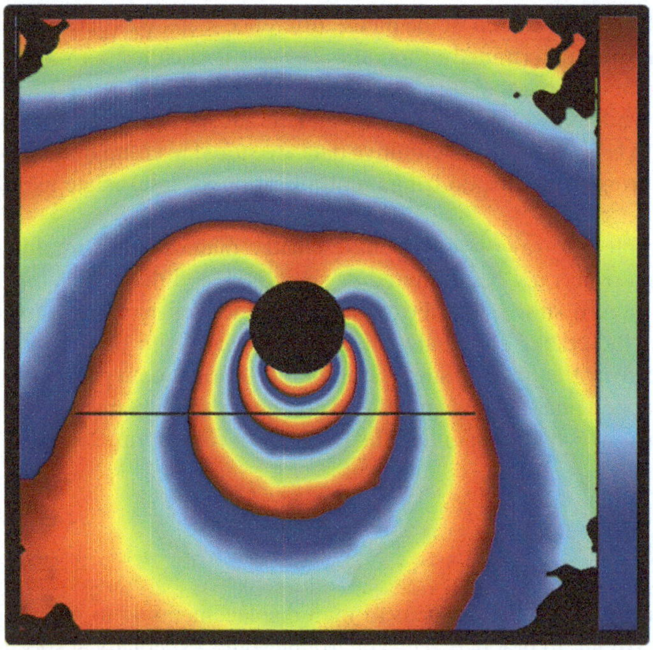

Fig. 21.29 False-color smoothed wrapped speckle interferometry phase change map showing in-plane displacements near a pin-loaded hole in a composite: Graphic by Dr. S. Hong

Fortunately, one needs no longer to write code to implement phase unwrapping. Certain software packages, such as MATLAB®, contain an unwrap function that seems to be well constructed and dependable. Figure 21.30 is a colorized example of an unwrapped change of phase difference map for the specimen of Fig. 21.28. Notice that position of the crack tip is easily seen; in practice, however, it is located quantitatively by scanning across the unwrapped phase map and looking for the first sign of a sharp discontinuity.

Fig. 21.30 Colorized unwrapped phase change map for an aluminum CT specimen with a crack: Graphic by X. Ding

21.12.4 Converting Phase Data into Displacements and Strains

After the change of phase difference map is obtained, conversion of the phase data into specimen displacements is quite easily implemented. First, phase difference change $\Delta\phi$ is converted to physical change of path length difference Δd for each pixel in the array using the well known relationship $\Delta d = (\lambda\ \Delta\phi)/2\pi$. A *sensitivity vector* that depends on the geometry of the setup is then used to calculate the specimen surface displacements. General instructions as to how to determine the sensitivity vector lie outside the purview of this article. For most cases encountered, the experimental setup is arranged to measure one or more specific displacement components, e.g. in-plane or out-of-plane, and the sensitivity can be worked out using trigonometry.

Consider, for example, the symmetrical arrangement shown in Fig. 21.20 and assume that the in-plane displacement Δd_x is sought as a function of the *(x,y)* coordinates in the specimen. Suppose that viewing is along the normal and that θ is the angle of illumination of the specimen with respect to its surface normal, which is the *z*-axis. Assume also that the setup is such that these angles are constant over the extent of the field. Incorporate the expression given above for path length change in terms of the wavelength λ of light used and of $\Delta\phi(x,y)$, which is the measured change of phase difference. The specimen displacement will be found to be,

$$\Delta d_x\,(x,y) = \frac{\Delta\phi\,(x,y)\,\lambda}{4\pi\sin\theta} \tag{21.17}$$

The apparently extra factor of 2 originates in the fact that both illuminating beams undergo equal but opposite path length changes. A calculation for out-of-plane displacement appears above as Equation 21.9, which is written in terms of fringe order, but the idea is the same.

Maps of strain components are usually required in experiments that support engineering design or testing, so differentiation of the displacement data is required. The difficulties inherent in differentiating experimental data are well known and have been alluded to at the head of this paper. Fortunately, the displacement data from speckle interferometry are usually dense enough so that, if appropriate data smoothing and noise removal have been executed during the acquisition and processing of the phase data, differentiation to obtain strain can be accomplished with success. In fact, given the favorable data density, the discrete differentiation will involve a spread of only three to five pixels, giving very small effective gage lengths.

Upon successful calculation of the strain field, the results can be plotted in various ways. Figure 21.31 is a false color example showing the normal strains ε_{yy} in a composite plate that was loaded in the *y*-direction (here horizontal) through a snug pin. Figure 21.32 shows the resulting strain concentration profile along the central load axis for the same composite plate. The visible discontinuity in the strain profile is caused by a fiducial mark on the specimen.

Fig. 21.31 False color strain map from speckle inteferometry for a pin-loaded composite plat: Graphic by F Lanza

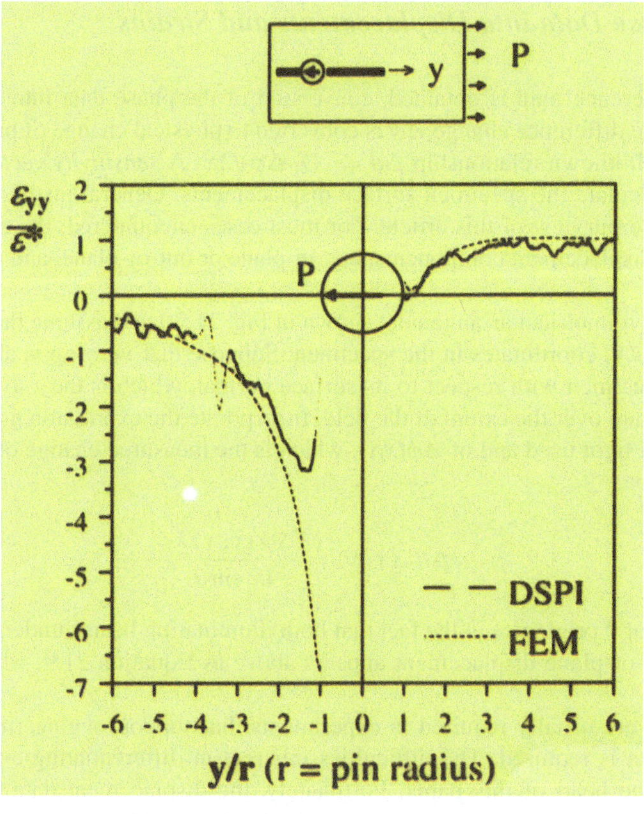

Fig. 21.32 Strain concentration factors along load axis for a pin-loaded composite plate: Graph by F. Lanza

21.13 Simultaneous Interferometries

A little used but potentially useful idea is that more than one interferometric technique can be implemented at once to gain complementary data about a specimen. Lasers having different wavelengths are used for illumination, and corresponding color filters separate the different types of data during recording. As an example, the author and colleagues have used Argon and HeNe lasers to perform simultaneous speckle interferometry and shearography NDI studies.

21.14 Concluding Remarks

Coherent and non-coherent speckle techniques are, by virtue of their ranges of sensitivities and densities of data, admirably suited for measurements of displacements, velocities, vibrations, and strains within the broad domain of experimental mechanics. Advances in apparatus and computational flexibility have eased implementation of these techniques. Nowadays, one can execute basic speckle interferometry with a laser, a couple of lenses, a piece of glass, a web camera, and standard photo processing software. At the least, practitioners should acquaint themselves with the capabilities of these techniques in order that informed choices of methodology can be made with confidence.

A.1 Appendix

Dr. S. Hong has kindly prepared for our use a short well-annotated MATLAB® routine that performs a complete phase-stepping interferometry analysis. The program was written for digital speckle interferometry, but it can be made to serve for other types of interferometry. The code initializes the input and output devices, sets the phase stepping increment, acquires before-load and after-load intensity maps at each of four phase steps, calculates the before- and after-load phase difference maps using the four-step algorithm, subtracts the two phase difference maps to obtain the change-of-phase-difference array, wraps those results back to the range $-\pi$ to $+\pi$, smooths the wrapped map to remove random noise, and finally unwraps it.

```
%=================================================================
% A MATLAB script for phase-shifting digital speckle pattern interferometry.
% Written by Dr. Soonsung Hong
% The script assumes that the optical system consist of a IEEE-1394 (firewire) digital camera
and an analog output device for phase shifting.
% MATLAB Image Acquisition Toolbox and Data Acquisition Toolbox are used.
%=================================================================
% Initialize a digital camera (Image Acquisition Toolbox)
vidobj = videoinput('dcam',1);
% Initialize an analog output device (Data Acquisition Toolbox)
ao = analogoutput('nidaq','Dev1');
ch = addchannel(ao,0);
% Define the voltage step assuming the increment of 1.5 volt
% introduces a 2*pi phase shift.
vstep=1.5/4;
% Acquire four phase-shifted speckle patterns in the undeformed state.
for ii=1:4,
    putsample(ao,vstep*(ii-1)); %Send a voltage output to PZT controller
    pause(0.1); %Wait for 0.1 second
    i0(:,:,ii)=double(getsnapshot(vidobj)); %Capture and store an image
end
putsample(ao,0);
pause %Wait for a loading to apply. Press any key to continue.
% Acquire four phase-shifted speckle patterns in the deformed state.
for ii=1:4,
    putsample(ao,vstep*(ii-1)); %Send a voltage output to PZT controller
    pause(0.1); %Wait for 0.1 second
    i1(:,:,ii)=double(getsnapshot(vidobj)); %Capture and store an image
end
putsample(ao,0);
%Calculate phase maps using the four-step algorithm
```

```
p0=atan2(i0(:,:,4)-i0(:,:,2),i0(:,:,1)-i0(:,:,3));
p1=atan2(i1(:,:,4)-i1(:,:,2),i1(:,:,1)-i1(:,:,3));
%Calculate the phase-change map and wrap back to -pi to +pi
dp=atan2(sin(p1-p0),cos(p1-p0));
%Remove random noise in the phase-change map
n=5; % Size of the smoothing window
sindp=conv2(sin(dp),ones(n)/n^2,'same');
cosdp=conv2(cos(dp),ones(n)/n^2,'same');
sdp=atan2(sindp,cosdp); %the smoothed wrapped map
%Unwrap the smooth phase-change map
usdp=unwrap(unwrap(sdp,[],1),[],2);
```

Chapter 22
Accurate Reconstruction of High-Gradient Strain Field in Digital Image Correlation: A Local Hermite Scheme

Xin Li, Jiaqing Zhao, Jianguang Shuai, Zhengming Zhang, and Xinxin Wu

Abstract High-gradient strain field measurement is attracting more attention from both engineering and scientific domains, where digital image correlation (DIC) is widely adopted to recover the dramatically changing displacement fields of stressed samples. Accurate reconstruction of reliable strain field from noisy displacement field is one of the key issues, since the strain calculation is highly sensitive to the displacement noise. The formally proposed improved Hermite finite element smoothing method [1] (IHFESM) and fast Hermite element method (FHEM) [2] are effective global Hermite method based on Tikhonov regularization. However, complex meshing process and high computation burden are inevitable downsides. In this work, we present a local version of FHEM, i.e. the local Hermite method, in which the Hermite element is taken as a smoothing window centered on one displacement data, and the whole field is handled point-wisely. The smoothed displacement and strain are consequently computed from the regularized local surface as in FHEM. The proposed method can deal with arbitrarily shaped region of interest, and is especially suitable for high gradient strain measurement due to the rich high gradient components contained in C^2 continuous Hermite element. Experiments show that the local Hermite method is more accurate than global Hermite method such as FHEM and IHFESM.

Keywords Digital image correlation · High-gradient strain · Displacement smoothing · Hermite element

22.1 Experiments and Results

Figure 22.1 shows the basic principle of the proposed local Hermite method. The comparison is made by smoothing the synthetic noisy displacement field shown in Fig. 22.2, which is created by adding the true sine-shaped displacement field with the zero-mean random Gaussian noise $N(0, \sigma)$, and σ is set to be 0.5% and 2% of the maximum displacement respectively. The result in Table 22.1 demonstrates the RMS error by the global and local Hermite method restrictively, in which the "C^1, $k = 2$" and "C^2, $k = 3$" mean the C^1 and C^2 continuous. To reduce the boundary effect, the data near the outer edges are omitted in RMS calculation. The RMS value indicates that the local Hermite method is not only able to solve the computation burden issue, but also provides a more reasonable field. Another experiment is done to test the smoothing result of proposed method under a high gradient strain field, which is displaced in Fig. 22.3. The theoretical displacement field and theoretical strain field are illustrated in Fig. 22.4 and the RMS values is in Table 22.2. It could not be deny the fact that the Local Hermite method performs better than the Global Hermite method.

X. Li · J. Zhao (✉) · Z. Zhang · X. Wu
Collaborative Innovation Center of Advanced Nuclear Energy Technology, The Key Laboratory of Advanced Reactor Engineering and Safety of MOE, Institute of Nuclear and New Energy Technology, Tsinghua University, Beijing, China
e-mail: jqzhao@mail.tsinghua.edu.cn

J. Shuai
Department of Mechanical Engineering, Tsinghua University, Beijing, China

L. Lamberti et al. (eds.), *Advancement of Optical Methods & Digital Image Correlation in Experimental Mechanics, Volume 3*, Conference Proceedings of the Society for Experimental Mechanics Series, https://doi.org/10.1007/978-3-319-97481-1_22

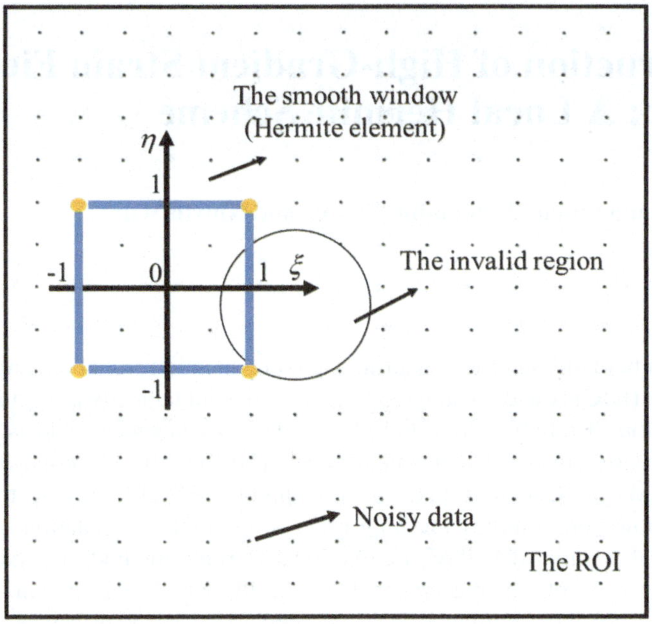

Fig. 22.1 Basic principle of the local Hermite method

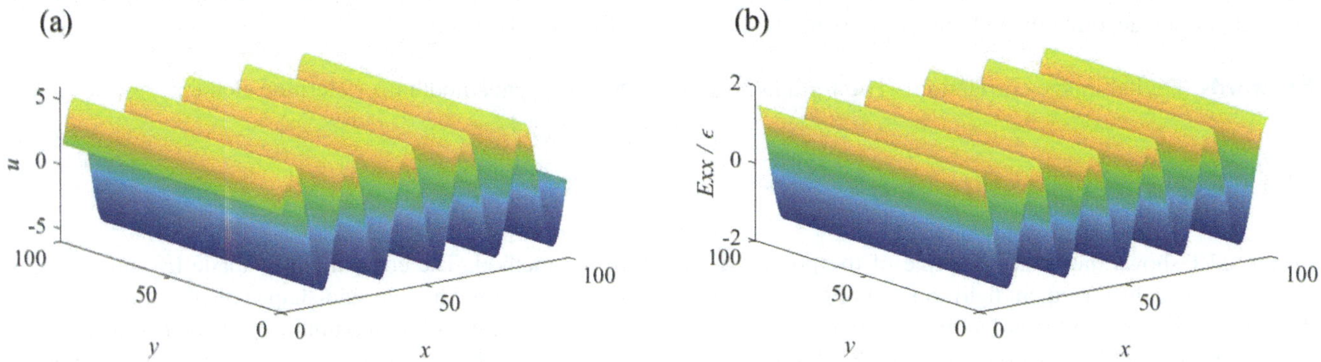

Fig. 22.2 The theoretical displacement field (**a**) and theoretical strain field E_{xx} (**b**)

Table 22.1 The RMS error of the strain (shown in Fig. 22.2b) results by global Hermite (C^1, k = 2), global Hermite (C^2, k = 3), local Hermite (C^1, k = 2) and local Hermite (C^2, k = 3)

Global Hermite method			Local Hermite method		
Parameter	C^1, k = 2	C^2, k = 3	Parameter	C^1, k = 2	C^2, k = 3
Sub-element			Window size		
(a) 0.5% noise					
8	0.88556	0.02372	7	0.00856	
9	0.15048	0.02529	9	0.00928	
10	0.12856	0.01337	11	0.01790	0.00457
11	0.07607	0.00849	13	0.03457	0.00483
			15		0.00474
			17		0.00650
(b) 2% noise					
8	0.89685	0.03323	7	0.03157	
9	0.14910	0.02405	9	0.02134	
10	0.12897	0.03611	11	0.02292	0.01519
11	0.07657	0.03232	13	0.03620	0.01414
			15		0.01491
			17		0.01375

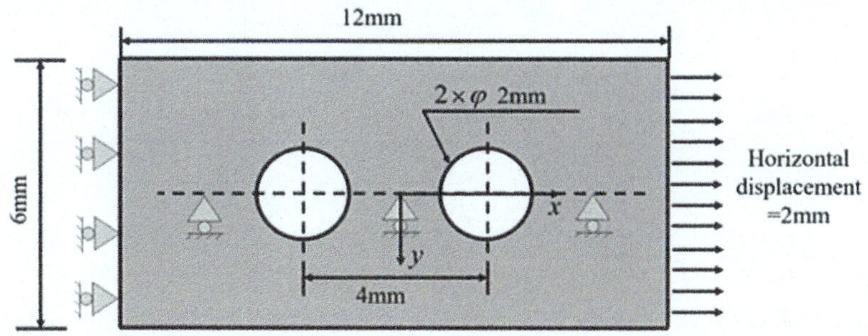

Fig. 22.3 Specimen geometry and the load applied onto the plate

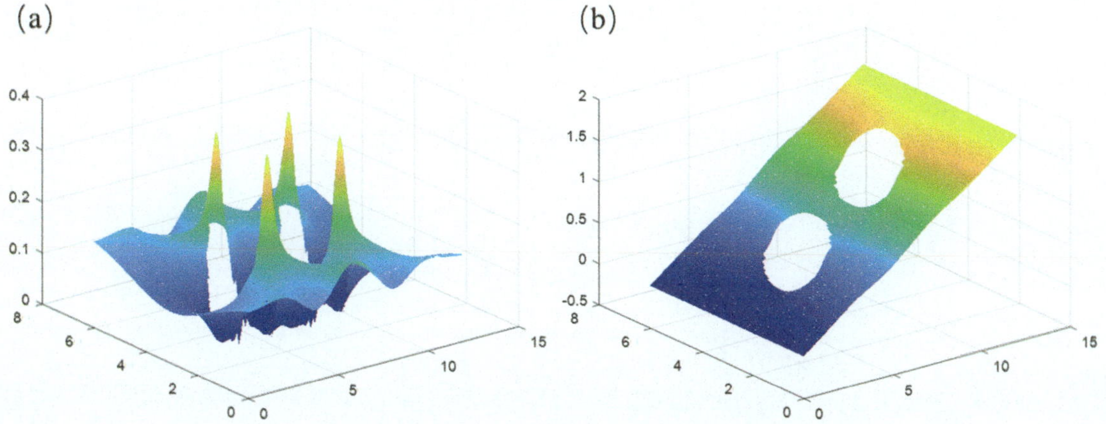

Fig. 22.4 The theoretical displacement field (**a**) and theoretical strain field (**b**) of the plate with two holes

Table 22.2 The RMS error of the strain (shown in Fig. 22.4b) results by global Hermite (C^1, k = 2), global Hermite (C^2, k = 3), local Hermite (C^1, k = 2) and local Hermite (C^2, k = 3)

Global Hermite method			Local Hermite method		
Parameter	C^1, k = 2	C^2, k = 3	Parameter	C^1, k = 2	C^2, k = 3
Sub-element			Window size		
(a) 0.5% noise					
8	0.01566	0.01010	7	0.01504	
9	0.00953	0.01046	9	0.00925	
10	0.01201	0.00993	11	0.00783	0.01029
11	0.01092	0.01027	13	0.00740	0.00815
			15		0.00757
			17		0.00747
(b) 2% noise					
8	0.02232	0.03323	7	0.08187	
9	0.01921	0.02405	9	0.03583	
10	0.02085	0.03611	11	0.02138	0.02722
11	0.02092	0.03232	13	0.01786	0.02557
			15		0.02110
			17		0.01919

References

1. Zhao, J., Zeng, P., Pan, B., Lei, L., et al.: Improved Hermite finite element smoothing method for full-field strain measurement over arbitrary region of interest in digital image correlation. Opt. Lasers Eng. **50**(11), 1662–1671 (2012)
2. Zhao, J., Song, Y., Wu, X.: Fast Hermite element method for smoothing and differentiating noisy displacement field in digital image correlation. Opt. Lasers Eng. **68**(0), 25–34 (2015)

Chapter 23
Development of a New Normalization Technique for Twelve Fringe Photoelasticity (TFP)

Ashutosh Pandey and K. Ramesh

Abstract The use of theoretical calibration table for fringe pattern demodulation in isochromatic images is revisited and a sensitivity analysis of material stress fringe value to generate theoretical calibration table is studied. The influence of filter parameter '*R*' in normalization is discussed and its nuances are brought out. A substitution for theoretical calibration table and a new normalization technique is proposed to alter the dynamic range over the entire model domain of the isochromatic image suitably.

Keywords Photoelasticity · Fringe pattern analysis · Theoretical calibration · Image processing · Normalization

23.1 Introduction

In TFP/RGB photoelasticity, isochromatic images recorded in white light are generally demodulated to get fringe order by using experimental calibration table [1]. Recently fringe demodulation has been explored by the use of fringe pattern normalization [2] in conjunction with theoretically generated calibration table (CT) [3]. To generate the theoretical CT, the procedure demands accurate evaluation of material stress fringe value (F_σ) for each of the channels. However, no sensitivity analysis has been reported in the literature. In this paper, the sensitivity of F_σ (N/mm/fringe-order) values of red and blue channels for theoretical CT is studied. One of the first steps in new fringe demodulation approach is that isochromatic image needs to be normalized. The normalization uses a frequency filter's parameter '*R*' to change the dynamic range in the image. The choice of '*R*' plays an important role in normalization and its influence for a standard problem of circular disc under diametral compression is brought out. A new normalization technique is proposed and a study of fringe order demodulation in circular disc under diametral compression with a normalized experimental calibration table is done and the results are compared with corresponding analytical solution.

23.2 Sensitivity Analysis of Theoretical Calibration Table

An isochromatic image of an epoxy C-specimen having $F_{\sigma\text{-red}} = 13.174$, $F_{\sigma\text{-green}} = 11.62$ and $F_{\sigma\text{-blue}} = 9.15$ is recorded till 12th fringe order (N) and a section till fourth fringe order is shown in Fig. 23.1a, which is normalized with $R = 5$ and is shown in Fig. 23.1b. Theoretical CT is generated with same F_σ values as application specimen (CT-1) and is shown in Fig. 23.1c. Since these F_σ values are experimentally calculated, there is a possibility of error in calculating F_σ. To study the effect of small errors in F_σ calculation on fringe order demodulation, an error of -2.5% in blue channel and $+2.5\%$ in red channel is introduced to generate a CT-2 which is shown in Fig. 23.1d. Intensity representation of the difference in grey values between CT-1 and CT-2 is shown in Fig. 23.1e. The plot of grey values along line AB is shown in Fig. 23.1f. Fringe order in normalized image (Fig. 23.1b) is demodulated by CT-1 and CT-2, the corresponding colour representation is shown in Fig. 23.1g, h respectively. The plot of difference of N_{CT} along line AB in Fig. 23.1g and 1 h with $N_{analytical}$ is shown in Fig. 23.1i which shows that the mean absolute deviation (MAD) of difference in fringe order by CT-2 is 0.14 which is 3.5 times greater than the MAD of 0.04 by CT-1.

A. Pandey · K. Ramesh (✉)
Department of Applied Mechanics, Indian Institute of Technology Madras, Chennai, India
e-mail: kramesh@iitm.ac.in

L. Lamberti et al. (eds.), *Advancement of Optical Methods & Digital Image Correlation in Experimental Mechanics, Volume 3*, Conference Proceedings of the Society for Experimental Mechanics Series, https://doi.org/10.1007/978-3-319-97481-1_23

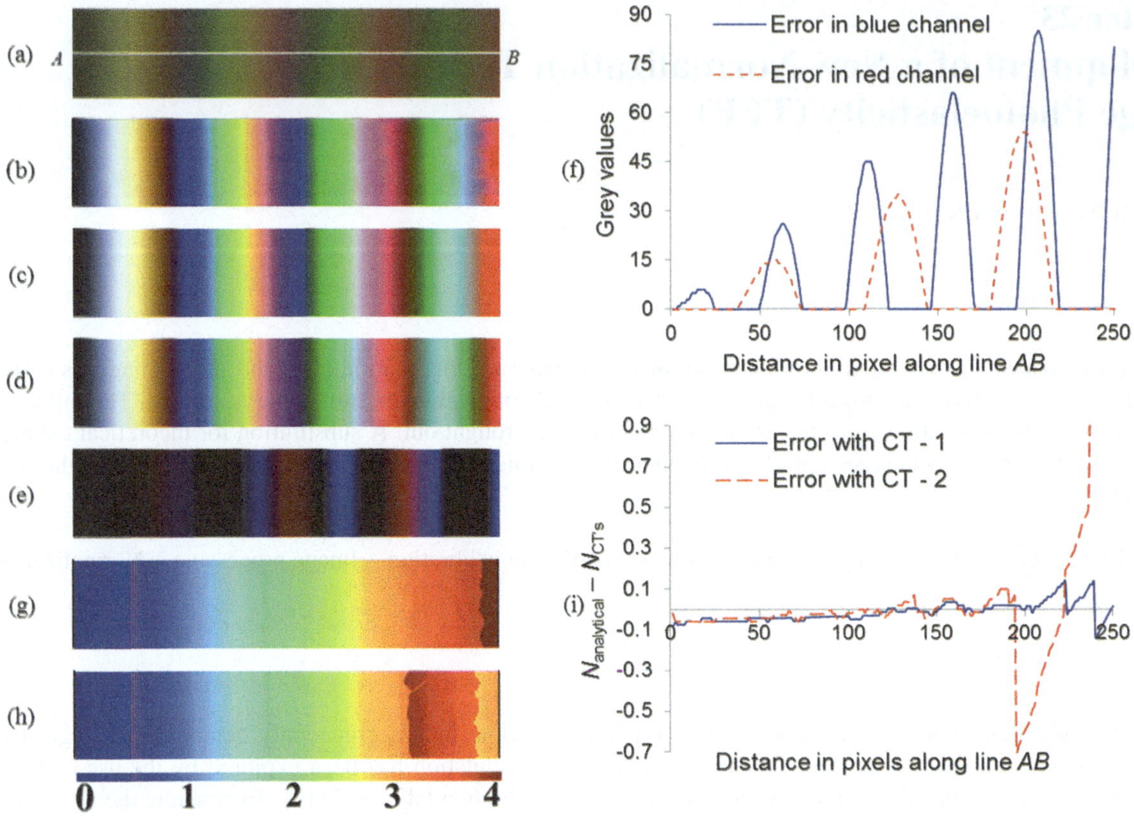

Fig. 23.1 (**a**) Cropped isochromatic image of epoxy C-specimen up to fourth fringe order, (**b**) Epoxy C-specimen image in 'a' normalized with $R = 5$, Intensity representation of theoretical calibration table: (**c**) with actual F_σ values (CT-1) (**d**) with error in red and blue channel F_σ values (CT-2), (**e**) Intensity representation of difference in grey scale values between CT-1 and CT-2, (**f**) Plot of grey values along line AB in 'e', Colour map of fringe order (N) by: (**g**) CT-1 (**h**) CT-2, (**i**) Error plot of fringe order by CT-1 and CT-2 with analytical solution along line AB

23.3 Influence of Filter Radius on Normalization and New Normalization Technique

In normalization, Phani et al. [4] demonstrated that a single R is insufficient to remap dynamic range in low and high fringe gradient zones. An epoxy circular disc (Fig. 23.2a) recorded with a load of 492 N, when normalized with $R = 2$, a loss of intensity values is seen at high fringe gradient zones (Fig. 23.2b) and when normalized with $R = 30$, noise intensities come at low fringe gradient zones (Fig. 23.2c). To handle this issue, a new normalization technique is proposed which uses a high pass (H_{HP}) and low pass (H_{LP}) filter, which are defined as follows:

$$H_{HP} = 1 - e^{\left(\frac{\sqrt{(u^2+v^2)}}{2R_1^2}\right)}, \quad \text{where } R_1 \text{ is the radius of the } H_{HP} \tag{23.1}$$

$$H_{LP} = e^{\left(-\frac{\sqrt{(u^2+v^2)}}{2R_2^2}\right)}, \quad \text{where } R_2 \text{ is the radius of the } H_{LP} \tag{23.2}$$

where u and v are the matrices defined for an isochromatic image of size $m \times n$ as follows.where R is the radius of the frequency filter, u and v are the matrices defined as

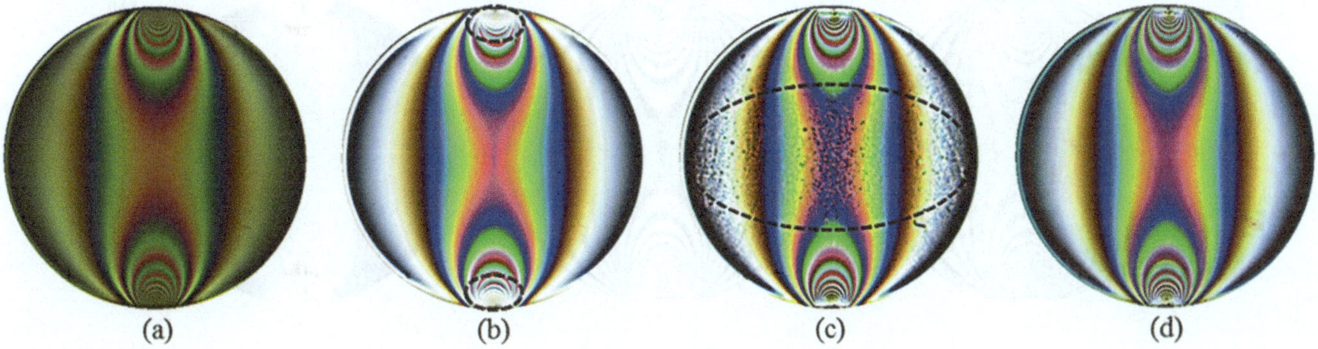

Fig. 23.2 (**a**) Isochromatic image of a circular disc with a load of 492 N, Image 'a' normalized with: (**b**) $R = 2$ (**c**) $R = 30$, (**d**) $R_1 = 30$ and $R_2 = 10$

$$u = \begin{bmatrix} 1-u_0 & 2-u_0 & \dots & n-u_0 \\ 1-u_0 & 2-u_0 & \dots & n-u_0 \\ \dots & \dots & \dots & \dots \\ 1-u_0 & 2-u_0 & \dots & n-u_0 \end{bmatrix}_{m\times n} \quad v = \begin{bmatrix} 1-v_0 & 1-v_0 & \dots & 1-v_0 \\ 2-v_0 & 2-v_0 & \dots & 2-v_0 \\ \dots & \dots & \dots & \dots \\ m-v_0 & m-v_0 & \dots & m-v_0 \end{bmatrix}_{m\times n}, \text{ where } u_0 = \frac{n}{2}+1 \text{ and } v_0 = \frac{m}{2}+1 \tag{23.3}$$

The circular disc in Fig. 23.2a is normalized with filter's radius $R_1 = 30$ and $R_2 = 10$ and shown in Fig. 23.2d, which clearly demonstrates that this algorithm has an ability to alter the dynamic range over the complete image reasonably better.

23.4 Whole-Field Fringe Order Evaluation with Normalized Experimental Calibration Table

An isochromatic image of circular disc (Fig. 23.3a) having diametral compression of 1373 N and $F_\sigma = 10.4$ N/mm/fringe-order for a wavelength of 546.1 nm is normalized with $R_1 = 30$ and $R_2 = 10$ and is shown in Fig. 23.3b. The calibration table is generated by normalizing the same epoxy isochromatic specimen (normalized CT) in Fig. 23.1a till 16th fringe order with $R_1 = 30$ and $R_2 = 10$. The colour map of fringe order obtained by using normalized CT is shown in Fig. 23.3c. The colour map of absolute error between fringe order in Fig. 23.3c and analytical fringe orders is shown in Fig. 23.3d. The plot of fringe order along lines *CD* and *EF* (Fig. 23.3e) shows a close match with analytical results.

23.5 Conclusion

The study of small changes in material stress fringe value shows that the theoretical calibration table is highly sensitive to this. Normalization algorithm is influenced by filter parameter '*R*' and a single value of '*R*' is not sufficient to alter the dynamic range in the image. The new normalization technique is capable of alter the dynamic range over the complete domain of the isochromatic image reasonably better. Use of experimentally recorded calibration table after normalization is found to extract up to 16th fringe orders.

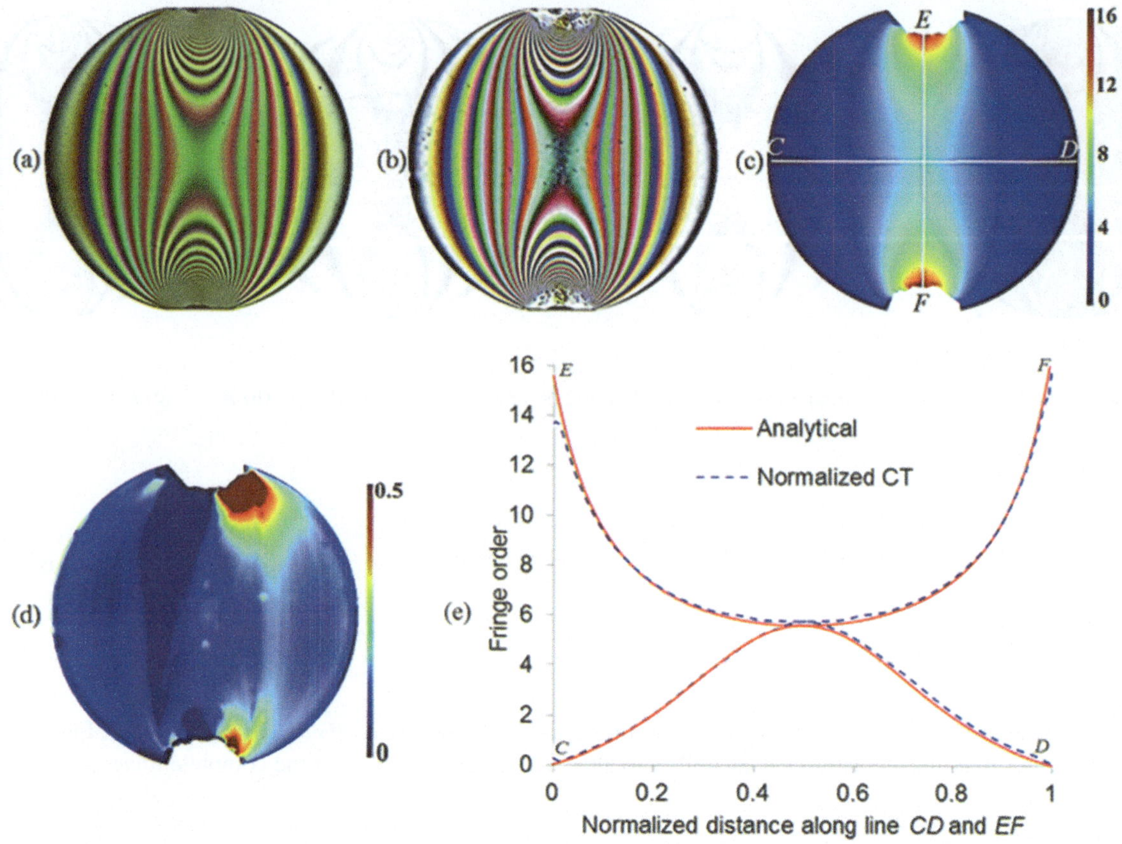

Fig. 23.3 (**a**) Isochromatic image of a circular disc with a load of 1373 N, (**b**) Image 'a' normalized with $R_1 = 30$ and $R_2 = 10$, (**c**) Colour map of fringe order obtained using normalized CT, (**d**) Colour map of absolute error between 'c' and analytical fringe orders, (**e**) The plot of fringe orders along line *CD* and *EF* in 'c'

References

1. Ramesh, K.: Digital photoelasticity. In: Rastogi, P. (ed.) Digital Optical Measurement Techniques and Applications, pp. 289–340. Artech House, Norwood MA (2015)
2. Larkin, K.G., Bone, D.J., Oldfield, M.A.: Natural demodulation of two-dimensional fringe patterns. I. General background of the spiral phase quadrature transform. J. Opt. Soc. Am. A. **18**(8), 1862–1870 (2001)
3. Swain, D., Thomas, B.P., Philip, J., Pillai, S.A.: Non-uniqueness of the color adaptation techniques in RGB photoelasticity. Exp. Mech. **55**, 1031–1045 (2015)
4. Phani Madhavi, C., Ramakrishnan, V., Ramesh, K.: Critical assessment of fringe pattern normalization for twelve fringe photoelasticity. In: Yoshida, S., Lamberti, L., Sciammarella, C.A. (eds.) Advancement of Optical Methods in Experimental Mechanics, pp. 295–299. Springer, Berlin (2017)

Chapter 24
On Performing Spatiotemporal Stereocorrelation at Very High Temperatures

M. Berny, T. Archer, F. Hild, A. Mavel, P. Beauchêne, V. Herb, and B. Lacombe

Abstract Global stereocorrelation has been deployed to measure 3D displacement fields of a ceramic matrix composite loaded at very high temperatures by a laser beam. The heat haze effect is analyzed thanks to an original calibration procedure.

Keywords Stereocorrelation · High temperature · Heat haze effect · Calibration target · Uncertainty quantification

24.1 Introduction

A multi-instrumented experiment at very high temperatures (600–1050–1500 °C) has been developed to monitor the displacement and strain fields of a coated ceramic matrix composite sample heated by a laser beam. A pair of optical cameras was installed to measure 3D displacement fields of the sample surface via stereocorrelation. The surface heated by the laser leads to a heat haze effect (due to variations of the refraction index) that biases the optical images and thus may invalidate the measurement of (small) 3D displacements of the sample by a regular stereocorrelation procedure. To estimate the artificial displacements induced by heat haze, a calibration target was specially designed for the experiment (Fig. 24.1), and set up so that it is seen by both cameras during the *whole* test.

A series of six inclined planes with a vertical offset of 3 mm allows for the calibration of the stereovision system with a *single* pair of images and a good initialization of the projection matrices of the stereovision system (14 corners at different heights within the field of view of the cameras). The speckled target is not heated during the experiment so that its displacement fluctuations are mainly attributed to the heat haze effect.

M. Berny
LMT (ENS Cachan/CNRS/University Paris-Saclay), Cachan, France

SAFRAN, Safran Ceramics, Le Haillan, France
e-mail: berny@lmt.ens-cachan.fr

T. Archer
LMT (ENS Cachan/CNRS/University Paris-Saclay), Cachan, France

SAFRAN, Safran Ceramics, Le Haillan, France

ONERA, The French Aerospace Lab, Palaiseau, France

F. Hild
LMT (ENS Cachan/CNRS/University Paris-Saclay), Cachan, France

A. Mavel · P. Beauchêne
ONERA, The French Aerospace Lab, Palaiseau, France

V. Herb · B. Lacombe
SAFRAN, Safran Ceramics, Le Haillan, France

L. Lamberti et al. (eds.), *Advancement of Optical Methods & Digital Image Correlation in Experimental Mechanics, Volume 3*,
Conference Proceedings of the Society for Experimental Mechanics Series, https://doi.org/10.1007/978-3-319-97481-1_24

Fig. 24.1 Top (**a**) and bottom (**b**) views of the 3D calibration target

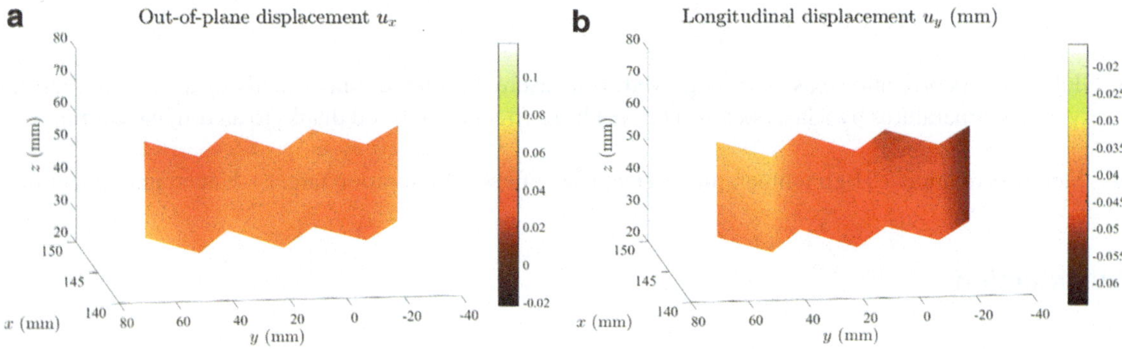

Fig. 24.2 Out-of-plane u_x (**a**) and longitudinal u_y (**b**) displacement fields of the calibration target surface for the highest applied temperature (T = 1050 °C)

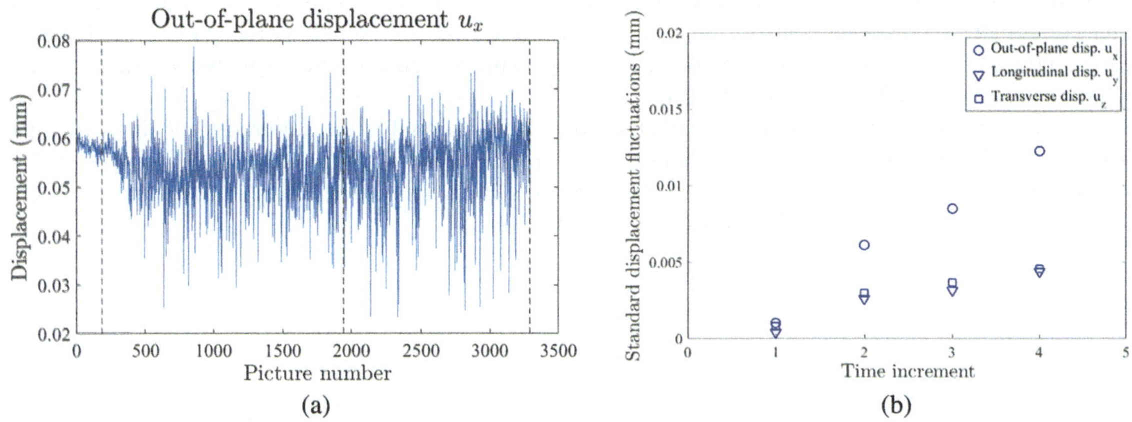

Fig. 24.3 (**a**) Temporal evolution of the out-of-plane (u_x) displacement of one evaluation point of the calibration target during the first and second temperature levels (the dotted vertical lines mark the ends of the acquisition at room temperature, at T = 650 °C then at T = 1050 °C). (**b**) Standard deviations of 3D displacements for the four steady states (at room temperature, T = 650 °C, T = 1050 °C and T = 1500 °C)

24.2 Results

The reported results are obtained with isogeometric stereocorrelation [1]. The calibration of the stereovision system is performed by using the NURBS model of the target surface thanks to CMM measurements. The optimization of the projection matrices reduces the global residuals from 5.7% to 4.4% of the dynamic range of the pictures. The 3D displacements (Fig. 24.4) are measured for the calibration target with a series of more than 5000 images. In this series, approximatively 200 images correspond to applied (thus known) horizontal displacements thanks to a micrometer screw at room temperature. When analyzing the measured displacements, the standard uncertainty is equal to 0.4 µm for the longitudinal displacement u_y, 0.8 µm for the transverse displacement u_z and 1.0 µm for the out-of-plane displacement u_x, which is consistent with the increase of uncertainties for out-of-plane measurements. The images corresponding to the three heating phases are then analyzed (Fig. 24.2). A temporal modulation (Fig. 24.3), whose amplitude increases with temperature, is observed when analyzing the displacements of the same point during time and characterizes the heat haze effect. The standard deviations of displacements increase (from the room temperature, referred as Time increment no. 1 in Fig. 24.3, to

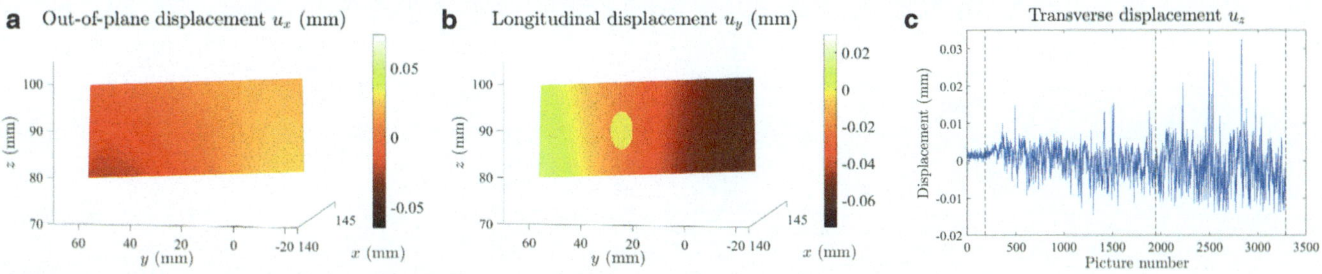

Fig. 24.4 Displacement fields (**a**, **b**) for the sample at T = 1050 °C. The masked areas correspond to the heated zone by the laser beam. (**c**) Temporal fluctuations of transverse displacement. The dashed vertical lines mark the ends of the acquisitions at room temperature, T = 650 °C, and T = 1050 °C

the final temperature level at 1500 °C, referred as Time increment no. 4 on Fig. 24.3) for the longitudinal displacement u_y from 0.4 to 4.7 μm, for the transverse displacement u_z from 0.8 to 4.8 μm, and for the out-of-plane displacement u_x from 1.0 to 13 μm, which is about ten times the fluctuations at room temperature.

The 3D surface displacements are then measured for the sample of interest (one NURBS patch of degree 1). A global correction of the surface was first performed within the frame of the calibration target. The 3D displacements (Fig. 24.4) are obtained considering cubic displacements of the control points for the longitudinal direction and linear displacements for the transverse direction. The longitudinal displacement field u_y is consistent with FE computations since the amplitude of elongation (i.e., 25.1 μm at 650 °C, 46.8 μm at 1050 °C, and 73.3 μm at 1500 °C for an evaluation point at the free edge of the beam compared to an evaluation point taken close to the grip) is greater than that induced by the heat haze (11.3 μm). The results for the transverse (u_z) and out-of-plane (u_x) displacement fields appear inconsistent (i.e., amplitudes far from the computed levels, such as a measured deflection of 36.2 μm for the highest temperature level, that is three times lower than the expected one), which can be understood thanks to the previous analysis that highlighted the heat haze effect and the temporal fluctuations (Figs. 24.2 and 24.3). The shape of measured fields appears consistent, thanks to a strong spatial regularization (only 24 kinematic degrees of freedom per frame) but displacement amplitudes are biased by artifacts (i.e., temporal fluctuations due to heat haze that increase up to 7 μm at T = 1500 °C) and cannot be attributed to the thermomechanical response of the material and/or control loop of the experiment.

24.3 Conclusion

The use of an original calibration target had two goals, namely, calibrate the stereovision system (with a single pair of images) and investigate experimental issues associated with very high temperatures. Spurious displacements due to the heat haze were observed when registering the images of the calibration target, which made out-of-plane displacements for the sample inconsistent. Consequently, the measurements carried out on the calibration target are very useful to fully characterize the heat haze effect. The real displacements for the sample require spatiotemporal approaches [2, 3] to be extended to stereocorrelation or integrated strategies [1, 4] to regularize measured displacement fields and filter out heat haze effects.

Acknowledgement This work was supported under PRC MECACOMP, French research project co-funded by DGAC and SAFRAN Group, piloted by SAFRAN Group and involving SAFRAN Group, ONERA and CNRS.

References

1. Dufour, J.-E., Hild, F., Roux, S.: Shape, displacement and mechanical properties from isogeometric multiview stereocorrelation. J. Strain Anal. **50**(7), 470–487 (2015)
2. Besnard, G., Leclerc, H., Roux, S., Hild, F.: Analysis of image series through digital image correlation. J. Strain Anal. **47**(4), 214–228 (2012)
3. Berny, M., Archer, T., Mavel, A., Beauchêne, P., Roux, S., Hild, F.: On the analysis of heat haze effects with space time DIC. Opt. Lasers Eng. **111**, 135–153 (2018)
4. Mathieu, F., Leclerc, H., Hild, F., Roux, S.: Estimation of elastoplastic parameters via weighted FEMU and Integrated-DIC. Exp. Mech. **55**, 105–119 (2015)

Chapter 25
Compression Tests on CFRP Analysed by Digital Image Correlation

C. Barile, C. Casavola, and G. Pappalettera

Abstract Carbon Fiber Reinforced Plastic materials (CFRP) are widely used for aeronautical and aerospace applications since they offer high specific mechanical properties. However, when damaged, their behavior is quite concerning and, if the damages are not visible on the surface their use could become highly dangerous if not properly monitored. Although traditional non-destructive techniques enable a posteriori detection of damage at successive stages of the life of these materials, it seems more difficult to monitor in-situ the evolution of internal damage, especially in opaque materials. In order to obtain more complete information about the damage mechanisms it is necessary to implement advanced strain analysis approach like those based upon full-field measurements. By using these techniques, in fact, it is possible to get information about the strain map all over the surface of the sample. In this paper, in particular, an approach based on the Digital Image Correlation was used and it was applied to analyze the behavior of CFRP specimens subjected to compression load with anti-buckling fixture.

Keywords Carbon Fibre Reinforced Plastic materials (CFRP) · Composite material · Compression tests · Digital Image Correlation (DIC) · Buckling

25.1 Introduction

Structures in composite materials are obtained by joining at least two materials, with very different physical and mechanical characteristics. The purpose of this matching is to obtain a final material with better characteristics than those referable to the single initial materials. The composite is generally made up by reinforcement and matrix, giving rise to a solid and continuous material, able to transmit and redistribute internal stress. We can state that composites are non-homogeneous and non-isotropic materials, where the individual constituents are bonded to each other in an insoluble way to obtain a final product that combines the best properties of the single components. The Fiber Reinforced Polymers (FRP) analyzed in this work are made up by a carbon fiber material (reinforcement) and by epoxy resin as a matrix.

The main carbon fiber properties [1–3] are: high mechanical strength, high elastic modulus, low density, low crack sensitivity, fatigue resistance and good ability to dampen vibrations. Moreover, the thermal expansion coefficients allow having structures with dimensional stability over a wide range of temperature variation.

Epoxy resins [4] represent a group of thermosetting polymeric materials that do not create reaction products when they cure (reticulate) and therefore have a low reticulation retention. They also have good adhesion to other materials, good chemical and environmental resistance, good mechanical properties and good electrical insulation properties. All these features, together with a remarkable weight reduction combined with high strength (high SWR: strength to weight ratio), make the possibility of combining those two materials very appealing in the aeronautic field.

Composites are non-homogeneous and anisotropic materials, so they respond differently to a given load, depending on the direction considered; therefore, a proper approach to material characterization, should employ full-field measurement techniques as, for example, those based on optical techniques or [5–8] Digital Image Correlation (DIC) [9].

The three-dimensional DIC technique uses two cameras to look at the object from two different directions, obtaining a binocular view that allows to determine the three displacement coordinates (X, Y, Z) for each generic point of the analyzed area [10]. Measurement requires the application of a speckle random pattern on the specimen test.

C. Barile · C. Casavola · G. Pappalettera (✉)
Dipartimento di Meccanica, Matematica e Management (DMMM), Politecnico di Bari, Bari, Italy
e-mail: claudia.barile@poliba.it; giovanni.pappalettera@poliba.it

L. Lamberti et al. (eds.), *Advancement of Optical Methods & Digital Image Correlation in Experimental Mechanics, Volume 3*, Conference Proceedings of the Society for Experimental Mechanics Series, https://doi.org/10.1007/978-3-319-97481-1_25

The 3D DIC technique can also be applied to non-flat surfaces that are displaced off the plane, as in the case of buckling structures; in fact, it provides information about both the shape of the body and the three-dimensional strain field. For this purpose, several images are captured by the two cameras and the speckle pattern for each deformed configuration is analyzed with respect to a reference one.

25.2 Materials and Methods

The material analyzed is a composite laminate made up of an epoxy resin mat with carbon fiber reinforcement. The size of the component subjected to uniaxial compression is 150 mm × 100 mm × 5 mm. The layout of the plies, according to the manufacturer's declarations is as it follows.

Type I layup: $[45F/0T/0T/45F/0T/0T/0F/0T/0T/0F/0T/0T/45F/0T]_S$ where F indicates the Fabric and T indicates the Tape.

The speckle pattern to be analyzed by means of the DIC method was introduced by spraying the specimen (Fig. 25.1). To this scope a matt white spray and a black matt spray were used.

Four electrical strain gauges, two for each side of the pieces, were applied on each specimen [9], in order to evaluate the local strain resulting from the application of compressive loads. In such a way it is possible to compare local behavior obtainable by strain gages with overall behavior obtained by DIC. The strain gauges used had a measuring base of $l_0 = 3$ mm, gage factor $k = 1.99$.

The strain gauges adopted are able to evaluate deformations in one direction (uniaxial). Strain gages with a resistance of 350 Ω were used because the composites are bad conductors.

Compression tests were performed by following the ASTM 7137 standard defined for the composite material. In particular this standard specifies the applied constrains, the environmental conditions as well as the compression rate.

Tests were performed on a SCHENCK servo-hydraulic machine. In this machine the lower clamp is movable while the upper crosshead is fixed and must be initially positioned with the highest precision in order to avoid torsional or flexural deformations that could introduce errors in the measurement process. The main setting for the tests are listed below:

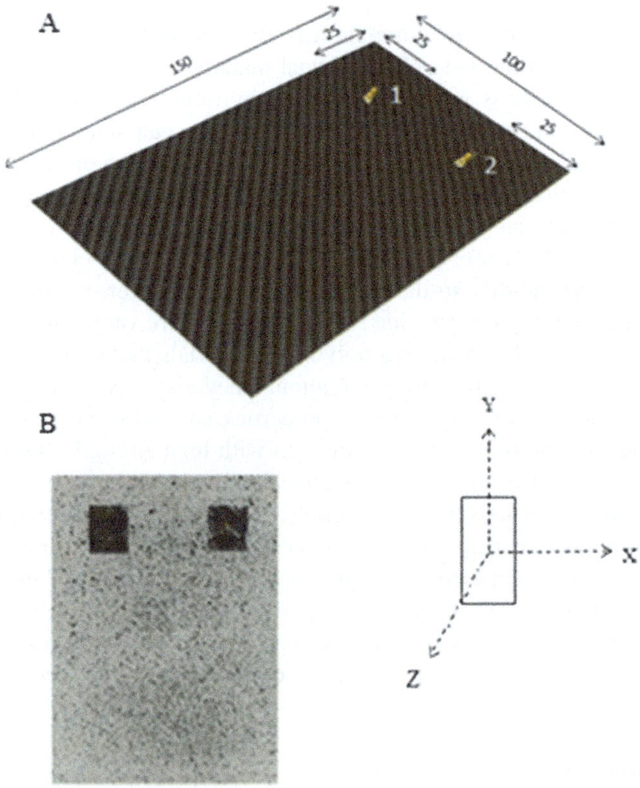

Fig. 25.1 Composite laminate (**a**) Position of the applied strain gages (**b**) real component sprayed for speckle pattern

- Monotonic test
- Maximum load 250 kN
- Control in "displacement rate" (1.25 mm/min until breakage [11])
- Acquisition frequency 10 Hz

For a greater accuracy of the tests and to measure the variables involved in the phenomenon of instability of the specimens, a measurement chain was required to complete the reading of displacements and deformations in a suitable manner.

In order to measure displacements and deformations of the whole surface of the specimens in 3D field, the Dynamics Q400 system with ISTRA 4D software was used.

It includes two Manta industrial cameras, equipped with Ricoh 16 mm lenses, fixed on the same support in order to avoid vibrations. System also includes a 4-channel data acquisition, a synchronization unit and a light source.

Lighting is used to ensure that the speckle pattern on the specimen, once acquired, has a wide grayscale span range; this means that the background is suitably clear (white) and dark (dark) avoiding saturation in areas subjected to the DIC analysis, but also ensuring uniformity of luminance in these same areas.

An optic with a reduced focal length was used for capturing test specimens large enough to monitor the whole component.

The cameras were coupled with a National Instruments® NIDAQ 9171 acquisition card which allows analog to digital conversion of the acquired signal. To synchronize acquisition with the start of the test, a trigger signal exiting from the loading machine was sent to the acquisition system by a BNC cable (Fig. 25.2).

Settings for the DIC system are listed below:

- Distance between cameras: l = 40 cm
- Distance between lens and piece: d = 50 cm
- Inclination angle of the cameras respect to the perpendicular axis of the specimen: 21.8°
- Diaphragm opening: f/8 (Fig. 25.3)

The adopted compressive test fixture is shown in Fig. 25.4. It utilizes adjustable retention plates to support the specimen edges and to inhibit buckling when the specimen is loaded.

The fixture consists of one base plate, two side plates and one top plate. The side supports provide no restraint to local out-of-plane rotation.

The top and bottom supports provide no clamp-up but offer some rotational restraint due to the fixture geometry (the slide plates have a squared geometry and overlap the specimen by 8 mm). The fixture is adjustable to accommodate small variations in specimen length, width and thickness. The top plate and slide plates are not directly attached to the lower portion of the fixture and they slip over the top edge of the specimen. The side plates are short enough for ensuring a gap between the side rails and the top plate for a proper test execution. Edge supports must be co-planar.

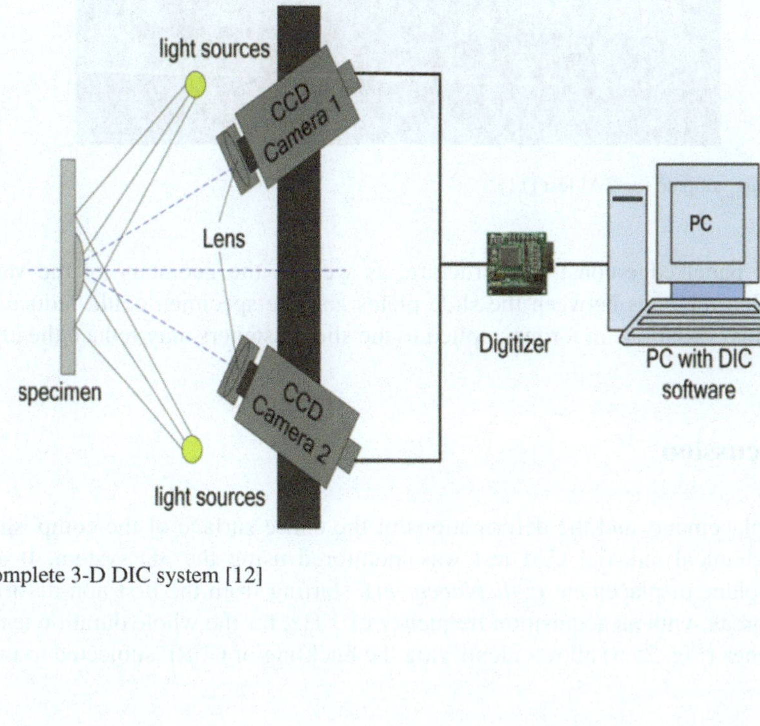

Fig. 25.2 Schematic of a complete 3-D DIC system [12]

Fig. 25.3 Overview of experimental set-up

Fig. 25.4 Edge-constraint structure adopted for CAI test [11]

The configuration of the panel edge-constraint structure, as well as the geometry of the various slide plates, could significantly affect test results: the gaps between the slide plates and the specimen could reduce the real edge support or introducing concentrated loads, variations in torque applied to the slide fasteners may reduce the effective edge support too.

25.3 Results and Discussion

In order to measure the displacements and the deformations of the entire surface of the composite specimen in the three-dimensional space, the mechanical uniaxial CAI test was monitored using the 3D system. It allows collection images representative of the out of plane displacement (*z-displacement*), starting from the first non-deformed configuration of the component until it came to break, with an acquisition frequency of 1 Hz, for the whole duration test.

The analysis of those frames (Fig. 25.5) allows identifying the buckling of CFRP subjected to uniaxial compression.

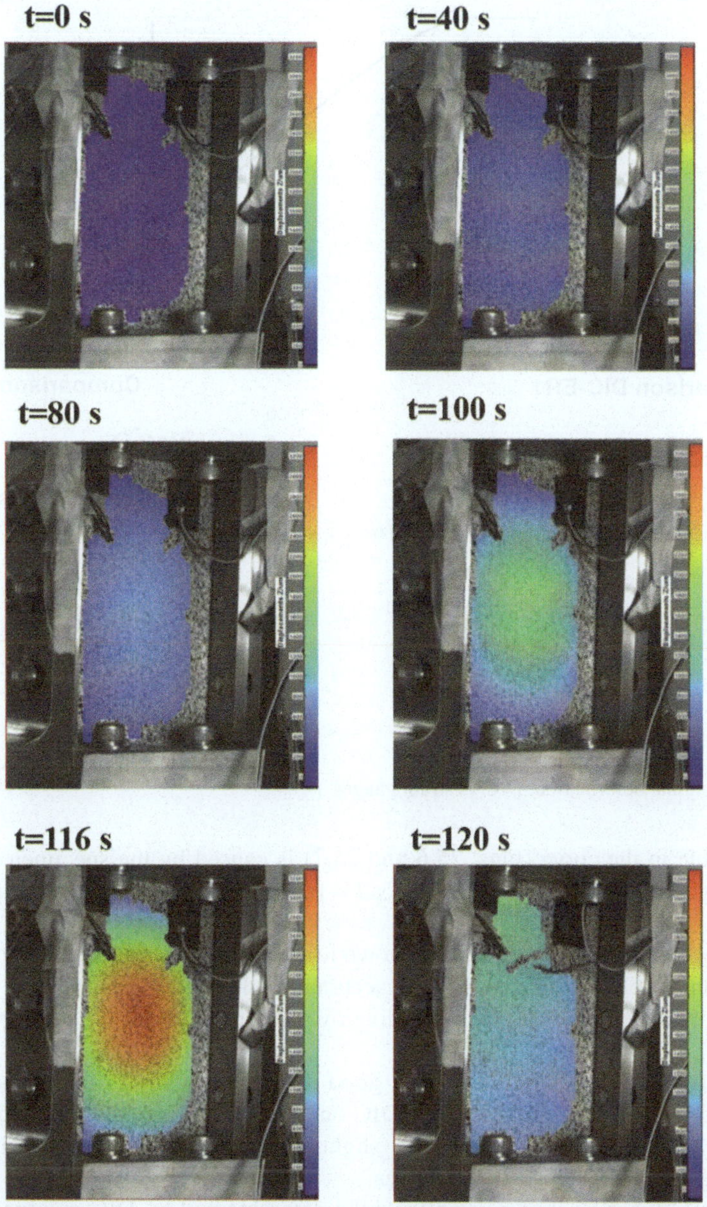

Fig. 25.5 DIC images at various steps: out of plane displacements of the composite specimen

The images are captured and correlated by setting a "high accuracy" mode. It carries out a correlation with a 3D residuum lower than 0.4 pixel, a facet size of 19 pixels and an accuracy of 0.1 pixels, locating 1180 grid points in the Region Of Interest (ROI) of the composite laminate.

Several considerations can be done by observing the captured images. In the first load steps, due to the adopted fixtures and the low displacement rate applied (1.25 mm/min) the component is subjected to very limited out of plane displacement (e.g. at step t = 40 s a z-displacement of 400 μm at the specimen center).

The component begins to buckle after 80 s from the beginning of the test with an out of plane displacement equal to 800 μm. After 116 s the out of plane displacement becomes more evident reaching 3200 μm; the specimen breakage occurs few seconds after this step, namely at 120 s. Temporal load law is reported in Fig. 25.6.

It is possible to observe that there is a first hyperbolic trend in the first 40 s until reaching a load of about −21 kN. At this point the first instability of the component is recorded because some out of plane displacements are observable. Successively there is a linear behavior, up to 80 s. From that point until 116 s a slight slope change occurs as a result of the buckling of the specimen, as it is inferable by observing DIC images.

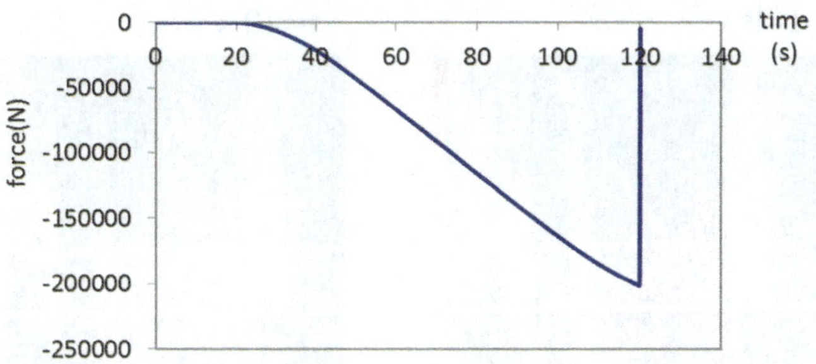

Fig. 25.6 Temporal load law of the machine

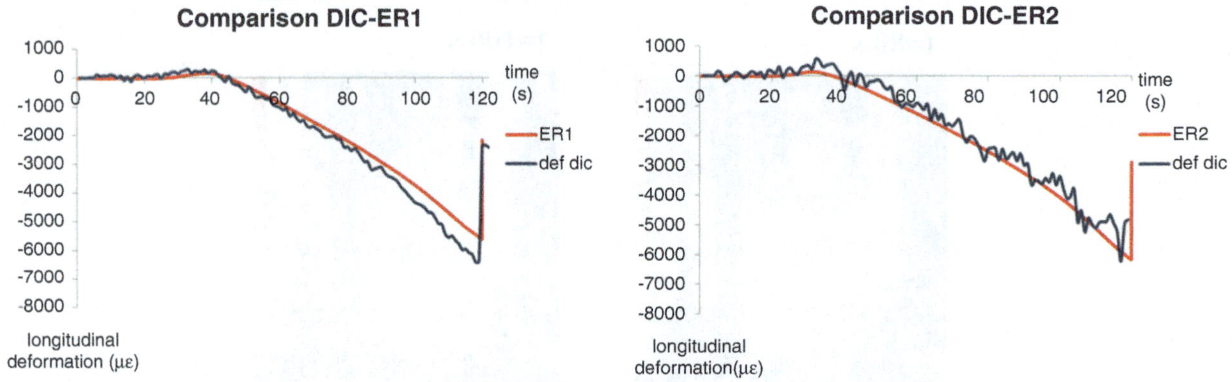

Fig. 25.7 Comparison of longitudinal deformation: DIC vs strain gauges

The drop down observable in the curve (Figs. 25.6 and 25.7) is caused by the specimen breakage with the consequent detachment of the strain gages as soon as the maximum load is reached. The break of the sample at $t = 120$ s corresponds to a load of -200 kN.

The use of two additional electrical strain gauges at known locations [11] had allowed to compare the longitudinal strains (ε_{yy}) obtained by both ER and DIC. In order to verify the accuracy of the evaluated strain ε_{yy}, the values of the strain gauges 1 and 2 were compared with the corresponding DIC reading over an area 3 mm long, corresponding to the gauge length of the strain gauges (Fig. 25.7).

The comparison of the two techniques shows a very good agreement with respect to the strain trends. In the case of ER1 lower deformations where recorded respect to the DIC ones. This attitude could be explained by considering a slight misalignment in the application of the strain gauges. A slight shift between the left and right side of the specimen was observed with respect to the center, in favor of the left area.

It should be observed that higher level of noise affects the data obtained by DIC measurements on the right side (ER2 location) as reported in Fig. 25.7. This could be justified by taking into account the high luminance reflected by the bounding structure.

In this case a "smoothing" operation is preferable.

In Fig. 25.8 the trend of longitudinal deformation along the cross section of the specimen is shown in correspondence of the step $t = 116$ s.

It can be noticed that the component has a normal negative strain ε_{yy} in the outer areas, according to a nearly symmetrical pattern with respect to the center of the specimen. At the center the deformations are positive thus the specimen stretches slightly.

There is a shift towards the top, in fact it happens that 15 mm far from the 0 location the specimen breaks.

In Fig. 25.9 the shear strain recorded after 116 s is reported. It was noticed that the maximum shear strain value can be traced from 15 to 20 mm far from the 0 location, and that is the point where the sample starts to break.

This may be understood in the framework of the continuum mechanics theory. Such deformation, in fact, is the sum of the mixed derivatives of the x e y displacements because of the angular deformation of the laminate that determines delamination effects.

The previously indicated asymmetry was also accounted by analyzing the out of plane z displacements (Fig. 25.10). In fact, despite of the typical gaussian trend, the lower part of the specimen has a different slope than the upper one. The

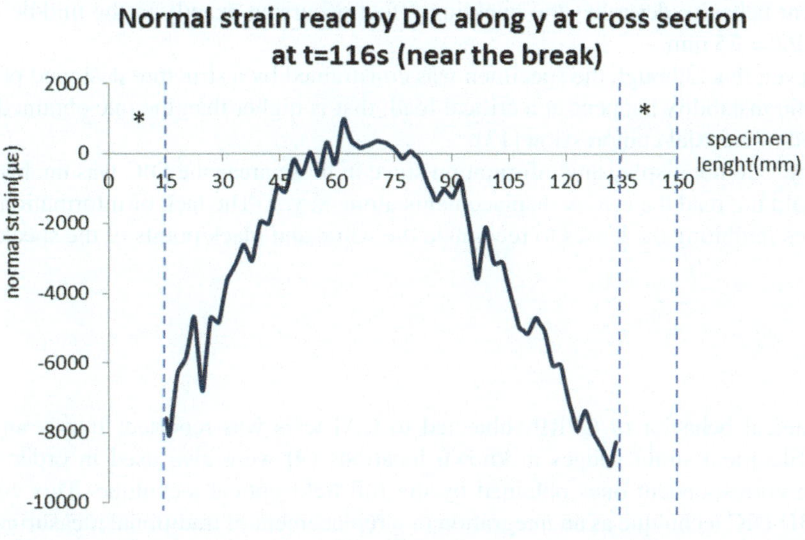

Fig. 25.8 Normal strain ε_{yy} at t = 116 s

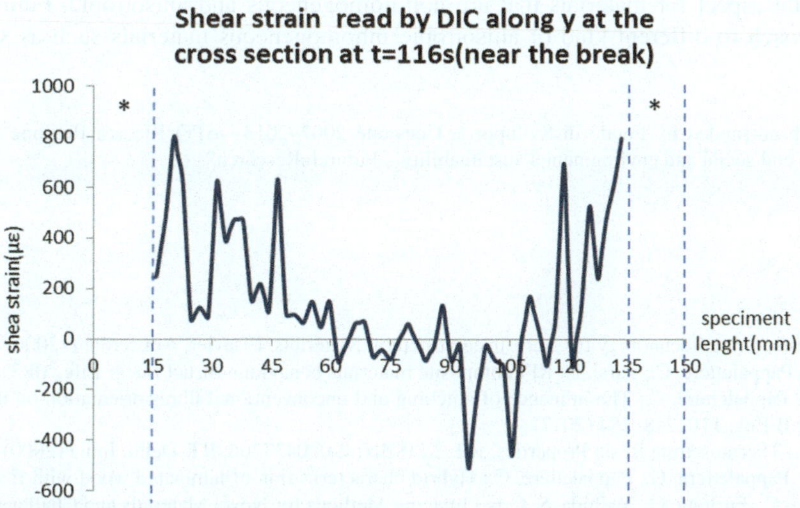

Fig. 25.9 Shear strain γ_{xy} at t = 116 s

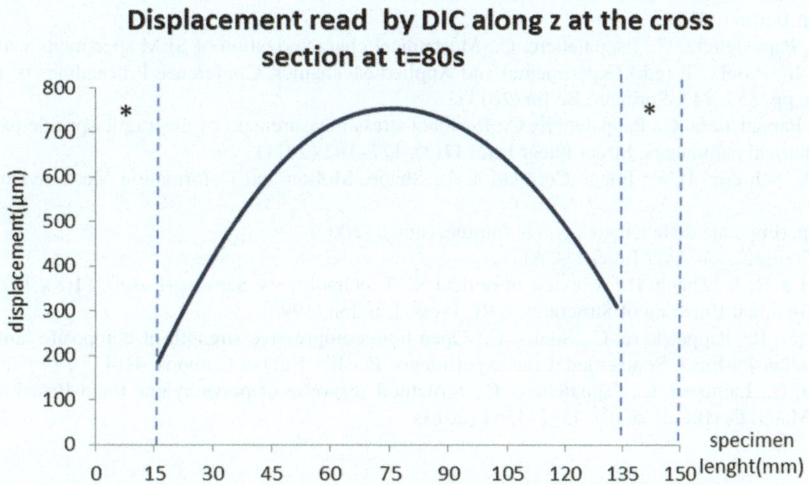

Fig. 25.10 Out of plane displacement w at t = 80s

out of plane displacement behavior describes the buckling effect occurring exactly in the middle of the specimen equal to w = 750 μm located at l/2 = 75 mm.

It can be stated, however, that although the specimen was constrained by a structure designed properly to avoid buckling, the phenomenon of elastic instability happens at a critical load, that is higher than the one obtained for a simply anisotropic supported plate subjected to uniaxial compression [13].

The areas indicated by * do not display any information since in those areas the DIC was unable to recognize the speckle pattern and therefore could not read the u, v, w displacements along x, y, z. The lack of information is due to the load fixture that creates shadow zones inhibiting the lenses to recognize the white and black points of the speckle pattern.

25.4 Conclusions

In this paper the mechanical behavior of CFRP subjected to CAI tests was reported. It was analyzed by Digital Image Correlation technique. Electrical strain gauges at known locations [9] were also used in order to compare local results obtained by ER and the correspondent ones obtained by the full-field optical technique. This comparison underlines the possibility to apply the 3D DIC technique as an integration or a replacement of traditional measuring instruments to calculate displacements and deformations on the whole surface of the specimen. This can allow also reduction of some experimental bias as those connected with ER misalignment. Moreover, DIC is able to capture asymmetries in the behavior of the sample, and this is a fundamental aspect for materials that are non-homogeneous and anisotropic. Future works can include the application of this approach to different kind of anisotropic inhomogeneous materials such as sandwich panels [14] and foam sandwich [15].

Acknowledgment Research co-funded by Fondo di Sviluppo e Coesione 2007–2013—APQ Ricerca Regione Puglia "Regional program to support smart specialization and social and environmental sustainability—FutureinResearch".

References

1. Campbell Jr., F.C.: Manufacturing Technology for Aerospace Structural Materials. Elsevier, Amsterdam (2006)
2. Barile, C., Casavola, C., Pappalettere, C., Tursi, F.: RFI composite materials behaviour. Struct Integr Life. **10**(3), 209–213 (2010)
3. Barile, C., Casavola, C., Pappalettere, C.: The influence of stitching and unconventional fibres orientation on the tensile properties of CFRP laminates. Compos. Part B Eng. **110**, 248–254 (2017)
4. Varma, I.K., Gupta, V.B.: Thermosetting Resin Properties, vol. 2.; ISBN: 0-080437206. IIT, Delhi, India (2000)
5. Barile, C., Casavola, C., Pappalettera, G., Pappalettere, C.: Hybrid characterization of laminated wood with ESPI and optimization methods. In: Jin, H., Sciammarella, C., Furlong, C., Yoshida, S. (eds.) Imaging Methods for Novel Materials and Challenging Applications, Conference Proceedings of the Society for Experimental Mechanics Series, vol. 3, pp. 75–83. Springer, Berlin (2013)
6. Barile, C., Casavola, C., Pappalettera, G., Pappalettere, C., Tursi, F.: Drilling speed effects on accuracy of HD residual stress measurements. In: Rossi, M., Sasso, M., Connesson, N., Singh, R., DeWald, A., Backman, D., Gloeckner, P. (eds.) Residual Stress, Thermomechanics & Infrared Imaging, Hybrid Techniques and Inverse Problems, Conference Proceedings of the Society for Experimental Mechanics Series, vol. 8, pp. 119–125. Springer, Berlin (2014)
7. Barile, C., Casavola, C., Pappalettera, G., Pappalettere, C.: Mechanical characterization of SLM specimens with speckle interferometry and numerical optimization. In: Proulx, T. (ed.) Experimental and Applied Mechanics, Conference Proceedings of the Society for Experimental Mechanics Series, vol. 6, pp. 837–843. Springer, Berlin (2011)
8. Barile, C., Casavola, C., Pappalettera, G., Pappalettere, C.: Residual stress measurement by electronic speckle pattern interferometry: A study of the influence of geometrical parameters. Struct Integr Life. **11**(3), 177–182 (2011)
9. Sutton, M.A., Orteu, J.J., Schreier, H.W.: Image Correlation for Shape, Motion and Deformation Measurements. Springer Science, Berlin (2009)
10. Ajovalasit, A.: Analisi sperimentale delle tensioni con la fotomeccanica (2009)
11. ASTM standard 7137 (Compression After Impact "CAI")
12. Zhu, Y.-K., Tian, G.-Y., Lu, R.-S., Zhang, H.: A review of optical NDT technologies. Sensors (Basel). **11**(8), 7773–7798 (2011)
13. Trahair, N.S.: Flexural Torsional Buckling of Structures. CRC Press, London (1993)
14. Casavola, C., Giannoccaro, R., Pappalettere, C., Soutis, C.: Open hole compressive strength of composite laminates and sandwich panels: Comparison between Budiansky-Fleck-Soutis model and experiments. Plastics Rubber Compos. **41**(4–5), 199–208 (2012)
15. Boccaccio, A., Casavola, C., Lamberti, L., Pappalettere, C.: Structural response of polyethylene foam-Based sandwich panels subjected to edgewise compression. Materials (Basel). **6**(10), 4545–4564 (2013)

Chapter 26
Evaluation of Residual Stress with Optical Methods

C. Pappalettere

Abstract Residual stress (RS) evaluation and management is a very important topic in mechanical design because its presence can severely affect final performances of the manufactured part especially under fatigue conditions. Due to the complex chain of intermediate steps that lead to the final manufactured component numerical simulations to predict RS are hard to be accomplished and the experimental approach is surely the most reliable one. Many different strategies have been developed along the years to perform RS measurements including hole-drilling (HD), X-Ray diffraction, neutron diffraction, Barkhausen noise etc. Also the adoption of optical techniques OT has been studied even in consideration of their common capability to provide full-field information with high accuracy and high sensitivity. Among OT the Electronic Speckle Pattern Interferometry is the technique which has experienced the greatest development in this field. In this paper the progresses made since the first experiments which date back 30 years ago are presented. The achievements occurred in these years are illustrated in terms of improvements of experimental set-up, developments of faster, simpler and more accurate analysis algorithms and increment of number of applications where the hole drilling combined with ESPI has been successfully exploited. Current state of the art is illustrated also with the indication of several different situations and materials successfully studied with this approach. Finally indications for future works and research activities on this topic are provided.

Keywords Optical methods · Residual stresses · ESPI · Hole drilling method · Interferometry

26.1 Introduction

Obtaining a final part is a complex process which can involve many steps. At the end of the production chain it is important to observe the presence of a possible residual stress state. This means that, without applying any external load, the material can exhibit the presence of stresses which are commonly referred, in literature, as residual stresses (RS). The presence of RS can be reconnected to the most relevant and widespread industrial manufacturing processes such as grinding, forming [1], rolling, casting, forging [2], welding [3–5], melting [6], cladding [7] and heat treatments [8]. Assessment of RS is a very relevant topic in consideration of the effects that the presence of a high level of these stresses can have. It should be considered, in fact, that they must be added to the applied stresses during the exercise and that they can contribute in reducing the lifetime of a component subjected to fatigue [9]. The attempt to follow the steps of the manufacturing chain in order to predict the final level of stress is a very complicated task also in consideration of the great difficulty to obtain detailed information of the relevant process parameters during the entire process. Experimental methods are, at the moment, the most adopted ones to the scope to get information about the residual stress state of the materials. Several different approaches were developed along the years each one displaying its proper drawbacks and advantages and their privileged fields of application. However, one of the most consolidated and widespread technique-both at industrial and research level –is the so called hole drilling method (HDM). This method, which is also nowadays ruled by an international standard, is based on the process of removing the stress state by drilling a hole inside the material. Stress removal results in a strain field around the hole that can be measured by using strain gages. Strain field measurement can be then used to retrieve information about the pre-existing stress state. Moreover, if the hole is drilled incrementally the methodology allows to get stress profile along the depth. Strain measurement, in this method, is performed by a rosette which, in the most common form, has three strain gages properly

C. Pappalettere (✉)
Dipartimento di Meccanica, Matematica e Management, Politecnico di Bari, Bari, Italy
e-mail: carmine.pappalettere@poliba.it

L. Lamberti et al. (eds.), *Advancement of Optical Methods & Digital Image Correlation in Experimental Mechanics, Volume 3*,
Conference Proceedings of the Society for Experimental Mechanics Series, https://doi.org/10.1007/978-3-319-97481-1_26

placed around the hole which give the minimum information allowing to solve the in-plane RS problem. The surface of the sample must be properly prepared (i.e. smoothed and degreased) before applying the rosette. This is a time consuming procedure which must be carried out with some care to avoid to alter the initial RS state. Also the relative placement between the drill bit axis and the center of the rosette must be managed with some care in order to avoid eccentricity errors that would introduce a bias in the measurement. In view of these considerations some efforts have been spent in the direction of studying alternative ways to measure the strain field.

In this framework several research works explored the feasibility of replacing strain gages by optical methods (OM). This class of experimental strain analysis methods constitutes, nowadays, a mature and well consolidated tool whose applications span from contour measurements [10] to crack monitoring [11] and corrosion monitoring [12]. The application of OM to the specific case of RS measurement displays several advantages. In fact, these methods do not require contact with the surface and this involves a reduction of the preparation time and it allows to save the cost of the strain gage rosette. The possibility of avoiding rosette attachment, moreover, extends the field of application of the HDM also to those materials with low Young's modulus where rosette reinforce action cannot be neglected. The use of moiré interferometry for strain determination in RS measurement was tested since 1983 by McDonach et al. [13]. This application, however, required grating attachment to the sample so that it shared several drawbacks with strain gages. In the same year the paper by Antonov [14] was published were the possibility of adopting holographic interferometry was analyzed. Also shearography was subjected to investigation in [15] where, however measurement principle for RS measurement is completely different because it does not relies on hole drilling but on indentation. Sherarography for RS measurement was also explored by [16] where also a comparison with the speckle based method is presented. Also the possibility to release residual stresses by using local heat treatment, combined with electronic speckle pattern interferometry was attempted in [17–19] and more recently repeated on aluminum alloys in [20, 21]. Among the OM analyzed to be employed in HDM residual stress measurement, at the current state of the art, ESPI (Electronic Speckle Pattern Interferometry) must be cited to be that which has undergone to the major developments along the years. The basic idea is to use the correlation fringe patterns obtained by the subtraction of a reference pattern, recorded before starting to drill the hole, and those obtained at each incremental drill step. Also for the case of the speckle interferometry we should say that it is, nowadays, a well consolidated approach which is successfully applied in many different applications in experimental mechanics [22–30].

Speckle pattern is obtained by the illumination of an optically rough surface by a coherent beam of light produced by a laser source. As a consequence of the random interference of scattered waves starting from each point of the illuminated surface a completely random pattern of illumination is created which is strictly correlated with the surface under analysis in the given state. Any surface modification introduces changes in the local roughness which introduce modifications in the recorded speckle pattern. If two speckle patterns corresponding to two different surface conditions (e.g. loaded and unloaded states) are recorded and subtracted each other, correlation fringes are displayed containing the information about the displacement field experienced by the sample along the direction of the sensitivity vector of the system. This approach can be easily transferred to the application of RS measurements by HDM in substitution of the application of a strain gage rosette. In principle, in fact, it is sufficient to illuminate the surface before drilling the hole to store information about the initial state of the sample. Successively new images can be recorded at the end of each drill increment. Subtraction of those images from the reference one allows determining the strain field, on the surface of the sample, generated as a consequence of the stress relaxation process. Starting from the very first studies on this approach, which date back to the last years of eighties, a lot of progresses occurred along the years. This paper intends to provide a concise overview of these advancements including improvements in terms of experimental equipment and developments of robust and faster calculation algorithms to end with a short and non-exhaustive list of applications where the technique was applied successfully out-of-laboratory. This last aspect can be considered a strong indication of the degree of maturity that this approach has achieved at this point of its development.

26.2 Pioneering Works

Among the first works where speckle interferometry was used as an experimental tool for stress analysis it is possible to cite [26] where digital subtraction of speckle pattern was used for full field metrology of a loaded sample and [27] where a speckle interferometer was used to analyze stress field arising as a consequence of mismatch in thermal expansion coefficients between two different ceramic materials. In the same period the very first studies about the possibility to use speckle interferometry for the problem of residual stress determination were published. One of the first and more relevant works on this topic was done by Nelson and McCrickerd in 1986 [31]. In their paper the case of a hole in a flat plate with uniform stress state along the thickness was considered. They explored different conditions of illuminations while the fringe analysis was only limited to simple visual processing of the recorded fringes. Almost in the same year, 1987, also the Italian group of Prof. Furgiuele, in Italy, presented a very complete study on the feasibility of applying speckle interferometry to residual stress determination [32]. The optical set-up adopted in [32] (Fig. 26.1) allowed to determine three components of

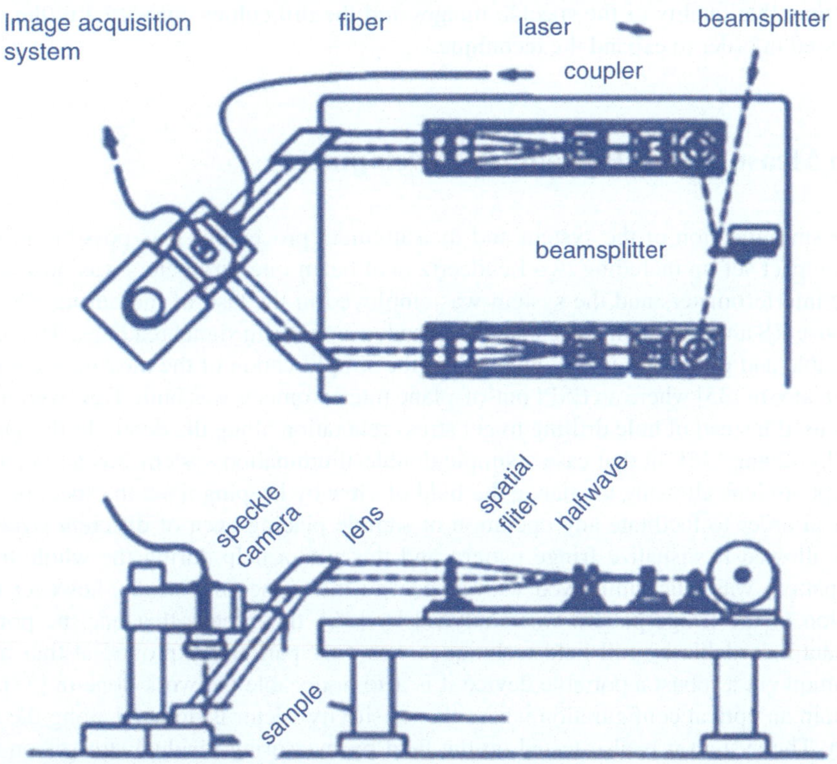

Fig. 26.1 Schematic of the HDM-ESPI Set-up for adopted in the pioneering work by Furgiuele et al. [32]

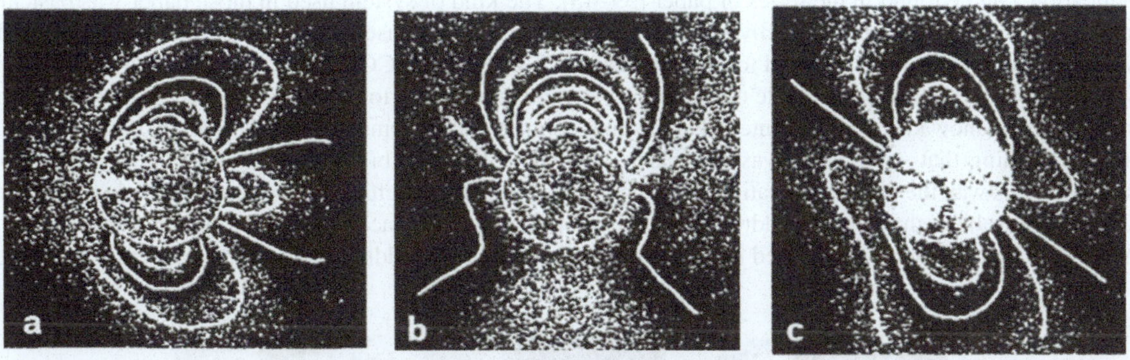

Fig. 26.2 Three different speckle correlation patterns obtained in the work by Furgiuele et al. [32]. Skeletonizing of the pattern is also shown. The three different patterns are referred to three different directions of illumination

the displacement in the plane of the specimen and two oblique to the same plane. In order to verify the reliability of the method the system was tested on a steel specimen where a well know stress state was introduced. Also in this case simple visual inspection was used in order to analyze fringes. Pattern were skeletonized manually (Fig. 26.2) by using a trackball in order to draw fringe edges and, successively, to assign them the proper fringe number. This allowed to determine the displacement just for the pixels along the fringe edges and to obtain stresses based upon calibration coefficients connecting the stress state to the displacement occurring as a consequence of the drilling procedure. The generic stress profile was hypothesized to be reconstructed starting by the superposition of eight different linear profiles. Also in this case the use of different directions of illumination was explored. It was found that the accuracy of the data was almost independent on the adopted configuration while, best results were obtained when data from all of the directions were combined. This was a strong indication that quality of the speckle fringes was a main issue which was only partially addressed by increasing the number of processed data. A follow up of their work was published, in italian language only, and it takes several years before being published in an international journal [33]. In this paper they specifically dealt with the problem to determine calibration constants to be used in conjunction with displacement fields obtainable by OM.

It was clear at that time that quality of the speckle images and the difficulties connected to the fringe analysis were the first issues to be addressed in order to expand the technique.

26.3 System and Measurement Procedures Developments

In the direction of the simplification of the system and measurement procedures it is possible to insert the work by [34] where a much more compact set-up including two Lendeertz dual beam interferometers was presented. Optical fibers were used to implement the interferometers and the system was employed in the task of measuring RS in CD-R disc. This is a practical problem because RS alters birefringence and this introduces errors in signal readings. This kind of system proved to be much more manageable and promising in the direction of the simplification of the measurement procedure. A simplified set up was also adopted later in [35] where an ESPI out-of-plane interferometer was built. However, in that case, the chemical etching procedure was used instead of hole drilling to get stress relaxation along the depth. In the same framework it should be inscribed the work by Zhang [36]. In that case a simple double-illumination system was adopted. The recording camera used a customized telephoto lens allowing to change the field of view by keeping fixed the specimen position. The working distance was designed in order to facilitate any operation of sample placing even of different sizes. Real time subtraction of the speckle images allowed to visualize fringe pattern and this was a help during the whole measurement procedure. Quality of the fringe pattern was much improved with respect to the pioneering works, however it should be underlined that only a cross section of the fringe pattern was analyzed by [36], this means that one the point of strength of using the great amount of data recordable by full-field techniques was only partially exploited at that time. In the direction of implementing the technique in a robust a portable device it is also noticeable the work done in [37]; in this paper a conical mirror was used to obtain an optical configuration where the sensitivity vector is directed along the radius for each point of the analyzed specimen. The system was also tested on the field by measuring residual stresses on a pressure vessel while the same interferometric configuration was adopted, in combination with indentation instead of hole drilling, in [38] even if a numerical model for quantitative evaluation of RS was missing. Other considerations, even in terms of system robustness and simplification, can be found in the series of paper [39–43]. The kind of system used in those paper was basically made a laser source whose output was split and delivered by two fibers; first fiber is used to illuminate the sample while the second one is used as a reference and it is directed towards the camera CCD matrix. CCD camera looks at the sample from a given angle giving the possibility to accommodate the drilling system perpendicular to the sample. This is a very simple and robust configuration which is nowadays implemented also in some commercial systems (Fig. 26.3) [44–46].

It worth underlining that some work was also carried on about the possibility of separating the ESPI set-up from the drilling set-up. In this way the drilling operation could be performed in the machine shop without requiring dedicated drilling system. The major problem that must be addressed, to this scope is the highly accurate repositioning of the sample after each drilled increment. This problem was studied in [47] and it was successfully addressed by using a kinematic mount.

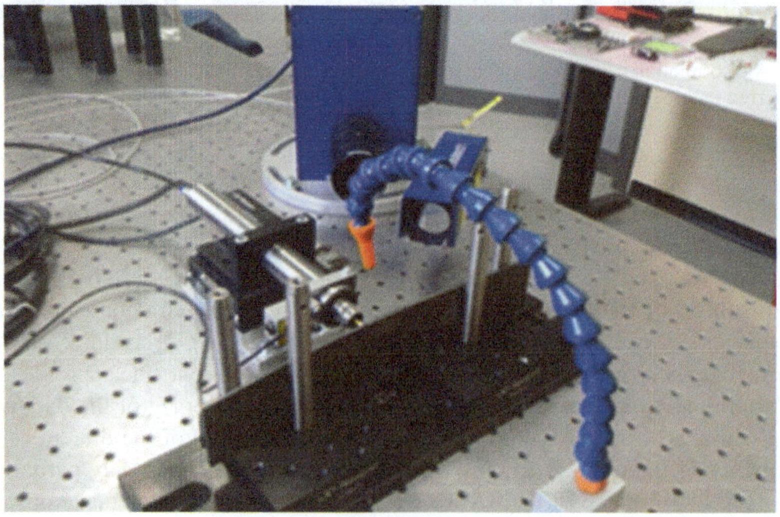

Fig. 26.3 Picture of the commercial system PRISM developed by Stresstech performing measurements on a Ti grade 5 sample at Politecnico di Bari

26.4 Analysis and Reliability Assessment Developments

A huge work in terms of data analysis improvement and consolidated reliability assessment of the method was done in the previously mentioned series of papers [39–42]. Management of the entire amount of information, provided by full-field technique, namely associated to each pixel, requires proper data management. In this direction the work by Steinzig et al. aimed to the implementation of a numerical solution allowing to determine surface deformation, generated by HD stress relaxation, in correspondence of whatever combination of elastic modulus and Poisson's ratio as well as hole diameter and single drill increments. This calculation allowed to determine the displacement corresponding to any in-plane stress state represented as a combination of the σ_x, σ_y and τ_{xy} components. Once that the displacement field is determined by ESPI then the stress profile can be obtained by applying least square technique to minimize the difference between the measured stress field and the calculated one for a given combination of the σ_x, σ_y and τ_{xy} components. Another aspect that should be conveniently taken into account during the data analysis as it was done in [39–42] is connected with rigid body motion corrections. In fact, differently by traditional strain gage approach, ESPI method measures displacements instead of strain. This means that possible rigid body motions that can occur during the measurement process can introduce a bias in the final results. A possible way to overcome this difficulty is by applying a derivative filter in order to obtain strain field. However, the typical noise associated with speckle measurements make this approach not recommendable. In [39–42] also this problem was addressed by generalizing the problem by adding terms which take into account also the effects of rigid body motion on the final displacement field. Accuracy and reliability assessments of the technique is another important aspect that was conveniently afforded by researchers in the last years. Main source errors can be associated with the uncertainty on the determination of the illumination and observation angles, as well as on the determination of the pixel size and so on. Also the influence of drilling parameter such as the choice of a given rotation speed of the drill bit was analyzed and studied in the case of aluminum and steel in [48] and, later, also for titanium [49–52]. In these papers also considerations about the definition of the area of analysis are reported (Fig. 26.4). In fact, for accurate measurements, information from pixels too much near to the hole edges or too far from them should be discarded. In the first case, in fact they can be affected by edge effects connected, for example to material heating during the drilling process. When too far from the center of the hole, instead, displacements can become too low to be detected properly [53–58].

Also the direction of the sensitivity vector, with respect to the principal stress direction, was found to have some relevance in final accuracy especially for the case of measurements of low level of stress (i.e. low displacements) [53]. Finally the problem of algorithm optimization for full data exploitation was analyzed in [59] while some extra considerations about the choice of the drilling strategy and its effects in terms of final accuracy can be found in [60].

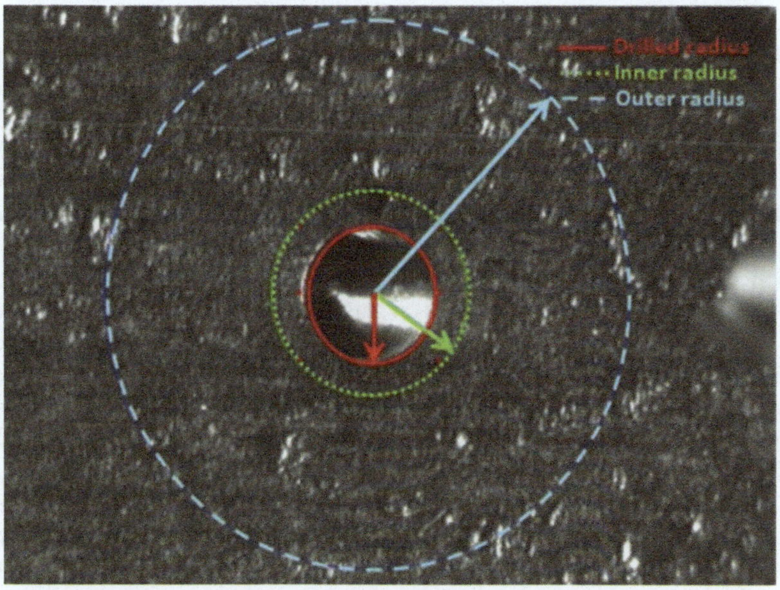

Fig. 26.4 Example of identification of the area of analysis by the indication of a circular crown centered at the center of the drilled hole. Only the pixels included between the inner radius and the outer radius are considered for the calculation

26.5 Recent Applications of the Technique

The increased easiness in using ESPI-HDM methods, the reduction in terms of initial costs required to set-up this kind of system, the development of software tools for fringe analysis and the increased performances of modern computers are all factors that facilitate the diffusion of this approach. It results that the number of publications on this topic in the last 10 years is increasing and, at the same time, it is increasing the range of different applications where this approach was successfully used. Nowadays, many applications of industrial relevance can be found; in particular connected to weld joint analysis, a field where the RS evaluation is a task of great importance. In [61] the technique was applied to evaluate residual stress generated by the friction hydro pillar processing, a process where welding is obtained by introducing a rod of welding material by exploiting frictional heating. The force used to push the road and its rotation speed needs to be optimized and this optimization process must also take into account RS issues. The ESPI-HDM approach demonstrated to be effective in RS determination and it put into evidence a radial symmetry of the stress field around the rod. It is also interesting to observe that, in the same paper the authors performed comparative measurements by using both strain gages and optical approaches obtaining very consistent data; at the same time, the authors observed that the measurement time required for measuring with ESPI-HDM can be estimated to be about 25% of the time required by strain gauges. This makes this approach appealing for industrial environment. In the same direction, it worth underlining that the specific set-up developed by the group of Viotti et al. is robust enough to be effectively used also outside laboratory. Another recent application of ESPI-HDM to RS evaluation in weld joint can be found in [62]. In that paper AlMg6 alloy argon arc welded shells were studied. Besides the previously mentioned advantages, connected with the use of the HDM ESPI technique, the authors underline that, for the specific case, employment of the optical method avoided removal of weld reinforcement. In this way, consequent redistribution of residual stresses is avoided as well and this goes in the direction of improving measurement reliability. Also in this case, it should be underlined that a robust set-up was employed, easily usable also outside laboratory. Another comparison between ESPI measurements and strain gages measurements for RS analysis can be found in [63]. In that paper three-layer submerged arc butt welded joints were analyzed and also in that case a good agreement among results obtained by the two different approaches was found. Moreover a finite element analysis was performed to simulate the butt welding process and to predict RS, also in this case a good agreement was found between RS results obtained HDM ESPI and numerical simulations. Another technological pathway which is assuming great importance at industrial level is connected with additive manufacturing [64, 65]. Also in this situation optimization of process parameters is a relevant task that must include also residual stress assessment. RS, in fact, can impact not only on the lifetime of the component but also on its final dimensional accuracy. RS management can avoid successive post-processing of the part allowing to exploit to the maximum the potentiality of additive manufacturing processes [66]. Fused Deposition Modelling is one of the most widespread additive manufacturing approach. Final part is built layer by layer by depositing a melted filament through a numerically guided heated nozzle [67]. Material under construction is constantly subjected to thermal cycles which introduce residual stresses that can also be released altering the shape of the final part. Measuring RS stress by ESPI HDM for this specific situation shows an additional advantage; in fact, in FDM, polymers are commonly employed. Due to their low Young's modulus, application of a strain gage rosette can act as a reinforcement so that measured strain values can be altered. The use of ESPI, being without contact, overcomes completely this issue. In [67, 68] RS measurement on ABS sample built by FDM is presented; to the scope an *ad hoc* orthotropic model based on the classical laminate theory was developed in order to calibrate coefficients. Other measurements on plastic materials were recently published in [69] where analyses on Babyblend (PC/ABS) and Makrolon (PC) were performed. In this case some deviations from the expected values were observed at the higher depths that can be reconnected to the low heat conductivity of such a kind of materials that limits heat dissipation during the drilling process. In the same paper also analyses on glass are presented. The surveys of recent researches on the use of ESPI methods for RS analysis must also include that works where alternative ways for stress relaxation are attempted. In [70] incremental grooving was used instead than drilling while in [71–72] cross slitting and dual-axis ESPI were introduced demonstrating that this configuration gives better results with respect to the determination of the shear stress profile. In [73] hole-drilling and ring-core techniques are used in combination. Recent works are also concerned with the improvement of calculation methods; in [74] the integral method for stress calculation in incremental hole drilling was modified so that the speckle images recorded at each drill increment are not subtracted from the initial speckle pattern but from the speckle pattern recorded on step before. By using this approach it can be observed that the matrix that rules the problem, that is to say that connecting measures strain and stresses, becomes diagonally dominant so that the equation system is better conditioned and error propagation is reduced.

26.6 Conclusions

The literature survey presented in this paper shows that many progresses occurred along the last thirty years in developing an approach which combines HDM and OM for RS measurements and in particular for that is concerning the adoption of Electronic Speckle Pattern Interferometry. Many technological progresses occurred along the years helped this progress. The appearance of high resolution cameras e better quality laser sources contributed to the possibility to manage data of higher quality that can guarantee analyses at a higher level of accuracy. At the same time the cost of the necessary equipment and size of the equipment itself has constantly reduced. This last aspect contributed to the development of portable and robust systems that can also be transported and used outside typical laboratory environment [75, 76]. Refinement of the calculation algorithms and higher performances processing units made the analysis process much faster and easy and, for some situations, a certain level of automation of several parts of the measuring process was implemented. This gives some indications that it is foreseeable the possibility that in the next future this technique can have some diffusion also at industrial level as it occurs, nowadays, for several different stress analysis methods [77, 78]. With regards to future research in this field we can expect that the number of investigated materials will increase more and more and it will comprises also non-isotropic materials. Some works is also expected to be done in terms of comparison with the emerging concurring techniques of displacement fields measurements based on Digital Image Correlation [79–82].

References

1. Lee, K.T., Park, C.S., Kim, H.Y.: Fatigue and buckling analysis of automotive components considering forming and welding effect. Int. J. Automot. Technol. **18**(1), 97–102 (2016)
2. Fu, Y., Li, W.Y., Yang, X.W., Ma, T.J., Vairis, A.: The effects of forging pressure and temperature field on residual stresses in linear friction welded Ti6Al4V joints. Adv. Manuf. **4**(4), 314–321 (2016)
3. Casavola, C., Lamberti, L., Pappalettere, C., Tattoli, F.: A comprehensive numerical stress—Strain analysis of laser beam butt-welded titanium compared with austenitic steel joints. J. Strain Anal. Eng. Des. **45**(7), 535–554 (2010)
4. Casavola, C., Pappalettere, C.: Discussion on local approaches for the fatigue design of welded joints. Int. J. Fatigue. **31**(1), 41–49 (2009)
5. Casavola, C., Pappalettere, C.: Application of WEL.FA.RE. method on aluminum alloy welded joints. In: Proceedings of the 2005 SEM Annual Conference and Exposition on Experimental and Applied Mechanics, pp. 1555–1562 (2005)
6. Casavola, C., Campanelli, S.L., Pappalettere, C.: Experimental analysis of residual stresses in the selective laser melting process. In: Society for Experimental Mechanics—11th International Congress and Exhibition on Experimental and Applied Mechanics 2008, vol. 3, pp. 1479–1486 (2008)
7. Alam, M.K., Edrisy, A., Urbanic, J., Pineault, J.: Microhardness and stress analysos pf laser-cladded AISI 420 martensitic stainless steel. J. Mater. Eng. Perform. **1**, 1–9 (2017)
8. Araghchi, M., Mansouri, H., Vafaei, R., Guo, Y.: A novel cryogenic treatment for reduction of residual stresses in 2024 aluminum alloy. Mater. Sci. Eng. A. **689**, 48–52 (2017)
9. Barile, C., Casavola, C., Pappalettera, G., Pappalettere, C.: Analysis of crack propagation in stainless steel by comparing acoustic emissions and infrared thermography data. Eng. Fail. Anal. **69**, 35–42 (2015)
10. Chen, F., Brown, G.M., Song, M.: Overview of 3-D shape measurement using optical methods. Opt. Eng. **39**(1), 10–22 (2000)
11. Tankam, P., Picart, P.: Use of digital color holography for crack investigation in electronic components. Opt. Lasers Eng. **49**(11), 1335–1342 (2011)
12. Casavola, C., Pappalardi, P., Pappalettera, G., Renna, G.: A fringe projection based approach for corrosion monitoring in metals. Exp. Tech. **42**(3), 1–7 (2018)
13. McDonach, A., McKelvie, P., MacKenzie, P., Walker, C.A.: Improved moiré interferometry and applications in fracture mechanics, residual stress and damaged composites. Exp. Tech. **7**(6), 20–24 (1983)
14. Antonov, A.: Development of the method and equipment for holographic inspection of residual-stresses in welded structures. Weld. Prod. **30**, 41–43 (1983)
15. Hung, Y.Y., Ho, H.P.: Shearography: an optical measurement technique and applications. Mater. Sci. Eng. R. Rep. **49**(3), 61–87 (2005)
16. Findeis, D., Gryzagoridis, J.: Determining residual stresses with the aid of optical interference techniques. In: Conference Proceedings of the Society for Experimental Mechanics Series, vol. 4, pp. 277–284 (2013)
17. Peckersky, M.J., Miller, R.F., Vikram, C.S.: Residual stress measurements with laser speckle correlation interferometry and local heat treating. Opt. Eng. **34**(10), 2964–2971 (1995)
18. Viotti, M.R., Albertazzi Jr., A., Kaufmann, G.H.: Measurement of residual stresses using local heating and a radial in-plane speckle interferometer. Opt. Eng. **44**(9), 093606 (2005)
19. Viotti, M.R., Sutério, R., Albertazzi Jr., A., Kaufmann, G.H.: Residual stress measurement using a radial in-plane speckle interferometer and laser annealing: Preliminary results. Opt. Lasers Eng. **42**(1), 71–84 (2004)
20. Barile, C., Casavola, C., Pappalettera, G., Pappalettere, C.: Feasibility of local stress relaxation by laser annealing and X-ray measurement. Strain. **49**(5), 393–398 (2013)
21. Barile, C., Casavola, C., Pappalettera, G., Pappalettere, C.: Preliminary analysis for a new approach to relieve residual stresses by laser heating. In: 11th IMEKO TC15 Youth Symposium on Experimental Solid Mechanics 2012, pp. 77–82 (2012)

22. Casavola C., Pappalettera G., Pappalettere C.: ESPI analysis of thermo-mechanical behavior of electronic components. In: Conference Proceedings of the Society for Experimental Mechanics Series, vol. 3, pp. 321–326 (2017)
23. Barile, C., Casavola, C., Pappalettera, G., Pappalettere, C.: Innovative mechanical characterization of materials by combining ESPI and numerical modelling. Int. J. Mech. **10**, 115–123 (2016)
24. Casavola, C., Pappalettera, G., Pappalettere, C.: Design of a double-illumination ESPI system for the measurement of very slow motions. In: Conference Proceedings of the Sfor Experimental Mechanics Series, vol. 3, pp. 97–102 (2015)
25. Casavola, C., Lamberti, L., Moramarco, V., Pappalettera, G., Pappalettere, C.: Experimental analysis of thermo-mechanical Behavior of electronic components with speckle interferometry. Strain. **49**(6), 497–506 (2013)
26. Barile, C., Casavola, C., Pappalettera, G., Pappalettere, C.: Hybrid characterization of laminated wood with ESPI and optimization methods. In: Conference Proceedings of the Society for Experimental Mechanics Seriers, vol. 3, pp. 75–83 (2013)
27. Barile, C., Casavola, C., Pappalettera, G., Pappalettere, C.: Mechanical characterization of SLM specimens with speckle interferometry and numerical optimization. In: Conference Proceedings of the Society for Experimental Mechanics Series, vol. 6, pp. 837–843 (2011)
28. Casavola, C., Pappalettera, G.: Strain field analysis in electronic components by ESPI: bad thermal contact and damage evaluation. J. Nondestruct. Eval. **37**(1), 11 (2018)
29. Chen, D.J., Chiang, F.P.: Computer-aided speckle interferometry using spectral amplitude fringes. Appl. Opt. **32**(2), 225–236 (1993)
30. Aswendt, P., Höfling, R., Totzauer, W.: Digital speckle pattern interferometry applied to thermal strain measurements of metal-ceramic compounds. Opt. Laser Technol. **22**(4), 278–282 (1990)
31. Nelson, D.V., McCrickerd, J.T.: Residual-stress determination through combined use of holographic interferometry and blind-hole drilling. Exp. Mech. **26**(4), 371–378 (1986)
32. Furgiuele, F.M., Pagnotta, L., Poggialini, A.: Application of holographic interferometry to hole-drilling technique for determining residual stresses. In: Proceeding of the XV National Conference AIAS, Pisa, Italy, pp. 607–620 (1987)
33. Furgiuele, F.M., Pagnotta, L., Poggialini, A.: Measuring residual stresses by hole-drilling and coherent optics techniques: a numerical calibration. J. Eng. Mater. Technol. **113**(1), 41–50 (1991)
34. Asundi, A., Zhang, J.: Industrial applications of residual stress determination using 2-D in-plane sensitive fibre ESPI and hole-drilling. In: Proceedings of SPIE, vol. 3740, pp. 78–81 (1999)
35. Lira, I.H., Vial, C., Robinson, K.: The ESPI measurement of the residual stress distribution in chemically etched cold-rolled metallic sheets. Meas. Sci. Technol. **8**(11), 150–157 (1997)
36. Zhang, J.: Two-dimensional in-plane electronic speckle pattern interferometer and its application to residual stress determination. Opt. Eng. **37**(8), 2402–2409 (1998)
37. Albertazzi, A., Kanda, C., Borges, M.R., Hrebabetzky, F.: A radial in-plane interferometer for ESPI measurement. In: Kujawinska, M., et al. (eds.) Laser Interferometry X: Technique and Analysis Proceedings of SPIE 4101, pp. 77–88 (2002)
38. Suterio R., Albertazzi A.G., Cavaco M.A.M.: Preliminary evaluation: the indentation method combined with a radial interferometer for residual stress measurement. In: Proceedings of the SEM Annual Conference Charlotte (NC) (2003)
39. Steinzig, M., Ponslet, E.: Residual stress measurement using the hole drilling method and laser speckle interferometry: part 1. Exp. Tech. **27**(3), 43–46 (2003)
40. Steinzig, M., Ponslet, E.: Residual stress measurement using the hole drilling method and laser speckle interferometry. Exp. Tech. **27**(4), 17–21 (2003)
41. Ponslet, E., Steinzig, M.: Residual stress measurement using the hole drilling method and laser speckle interferometry part III: analysis technique. Exp. Tech. **27**(5), 45–48 (2003)
42. Steinzig, M., Takahashi, T.: Residual stress measurement using the hole drilling method and laser speckle interferometry part IV: measurement accuracy. Exp. Tech. **27**(6), 59–63 (2003)
43. Casavola, C., Pappalettera, G., Pappalettere, C., Tursi, F.: Analysis of the effects of strain measurement errors on residual stresses measured by incremental hole-drilling method. J. Strain Anal. Eng. Des. **48**(5), 313–320 (2013)
44. http://www.stresstech.com/en-fi/products/espi-hole-drilling-equipment/prism-equipment/
45. Rickert, T.J., Gubbels, W.: ESPI hole-drilling of rings and holes using cylindrical hole analysis. In: Conference Proceedings of the Society for Experimental Mechanics Series, vol. 9, pp. 83–89 (2017)
46. Rickert, T.: Stress measurement repeatability in ESPI hole-drilling. In: Conference Proceedings of the Society for Experimental Mechanics Series, vol. 9, pp. 363–369 (2016)
47. Baldi, A., Jacquot, P.: Residual stressed investigations in composite samples by speckle interferometry and specimen repositioning. In: Proceedings SPIE 4933, Speckle Metrology, pp. 141–148 (2003)
48. Steinzig, M., Upshaw, D., Rasty, J.: Influence of drilling parameters on the accuracy of hole-drilling residual stress measurements. Exp. Mech. **54**(9), 1537–1543 (2014)
49. Barile, C., Casavola, C., Pappalettera, G., Pappalettere, C.: Analysis of the effects of process parameters in residual stress measurements on Titanium plates by HDM/ESPI. Measurement. **48**, 220–227 (2014)
50. Barile, C., Casavola, C., Pappalettera, G., Pappalettere, C.: Residual stress measurements by ESPI-HDM in titanium grade 5: Comparative measurements with different hole diameters. Ciencia e Tecnologia dos Materiais. **27**(2), 79–93 (2015)
51. Barile, C., Casavola, C., Pappalettera, G., Pappalettere, C.: Considerations on the choice of experimental parameters in residual stress measurements by hole-drilling and ESPI. Frattura ed Integrità Strutturale. **30**, 211–219 (2014)
52. Barile, C., Casavola, C., Pappalettera, G., Pappalettere, C., Tursi, F.: Drilling speed effects on accuracy of HD residual stress measurements. In: Conference Proceedings of the Society for Experimental Mechanics Series, vol. 8, pp. 119–125 (2013)
53. Barile, C., Casavola, C., Pappalettera, G., Pappalettere, C.: Consideration on temperature fields and internal radius of analysis in hdm+espi residual stress measurement. In: 13th IMEKO TC15 Youth Symposium on Experimental Solid Mechanics, pp. 11–14 (2014)
54. Barile, C., Casavola, C., Pappalettere, C., Pappalettera, G.: Experimental and numerical characterization of sintered materials with speckle interferometry and optimization methods. In: 10th IMEKO TC12 Youth Symposium on Experimental Solid Mechanics, pp. 35–36 (2012)
55. Barile, C., Casavola, C., Pappalettera, G., Pappalettere, C.: Remarks on residual stress measurement by hole-drilling and electronic speckle pattern interferometry. Sci. World J. **2014**, 1–7 (2014)

56. Barile, C., Casavola, C., Pappalettera, G., Pappalettere, C.: Overview of the effects of process parameters on the accuracy in residual stress measurements by using HD and ESPI. In: Conference Proceedings of the Society for Experimental Mechanics Series—Residual Stress, Thermomechanics & Infrared Imaging, Hybrid Techniques and Inverse Problems, vol. 9, pp.113–118 (2016)
57. Barile, C., Casavola, C., Pappalettera, G., Pappalettere, C.: Residual stress measurement by electronic speckle pattern interferometry; a study of the influence of geometrical parameters. Struct. Integr. Life. **11**(3), 177–182 (2011)
58. Barile, C., Casavola, C., Pappalettera, G., Pappalettere, C.: Residual stress measurement by electronic speckle patterni interferometry: a study of the influence of analysis parameters. Struct. Integr. Life. **12**(3), 159–163 (2012)
59. Baldi, A.: A new analytical approach for hole drilling residual stress analysis by full field method. J. Eng. Mater. Technol. **127**, 165–169 (2005)
60. Stefanescu, D., Truman, C.E., Smith, D.J., Whitehead, P.S.: Improvements in residual stress measurement by the incremental centre hole drilling technique. Exp. Mech. **46**(4), 417–427 (2006)
61. Viotti, M.R., Albertazzi, A.: Compact sensor combining digital speckle pattern interferometry and the hole-drilling technique to measure nonuniform residual stress field. Opt. Eng. **52**(10), 1–8 (2013)
62. Lobanov, L.M., Pivtorak, V.A., Savitsky, V.V., Tkachuk, G.I.: Technology and equipment for determination of residual stresses in welded structures based on the application of electron speckle-interferometry. Mater. Sci. Forum. **768–769**, 166–173 (2014)
63. Kim, K.S., Choi, S.B., Lee, J.H., Park, S.M., Kim, B.I., Lee, N.H., Lee, C.H., Woo, M.: A study on measurement of welding residual stress using ESPI system. Key Eng. Mater. **324-325**, 859–862 (2006)
64. Contuzzi, N., Campanelli, S.L., Casavola, C., Lamberti, L.: Manufacturing and characterization of 18Ni marage 300 lattice components by selective laser melting. Materials. **6**(8), 3451–3468 (2013)
65. Barile, C., Casavola, C., Pappalettere, C., Pappalettera, G.: Experimental and numerical characterization of sinterized materials with speckle interferometry and optimization methods (2011). In: 10th IMEKO TC15 Youth Symposium on Experimental Solid Mechanics, pp. 35–36 (2011)
66. Casavola, C., Cazzato, A., Moramarco, V., Pappalettera, G.: Preliminary study on residual stress in FDM parts. In: Conference Proceedings of the Society for Experimental Mechanics Series, vol. 9, pp.91–96 (2017)
67. Casavola, C., Cazzato, A., Moramarco, V., Pappalettera, G.: Residual stress measurement in fused deposition modelling parts. Polym. Test. **58**, 249–255 (2017)
68. Casavola, C., Cazzato, A., Moramarco, V., Pappalettere, C.: Orthotropic mechanical properties of fused deposition modelling parts described by classical laminate theory. Mater. Des. **90**, 453–458 (2016)
69. Laakkonen, M., Rickert, T., Suominen, L.: Stress measurements in glass and plastic by optical hole-drilling. Mater. Sci. Forum. **768–769**, 95–100 (2014)
70. Montay, G., Sicot, O., Maras, A., Rouhaud, E., Francois, M.: Two dimensions residual stresses analysis through incremental groove machining combined with electronic speckle pattern interferometry. Exp. Mech. **49**, 459–469 (2009)
71. Schajer, G.S., An, Y.: Residual stress determination using cross-slitting and dual-axis ESPI. Exp. Mech. **50**(2), 169–177 (2010)
72. Schajer, G.S., Steinzig, M.: Dual-axis hole-drilling ESPI residual stress measurements. J. Eng. Mater. Tech. Trans. ASME. **132**(1), 0110071–0110075 (2010)
73. Baldi, A.: Combining hole-drilling and ring-core techniques. In: Conference Proceedings of the Society for Experimental Mechanics Series, vol. 9, pp. 105–112 (2017)
74. Schajer G.S., Rickert T.J. Incremental computation technique for residual stress calculations using the integral method. In: Conference Proceedings of the society for Experimental Mechanics Series, vol. 6, pp. 185–191 (2011)
75. Albertazzi, A., Viotti, M.R., Buschinelli, P., Hoffmann, A., Kapp, W., Residual stress measurement and inner geometry inspection of pipelines by optical methods. In: Conference Proceedings of the Society for Experimental Mechanics Series, vol. 8, pp. 1–12 (2011)
76. Albertazzi, A.G., Viotti, M.R., Kapp, W.A.: A robust achromatic dspi interferometer for measurement in polar coordinates. Tech. Mess. **78**(11), 513–519 (2011)
77. Casavola, C., Pappalettere, C., Tursi, F.: Calibration of barkhausen noise for residual stress measurement. In: Conference Proceedings of the Society for Experimental Mechanics Series, vol. 4, pp. 255–266 (2013)
78. De Paula Dias, A.R., Nunes, R.M., De Lima, T.R.S., Clarke, T.G.R.: Evaluation of the residual stress state of 42crmo4 steel sheets in a production line. Mater. Res. **19**(1), 153–157 (2016)
79. Harrington, J.S., Schajer, G.S.: Measurement of structural stresses by hole-drilling and DIC. Exp. Mech. **57**(4), 1–9 (2017)
80. Baldi, A., Bertolino, F.: A low-cost residual stress measuring instrument. In: Conference Proceedings of the Society for Experimental Mechanics Series, vol. 9, pp. 113–119 (2017)
81. Harrington, J., Schajer, G.S.: Measurement of structural stresses by hole-drilling and DIC. In: Conference Proceedings of the Society for Experimental Mechanics, vol. 4, pp. 87–96 (2017)
82. Baldi, A.: Sensitivity analysis of i-DIC approach for residual stress measurement in orthotropic materials. In: Conference Proceedings of the Society for Experimental Mechanics Series, vol. 9, pp. 355–362 (2016)

Chapter 27
Elevated Temperature Optical Microscopy DIC

Kevin B. Connolly and W. Carter Ralph

Abstract When determining the microstructural performance of new materials, such as composites and additively manufactured parts, it is necessary to understand their full-scale behavior. Since non-homogeneous materials are being used more often in critical applications, understanding their microstructural response is becoming more important. Digital Image Correlation (DIC) has proven to be invaluable in characterizing the small-scale performance of non-homogenous materials, but presents a number of challenges for certain tests. One such area is determining the local strain behavior for non-homogeneous micro-scale materials as the part is heated. Using DIC on a sample at elevated temperatures must deal with optical distortions caused by uneven heating of the air above the part. Using DIC at a microstructural level is difficult due to the limited depth of field, which can cause a loss of focus and difficulty in generating an appropriate speckle pattern. To examine the microstructural thermal response both of these difficulties must be overcome simultaneously. Southern Research has developed methods and apparatus to allow optical DIC to be used over a range of temperatures and microscopic fields of view. This paper will address how each of the problems was overcome in the apparatus design and testing method as well as show results validating the technique.

Keywords Strain measurement · Digital image correlation · Thermal strain · Optical microscopy · Surface preparation

27.1 Introduction

Non-homogeneous materials are becoming more common in many different engineering disciplines. These aspects may include manufacturing-induced variations from 3D printing, the material mismatches inherent to microelectronic structures, and the differences between the fiber and matrix for composite materials. The differences in materials make defining the local thermal response difficult. Direct measurement of the displacement within these structures is difficult due to their small scale and temperature-dependent response. This same information can be crucial in performing quality assurance checks, or determining whether the material is appropriate for a desired application.

Interferometry can be used to measure full-field displacement on this scale, but has significant disadvantages to its use. Specimen preparation is labor intensive, requiring careful surface preparation and test setup [1]. The technique requires that one grating be carefully applied to the specimen surface, which is then viewed through another grating, and displacement can only be measured in one direction at a time with a linear grating, and in two directions with a cross-line grating. The grating pattern may be projected onto the specimen surface to reduce surface preparation, but this comes at the cost of lower resolution, greater susceptibility to external vibration, and more restrictive test setup due to the fact that the specimen must be located close to the grating surface. Laser interferometry techniques offer higher resolution with reduced surface preparation, but come at the cost of high sensitivity to environmental vibrations and air fluctuations. All of these techniques require specialized equipment and training.

Digital image correlation (DIC) is a newer technique that offers several distinct advantages. DIC was developed by Michael Sutton and his colleagues at the University of South Carolina in the 1980s [2], and has become widely used in the last ten years due to the advances in digital imaging and processing. Displacement and strain are computed by comparing deformed images to a reference image and measuring the change in location of common features in both images. The software algorithms can locate the position of the features with sub-pixel resolution.

K. B. Connolly (✉) · W. C. Ralph
Southern Research Institute, Birmingham, AL, USA
e-mail: kconnolly@southernresearch.org

L. Lamberti et al. (eds.), *Advancement of Optical Methods & Digital Image Correlation in Experimental Mechanics, Volume 3*, Conference Proceedings of the Society for Experimental Mechanics Series, https://doi.org/10.1007/978-3-319-97481-1_27

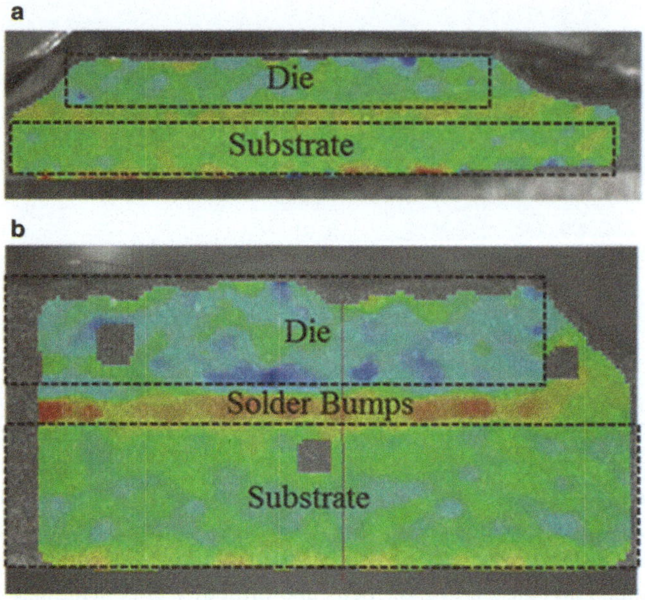

Fig. 27.1 Thermally induced strain measured on a microprocessor at 10 and 4 mm fields of view. (**a**) 10 mm field of view, (**b**) 4 mm field of view

Displacement can be measured simultaneously in two dimensions (and 3 axes) with a single camera, or in three dimensions (and 6 axes) over the entire field of view with two or more cameras positioned at an angle from each other. Surface features are either intrinsic to the material or applied to the surface. Surface preparation is typically less labor-intensive than for interferometry because the surface does not need to be flat, as long as the regions of interest are visible and in focus. The technique is scalable to any size that can be digitally imaged, and has been applied across the meter and nanometer scales.

Application of this technique to the micrometer scale, which is ideal for various 3D printed and composite structures, presents several technical challenges. While depth of focus is not a practical problem for most macro-scale camera lenses or for scanning electron microscopy [3], the depth of focus of typical microscope lenses is very small, so even small out-of-plane motion can result in unusable images. Surface features of the proper size are difficult to apply: paint speckles are too large, microspheres are too tall, fine powders clump due to van der Waals forces, and focused ion beam features are too small. In addition, thermal problems arise: uneven heating of the air causes optical distortions, the necessarily close proximity of the microscope lens to the specimen can cause the lens to be heated to damaging temperatures, and window panes can severely degrade image contrast.

The present study has developed from work that was published in 2013 covering DIC at both high and low temperatures to measure thermally induced strain on electronic assemblies under reflow and temperature cycling conditions [4]. That work has been extended to continually smaller scales until the test apparatus and methods could no longer produce acceptable results. At fields of view less than 10 mm (Fig. 27.1a), higher magnification camera lenses had to be used. This setup produced usable images down to a 4 mm wide field of view, at which scale individual microelectronic features were large enough to distinguish. Figure 27.1b shows a 4 mm wide image of a portion of a microelectronics package overlaid with the strain caused by a temperature profile, and the strain in the first level interconnect can begin to be seen.

The optics of traditional camera lenses broke down at this scale, and an optical microscope was employed. A number of issues immediately surfaced preventing sufficiently good quality images for DIC measurement at this scale: the paint speckling technique resulted in speckle features that were too large for the size of the images, the microscope lens was heated dangerously, the window pane greatly reduced image contrast, and out of plane thermal expansion caused loss of focus. No published studies could be found that addressed these issues.

These technical challenges have been systematically addressed by the authors in a number of ways, which are discussed in this paper. A new apparatus has been designed, and the basic features and their iterations are described; specimen preparation methods for providing high contrast surface features are illustrated; and validation testing up and 170 °C are presented.

27.2 Apparatus

The heating apparatus developed for board-level DIC measurements was hampered by several problems when used with an optical microscope. The board-level setup consists of a modified Bomir batch reflow oven with a 28 × 28 cm double-pane glass window. The oven uses convection heating to simulate factory reflow conditions and has the added benefit of creating turbulent air flow that reduces refractive distortions caused by regions of unevenly-heated air. The convection vents and specimen stage were re-designed to produce more even air flow distribution in both the vertical and horizontal directions across the viewing area of the oven. The two-pane window was re-designed to eliminate refractive distortions caused by air recirculation patterns that developed between the two panes. An external fan was used to blow ambient air across the outer surface of the window and break up pockets of stagnant hot air between the window panes and the cameras. A stereo pair of 5 megapixel digital cameras was placed outside the oven approximately 30 cm (depending on the lenses and the size of the specimen) from the specimen surface to produce the DIC images. Thermocouples were mounted directly to the specimen, often in multiple locations within the field of view. This setup, illustrated schematically in Fig. 27.2, has been proven capable of producing good results on centimeter-scale parts from −55 to 300 °C.

The primary problems with using this setup with an optical microscope were that the specimen quickly went out of focus due to the thermal expansion of the metal oven structure and the specimen, causing the specimen's viewing surface to move towards the lens. This is not a problem for typical camera lenses due to the large depth of focus that is much greater than the thermal expansion of both the oven and the specimen. In addition, the window, even with a single pane, greatly degraded the image quality of the microscope images due to the refraction of light through the glass and the lens, which was also not a problem for magnifications less than or equal to 4 mm field of view. Finally, the small focal length required the microscope lenses to be very close to the heated specimen and resulted in unacceptable heating of the lens, which would have exceeded the manufacturer's specifications.

A new heating apparatus was designed in which the specimen is inverted on a low thermal expansion stage. The microscope is also inverted to accommodate this setup. The stage has a viewing port machined into it and rests on a vented thermal insulator, which in turn rests on a stand set above the microscope. A resistance heater is placed around the specimen on the stage. This setup is illustrated schematically in Fig. 27.3. This design has several specific advantages. The viewing surface does not go out of focus because the thermal expansion of the stage material is negligible and all expansion of the specimen occurs away from the viewing plane. The heat rises away from the viewing plane and the microscope lens, keeping the lens at ambient temperature and producing less refractive distortions in the air between the lens and the viewing plane. Finally, the specimen is resting on the stage, supported only by its weight, and is free to move in all directions in response to its thermal expansion. A thermocouple is adhered to the specimen for temperature readings.

This microscope apparatus design has remained largely unchanged as various quality improvements were made to improve design shortcomings. The apparatus superstructure has been improved to reduce motion of the specimen relative to the camera. The microscope has also been joined to the superstructure so that most of the movement of the specimen and lens occur in sync, further reducing the occurrence of motion blur. Foam insulation under the expansion stage has been replaced by

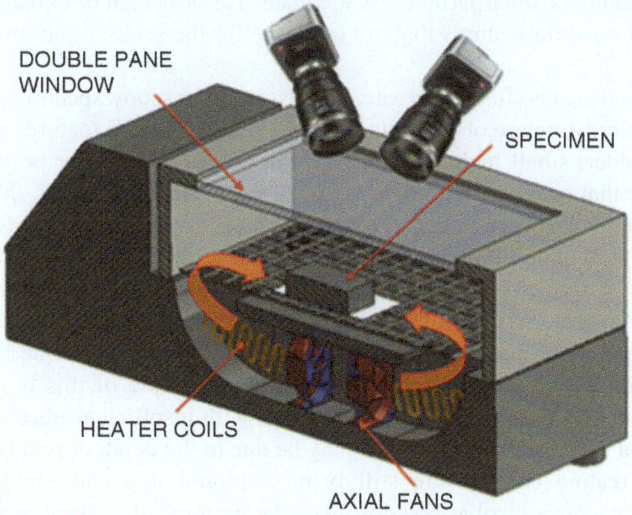

Fig. 27.2 Schematic of board-level DIC apparatus

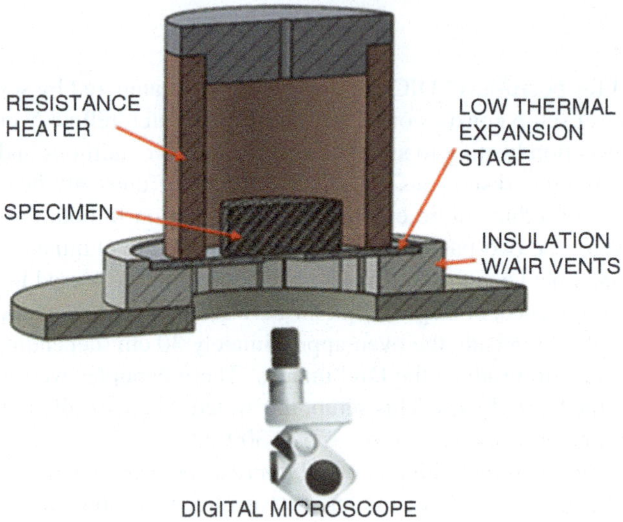

Fig. 27.3 Schematic of prototype microscopy apparatus

ceramic bearings allowing for greater airflow under the stage, further disrupting heat distortions and keeping the microscope lens cooler. A translation stand was added under the insulation layer to allow for precision control of specimen placement. Finally, the lens was replaced with a telecentric lens to improve the depth of field that will remain in focus.

27.3 Specimen Preparation

Preparation of the specimen surface is required to produce good results. DIC requires high-contrast surface features in order to track the displacement in the images. These features need to be a certain number of pixels in size and remain consistently identifiable throughout the test. On rare occasions, the surface to be measured has sufficient contrast that it can be measured as-is. An example of this is shown in Fig. 27.4, in which an as-manufactured second level interconnect copper pad from a microelectronics package is photographed at 1000× magnification and shows high contrast surface features. It is more often the case that the specimen must be prepared in order to add or reveal high contrast features. Typical specimen preparation in centimeter scale specimens is typically done using spray paint to produce a field of randomly shaped and sized speckles, and a variety of techniques can be used to produce acceptable sizes of speckles. Paint droplets are too large for millimeter scale specimens, even when applied with an air brush, and particles can be adhered to the surface instead. For objects that are approximately 3 mm or less in width, powder particles of adequate size (less than or equal to 10 μm) tend to clump together due to van der Waals forces and result in features that are too large for the scale of the image and the depth of focus of the lens.

Two techniques have been used successfully to create contrast in microscopy specimens. The first is to finely polish the specimen surface to reveal the microstructure of the materials. The microstructure features may include the grain or dendritic structures of metals, such as solder; small inclusions or voids in metals, silicon, or polymers; filament ends or edges in composites; or any other feature that produces sufficient contrast. An example of this is shown in Fig. 27.5, where a pair of first-level interconnect solder bumps has been polished to reveal the solder dendrites, voids in the copper traces and silicon, along with features in the polymeric underfill and substrate resin. All of these features produced sufficient contrast for DIC software to track.

At lower magnifications, or on materials with insufficient intrinsic features to produce surface contrast, the specimen surface can be less finely polished to produce surface features. An alternative technique is to finely polish the surface and then "de-polish" the surface with a coarser compound or paper. An example of this is shown in Fig. 27.6, where small scratches produced contrast in regions of the solder microstructure with insufficient intrinsic features. This technique tends to produce higher levels of signal noise in the data, which may be due to the depth or relatively uniform nature of the marks.

Additional techniques for creating contrast are still being explored to create smaller trackable features at higher magnification. Most of these techniques involve powder mixes being applied to the specimen surface. Concern still exist to make sure that contrast material stays adhered to an inverted surface. The inverse of this problem also exists if the surface coating is too thick it will effect strain measurements.

Fig. 27.4 As-manufactured surface features on a second level copper pad

Fig. 27.5 Polished first level solder bumps viewed at 1000× magnification

Fig. 27.6 Polished and "de-polished" first level solder bump viewed at 1000× magnification

27.4 Results

Tests using the inverted microscope apparatus have been performed over a range of temperatures up to 180 °C, over a range of magnifications, and on a variety of microelectronics structures and homogeneous metal plates.

Higher magnification tests are more difficult due to the fact that the depth of focus decreases as magnification increases. When focus is lost, re-focusing cannot correct the problem because this changes the contrast of many of the features. The loss of focus is thought to be the result of bending, warpage, and/or uneven out-of-plane thermal expansion of the specimen. This effect was mitigated by reducing vibrations, improving the level of the apparatus, and replacing the foam insulator with ceramic bearings.

For purposes of illustration of the capability, two tests are presented below. The first test is of a second level solder ball at 200× magnification up to 170 °C. The second test is a calibration using an aluminum specimen at similar magnification to illustrate the validation of an improved system.

The second level interconnect test was performed on a FCBGA package mounted to a 1.6 mm thick board. The ball pitch is approximately 1.2 mm and the solder balls are approximately 1 mm wide. The part was cross-sectioned near the die and polished down to a 1 μm diamond paste finish. The solder ball that was imaged was the second from the last at the outer edge, and is shown in Fig. 27.7 at 200× magnification, producing an image that is approximately 1.4 mm wide. The specimen was photographed every 20 °C from 30 to 170 °C. The response of the specimen was essentially uniform expansion, as shown in Fig. 27.8, with significant rigid body motion in the in-plane direction due to the expansion of the package (which was removed from the data using a correction), but negligible shearing across the solder joint by the substrate and board. The unit thermal expansion across (vertically, as shown in the figure) the solder ball was measured as a function of temperature, and the data are plotted in Fig. 27.9, showing general agreement with typical values. However, these results proved difficult to reproduce leading to the need for a validation test of the apparatus.

Validation specimens were made from aluminum, copper, and silicon. Of these materials the aluminum produced the best contrast when polished, this is shown in Fig. 27.10. Testing with the aluminum calibration specimen continued to show high amounts of variation from test to test, in some cases resulting in twice the thermal expansion than is correct for this material. After examining a number of possible causes for this error it was determined that the digital camera in the microscope was producing images that resulted in strain calculation errors. At this point a new microscope was purchased with the proper type of camera to ensure that the exposure time, gain, and white balance could be precisely controlled and would not change over the course of the test. A telecentric lens was also introduced to reduce focus loss issues.

Controlling the camera settings and light intensity allowed for repeatable calibrations runs using the improved apparatus. These calibration runs reported a uniform thermal expansion of approximately 22.5 $\mu\varepsilon$/°C, 5% lower than the expected 23.6 $\mu\varepsilon$/°C for 6061 aluminum. The temperature versus thermal stain plot is shown in Fig. 27.11. The response also appeared to differ on the heat up versus the cool down portion of the test. Thermal examination of the specimen using multiple thermocouples revealed a discrepancy between the temperature of the specimen top, where the temperature was being measured during testing, and the bottom of the specimen, which is imaged during the test. This discrepancy in temperature

Fig. 27.7 Pre-test second level interconnect specimen at 200× magnification

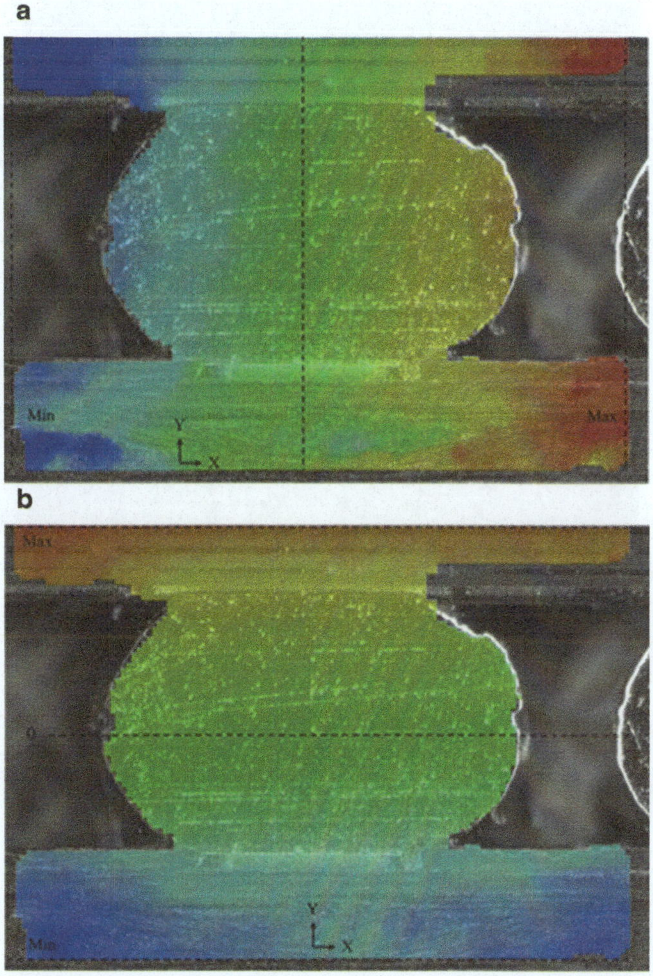

Fig. 27.8 Relative displacement of second level interconnect specimen at 170 °C. (**a**) X displacement, (**b**) Y displacement

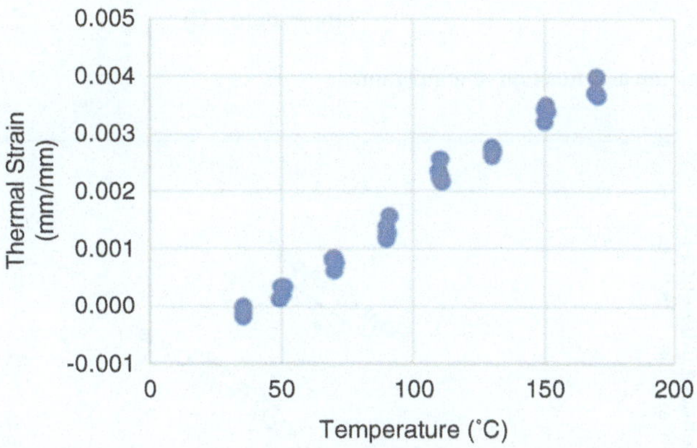

Fig. 27.9 Thermal strain of second level interconnect as a function of temperature

appears to be dependent on the temperature of the apparatus and whether the measurement was taken during a heat up or cool down cycle, this calibration is shown in Fig. 27.12. Using the temperature calibration to adjust the thermal expansion curves produced the expected 23.6 $\mu\varepsilon$/°C thermal expansion for the aluminum sample.

Fig. 27.10 As-polished surface features on an aluminum calibration sample

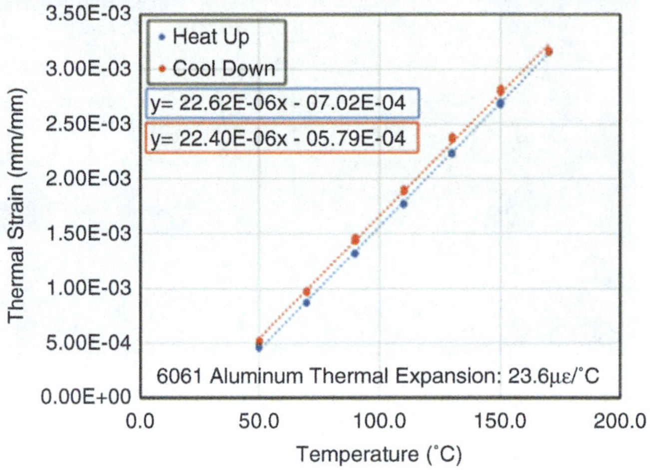

Fig. 27.11 Thermal strain of the aluminum as a function of temperature

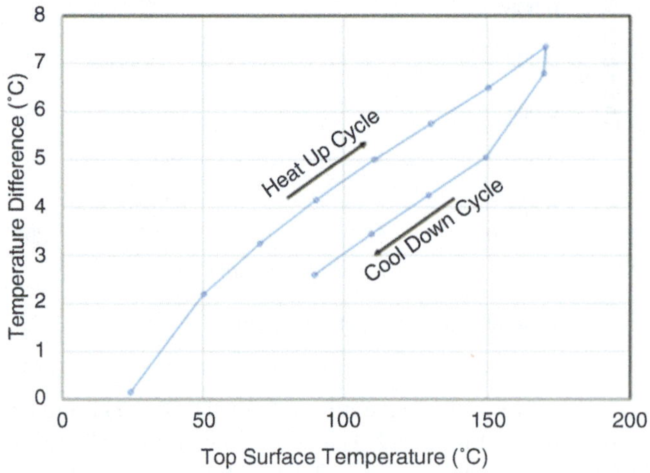

Fig. 27.12 Temperature difference between top surface and view surface

27.5 Conclusions

The test apparatus has undergone a number of changes to improve test stability and produce more accurate strain measurements. The current apparatus and specimen preparation method has been has been validated by measurement of the thermal expansion of a uniform material that compares well to known values. This setup operates between room temperature and 170 °C at approximately 200× magnification. It would likely continue to work well at 1000× magnification if the contrast features could be sized appropriately.

Additional specimen preparation techniques would greatly improve the types of material that can be examined with this system. Until a universal speckling procedure can be developed for these size scale additional material will be polished to find those with naturally occurring contrast features. There is hope to identify 3D printed metals with the appropriate microstructure to allow for DIC measurements.

Improvements will continue to be made on this system to aid test repeatability. As additional specimens are identified, different techniques will be examined to determine the temperature calibration necessary to correct the thermal expansion measurements.

The results produced by this testing continue to improve and validation techniques are beginning to take form. Results from additional material should improve the confidence of this testing method. Even with the current limitations this approach is ready to provide useful data regarding relative strain levels within a non-homogeneous material.

References

1. Chu, T.C., Ranson, W.F., Sutton, M.A., Peters, W.H.: Applications of digital image correlation techniques to experimental mechanics. Exp. Mech. **25**(3), 232–244 (1985)
2. Kammers, A.D., Daly, S.: Digital image correlation under scanning electron microscopy: methodology and validation. Exp. Mech. **53**(9), 1743–1761 (2013)
3. Ralph, W.C., Raiser, G.F.: Assembly level digital image correlation under reflow and thermal cycling conditions. In: Proceedings of the 63rd Electronic Components & Technology Conference, Las Vegas, NV, pp. 2223–2227, 28–31 May 2013

Chapter 28
A Digital Laser Speckle Technique for Generating Slope Contours of Bent Plate

Austin Giordano and Fu-Pen Chiang

Abstract A digital method is proposed whereby spatial speckles created when an optically rough surface is illuminated by a diffuse coherent laser beam are used to generate slope contour fringes. This is done by photographing the speckles contained in a parallel plane in front of the plate before and after deformation. The resulting photographs are then Fast Fourier Transformed twice with the digital image correlation software CASI to produce slope contours.

Keywords Laser speckle · Digital speckle photography · Digital image correlation · Stress analysis · Plate bending

28.1 Introduction

When a coherent laser beam impinges upon an optically rough surface of a specimen the reflected wavelets from every point of the surface mutually interfere into a volumetric speckle pattern. These speckles move when the surface of the specimen move and they tilt when the surface of the specimen tilts in much the same way as a mirror tilts an impinging light beam. These speckles can be used to measure the deformation of the specimen. If a camera is focused at the surface of the specimen the speckles act as if they are in-plane gages; but when the camera focuses at a plane away from the surface of the specimen, the movement of the speckles can be used to measure the slope of the surface. A detailed description of using laser speckle for deformation measurement is given by Chiang [1]. Laser speckle has been used for metrology and experimental mechanics for some time, but most of the work has been done before digitization. In this paper, we propose a digital technique for slope measurement. Defocused speckles have been used by Juang and Chiang [2] to generate slope contours and by Keene and Chiang [3] to measure vibrational modes. Many experimental methods for solving plate bending problems are methods of deflection measurement [4, 5]; however, in the case of thin plate bending the more advantageous approach is to generate the slope contour thus eliminating the need to differentiate twice. This would reduce the errors associated with numerical differentiation.

In this paper, a whole-field technique of slope measurement is described in which the specimen surface can be of almost any texture. This technique is based on the usage of speckles that illuminate a surface, which is created by a beam of coherent light from a laser that is passed through a microscope objective and a pinhole aperture. With this method, no surface preparation is necessary if sufficient light is reflected back from the surface. Otherwise, a thin coating of white paint sprayed onto the surface is all that is needed.

28.2 Experimental Procedures

The system used for the data acquisition is as shown in Fig. 28.1a. The specimen is illuminated by an expanded laser beam, and the image of the object is captured by a digital camera. The light intensity of each speckle pattern is digitized into an array of 6000 × 4000 pixels. The image processing software ImageJ [6] is then used to convert the JPG image files into 8-bit TIFF files. The 8-bit TIFF files are then loaded into an image correlation software based on Fourier transform [7] hereby referred as CASI (Computer Aided Speckle Photography), and processed. Basic processes of the technique involve

A. Giordano (✉) · F.-P. Chiang
Department of Mechanical Engineering, State University of New York at Stony Brook, Stony Brook, NY, USA
e-mail: austin.giordano@stonybrook.edu

L. Lamberti et al. (eds.), *Advancement of Optical Methods & Digital Image Correlation in Experimental Mechanics, Volume 3*,
Conference Proceedings of the Society for Experimental Mechanics Series, https://doi.org/10.1007/978-3-319-97481-1_28

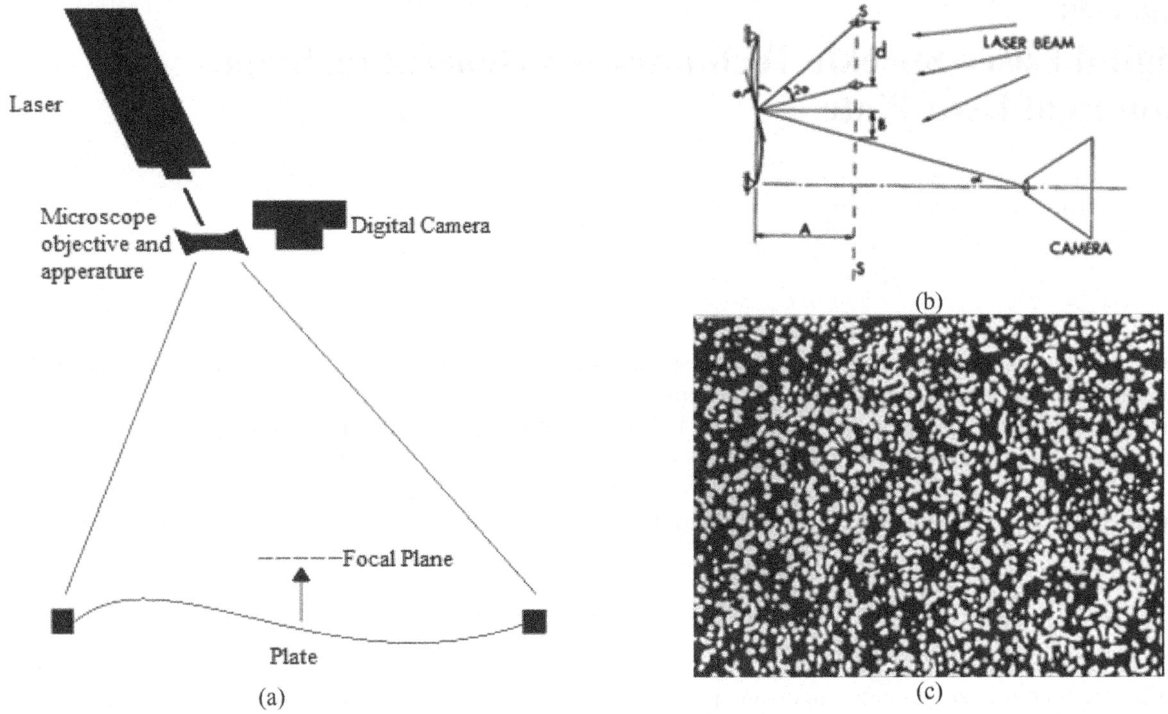

Fig. 28.1 (**a**) Schematic of the data-acquisition system. (**b**) Arrangement for recording spatial speckles for plate bending. (**c**) Sample of black and white speckles

data acquisition and image processing. In the data acquisition stage, two speckle patterns of the specimen, one before and one after the deformation, are captured by the camera and stored on an SD card, which is then registered into the computer. The image processing stage consists of four steps with ImageJ and CASI. First, the JPG image files are converted into 8-bit TIFF files which are then input into CASI. The three steps of the process that CASI is necessary for was described by Chen and Chiang [7]. First, an equivalent double-exposure speckle pattern is obtained by superimposing the two digital images, and it is segmented into a series of small subimages. Second, a fast-Fourier transform (FFT) is applied to each subimage and "compared" with a correlation calculation. Last, a second FFT is applied to the resulting value giving rise to an impulse function whose position vector is nothing but the displacement vector experienced collectively by all the speckles contained within the subimage. in the subimage. Repeating the last two steps to all other subimages, we can resolve the full-field 2-D displacements. Upon completion of this process, the slope fields are generated.

28.3 Principle of the Method

Fringes are formed when coherent waves interfere. When the number of point sources increases the interference pattern takes the form of randomly distributed light and dark spots of varying size. These light and dark spots are speckles. The speckles in space will move when the specimen surface is moved. Under the classical assumptions of the thin plate theory, the stress field can be determined once the surface curvature is obtained. The present technique produces slope contours from which curvature can be calculated. By selecting an arbitrary plane in space parallel to the surface of the plate the speckles contained in the plane will move according to the local surface tilt. The displacement of the speckles d can be related to the surface slope ϕ. From Fig. 28.1b it can be shown that [2]

$$\mathrm{d} = \frac{\mathrm{A}\left[\tan\left(2\phi+\alpha\right)-\tan\left(\alpha\right)\right]}{\frac{\tan 2\phi\left(1+\tan^2\alpha\right)}{1-\tan\alpha\,\tan 2\phi}}. \tag{28.1}$$

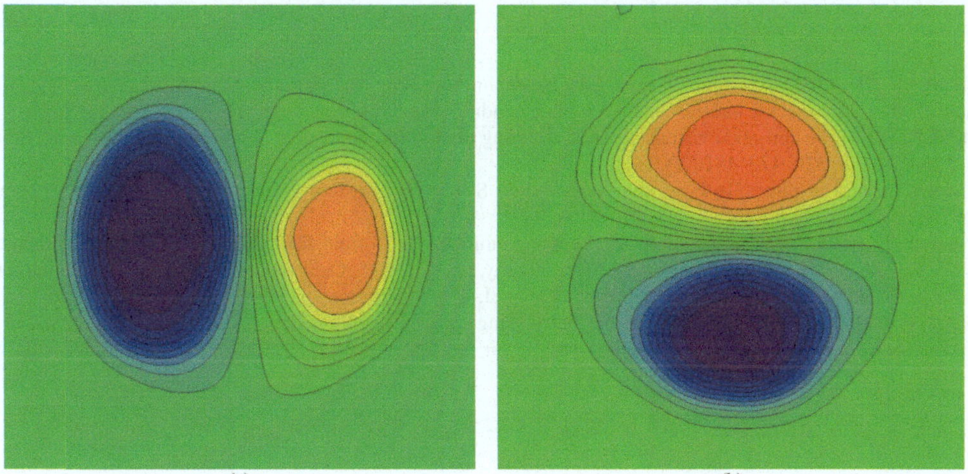

Fig. 28.2 Fringe patterns of partial slope contours of a clamped circular polycarbonate plate under concentrated central load (a) ∂w/∂x-field (b) ∂w/∂y-field

From Eq. (28.1) we can obtain the slope contours as follows depending on which direction we select to display the result:

$$\frac{\partial w}{\partial x} = \frac{\partial w}{\partial y} = \phi = \frac{\mathrm{d}}{2\mathrm{A}}. \tag{28.2}$$

28.4 Results

In this experiment, we tested the deformation of a circular disk under a central transverse load using the proposed laser-speckle method. The plate was made of polycarbonate (diameter = 10.16 cm, thickness = 6.35 mm) and loaded by a central deflection of 12.7 μm. A 15 mW expanded He-Ne laser beam ($\lambda = 0.6328$ μm) was used to illuminate the object from 2.2 m away. The digital camera was situated 2.1 m away from the plate and focused on a parallel plane 17.5 cm in front of the specimen surface. Figure 28.2 shows a set of slope fringes obtained by processing a before and after loading image with CASI. The slope fringes displayed in Fig. 28.2 are in good agreement with previous experiments.

28.5 Conclusion and Discussion

A digital method has been proposed to use laser speckle generate slope contours of a bent plate. By focusing the digital camera on a parallel plane in front of the plate and capturing images before and after loading the plate. The resulting photographs are then Fast Fourier Transformed twice with the digital image correlation software CASI to produces the slope contours. The digital method is analogous to Ligtenberg's [8] reflection moiré method in which a grating and a model plate with a mirrored surface are used. Indeed, the proposed method can be called an equivalent reflection moiré method with a grating of continuously variable pitch and orientation. Unlike the reflection moiré method which can only be applied to mirrored specimens, the speckle method can be applied to almost any material. An added benefit of having this technique being digital is that the results provide a direction as well as a magnitude to the contours produced by CASI. Experience has shown that a thin coat of white paint usually improves a great deal the quality of the fringe pattern.

Acknowledgement We gratefully acknowledge the support of Dr. Rajapakse, and the financial support of the Office of Naval Research through award NO.-N00014-17-1-2873.

References

1. Chiang, F.P.: A family of 2D and 3D experimental stress analysis technique using laser speckle. Solid Mech. Arch. **3**(1), 1–32 (1978)
2. Chiang, F.P., Juang, R.M.: Laser speckle interferometry for plate bending problems. Appl. Opt. **15**, 2199–2204 (1976)
3. Keene, L., Chiang, F.P.: Real-time anti-node visualization of vibrating distributed systems in noisy environments using defocused laser speckle contrast analysis. J. Sound Vib. **320**, 472–481 (2009)
4. Theocaris, P.S.: Moire method in plates. In Proc. Int. Assoc. Shell Struct. Symposium, Warsaw (North Holland, Amsterdam), pp. 877–889 (1963)
5. Havanesian, J., Varner, J.: Method for determining the bending moments in normal loaded thin plates by holographic interferometry. In: Robertson, E.R., Harvey, J.M. (eds.) Engineering Uses of Holography, pp. 173–185. Cambridge University Press, London (1970)
6. Schneider, C.A., Rasband, W.S., Eliceiri, K.W.: NIH Image to ImageJ: 25 years of image analysis. Nat. Methods. **9**, 671–675 (2012)
7. Chen, D.J., Chiang, F.P.: Computer-aided speckle interferometry using spectral amplitude fringes. Appl. Opt. **32**, 225–236 (1993)
8. Ligtenberg, F.K.: The moiré method: a new experimental method for the determination of moments in small slab models. SESA Proc. **12**(2), 83–98 (1955)

Chapter 29
Deflectometry on Curved Surfaces

Y. Surrel and F. Pierron

Abstract This paper presents an application of deflectometry to measure the deformation of a thin cylindrical shell in bending. The principle of the calibration method is briefly outlined. Then, the experimental set-up is presented, followed by comparison of slopes, deflections and curvatures with results from a finite element model of the test. The results are satisfactory though limited by the quality of the reflective surface of the test samples. Future work will focus on more complex shapes to extend the technique.

Keywords Deflectometry · Full-field measurements · Slopes · Bending · Curvatures

29.1 Introduction

Deflectometry is a full-field optical measurement technique where the shape of an object is deduced from the image of a regular grid distorted by the reflection on this object [1]. Deflectometry is mainly used for shape measurement of naturally specularly reflective objects [1, 2] but a small community in experimental mechanics has been using this technique, starting with the seminal paper by Ligtenberg [3] and followed by other authors over the years [4–8], sometime under a different name [9]. However, all these applications were on flat surfaces as the calibration process for arbitrary curves surfaces is much more onerous. Sciamarella et al. [10] performed measurements on a curved panel but in a point-wise manner rather than full-field. The objective of this paper is to show feasibility for the measurement of slopes and curvatures on a simple cylindrical shell under bending and compare the results against a finite element simulation. This is an effort to revive this somewhat underrated technique that has a lot of potential for applications such as vibrations where deformations are very small and techniques like digital image correlation lack sensitivity, while at the same time providing a rather inexpensive alternative to interferometry.

29.2 Objectives and Methodology

As in many other fringe-based techniques (interferometry, fringe projection etc.), the standard technique is nowadays to use phase-stepping to make the measurements [11]. The set-up is shown schematically in Fig. 29.1. It consists of a display screen, a specularly reflective object and a camera. The x and y unwrapped phase fields obtained for two directions of lines (vertical and horizontal) on the display screen are directly proportional to the coordinates of the screen pixels seen by the camera sensor pixels, so in the sequel we refer to the acquired fields as 'coordinate fields' instead of 'phase fields'. Deflectometry is known to have an excellent sensitivity to slope changes, and thus is well adapted for bending tests on reflective surfaces. However, obtaining the slopes from the acquired screen coordinate fields requires to know the shape of the surface, as the

Y. Surrel
Engineering and the Environment, University of Southampton, Southampton, UK

Yves Surrel Expertise and Consultancy, St-Étienne, France

F. Pierron (✉)
Engineering and the Environment, University of Southampton, Southampton, UK
e-mail: F.Pierron@soton.ac.uk

L. Lamberti et al. (eds.), *Advancement of Optical Methods & Digital Image Correlation in Experimental Mechanics, Volume 3*, Conference Proceedings of the Society for Experimental Mechanics Series, https://doi.org/10.1007/978-3-319-97481-1_29

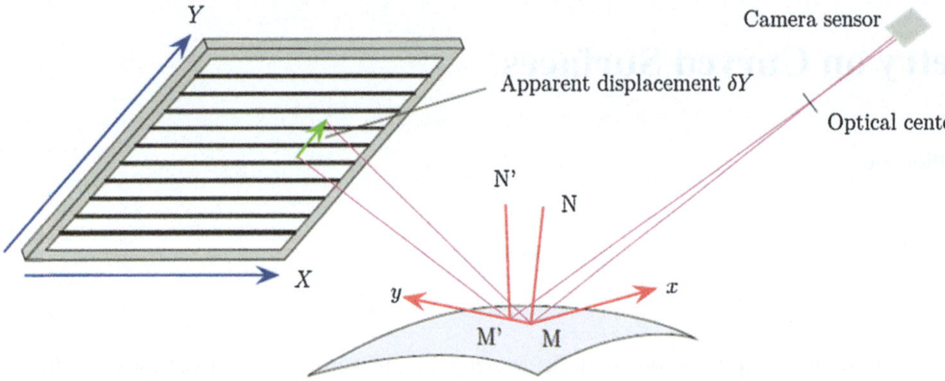

Fig. 29.1 Schematic of a deflectometry setup; two rays corresponding to adjacent sensor pixels are displayed

ray tracing from the sensor to the display screen requires the three coordinates in space of the object point where each ray impinges (e.g. M and M′ in Fig. 29.1).

The methodology we followed to obtain the object shape and thus the object slopes was:

- Identification of the camera intrinsic parameters, using different oblique views of the same plane cross grid of pitch $p = 0.67$ mm, following the method indicated in [12], based on a pin-hole camera model. The perspective distortion characteristics of the images allow for the distance from the projection centre to the sensor and the position of the optical axis in the sensor plane to be determined.
- Identification of the screen position in the camera frame of reference: coordinate fields of the screen observed by reflection on a plane mirror placed at different arbitrary orientations are acquired. The key idea in this procedure is the fact that the geometrical transform from one screen image to the other is a pure rotation (as a product of two plane symmetries), the axis and angle of which can be determined. As the rotation axis is the edge of the wedge resulting from the two positions of the flat mirror, it provides the location in space of one line belonging to both mirror planes. With three different mirror orientations (1, 2, 3), three couples of images (1, 2), (2, 3) and (1, 3) can be processed, giving two lines for each mirror orientation, allowing the mirror positions in space to be fully determined. Once these positions are obtained, the screen position can be derived by reflecting the corresponding screen images. Due to noise and imperfections, the three identified mirror positions will provide a slightly different screen position, so some sort of optimized averaging has to be performed. Of course, redundancy can be increased with additional mirror orientations.
- Determination of the object shape by a phase-locked iterative method: starting from the assumption of a plane object, the screen phase fields are reconstructed. From the discrepancy between the acquired and reconstructed fields, an approximate new shape can be calculated, and new screen coordinate fields can be reconstructed. This process is iterated until the shape stabilizes.

The tested object was a rectangular ($w = 50$ mm, $h = 78.4$ mm) piece of aluminium, of thickness $e = 0.7$ mm, formed by rolling to a cylindrical shape of curvature $R = 105$ mm. It was pinned at three corners, the fourth one remaining free. The load was applied from the rear (concave surface) at the centre with a simple screw driven system equipped with a load cell. The camera was an Allied Vision Manta G504 with 2452 × 2054 pixels 3.45 μm in size, paired with a Nikkor 50 mm lens. Views of the experimental set-up are provided in Fig. 29.2.

29.3 Results

The two slope components are provided in Figs. 29.3 and 29.4 and compared to the results from a finite element model of the test. The tangential and axial directions are provided in Fig. 29.2 t(t and z axes respectively). The model assumed a Young's modulus of 70 GPa and a Poisson's ratio of 0.3 and used the same dimensions as the test specimen. ANSYS SHELL181 shell elements were used with a mesh of 100 × 100 elements. The boundary conditions were pinned nodes at all corners except the bottom right one, and a 12 N load was applied to the centre of the shell. Figure 29.5 shows the out of plane deflection while Figs. 29.6 and 29.7 show the curvatures. The comparison is rather good even though the difference maps clearly show a low spatial frequency bias. The origin of this is believed to be because the shell was not a perfect cylinder. Deviations

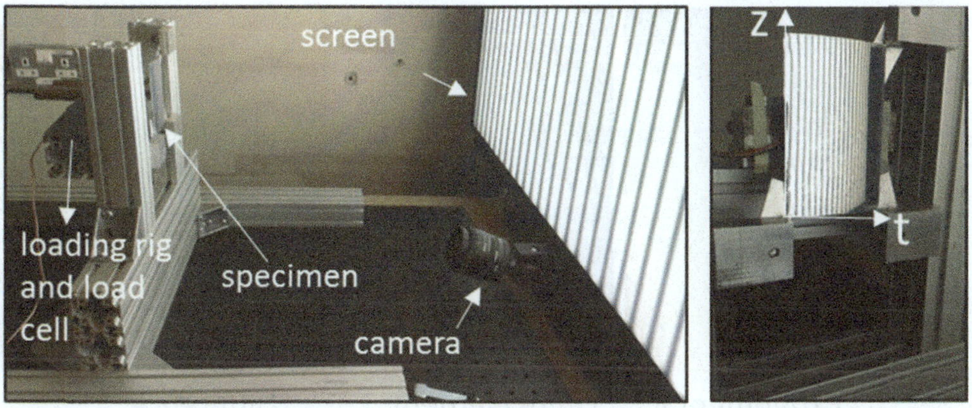

Fig. 29.2 Mechanical setup and cylindrical test specimen

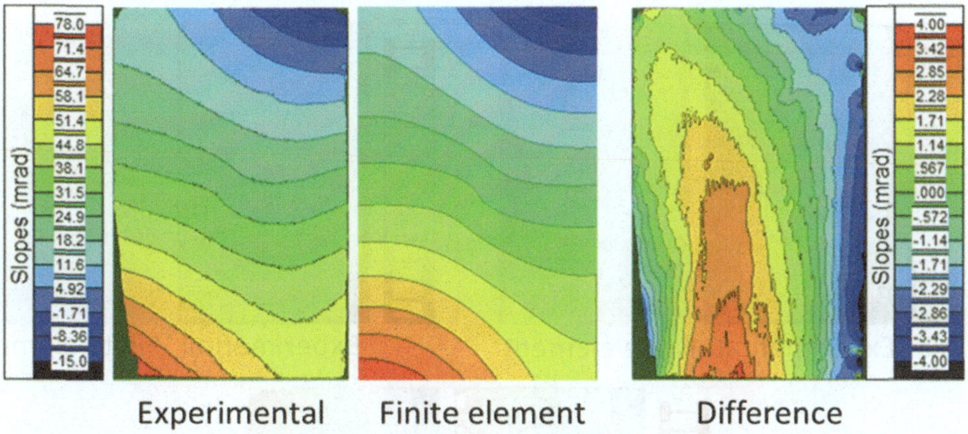

Fig. 29.3 Comparison between experimental and simulated $\frac{1}{R}\frac{dw}{d\theta}$ slopes

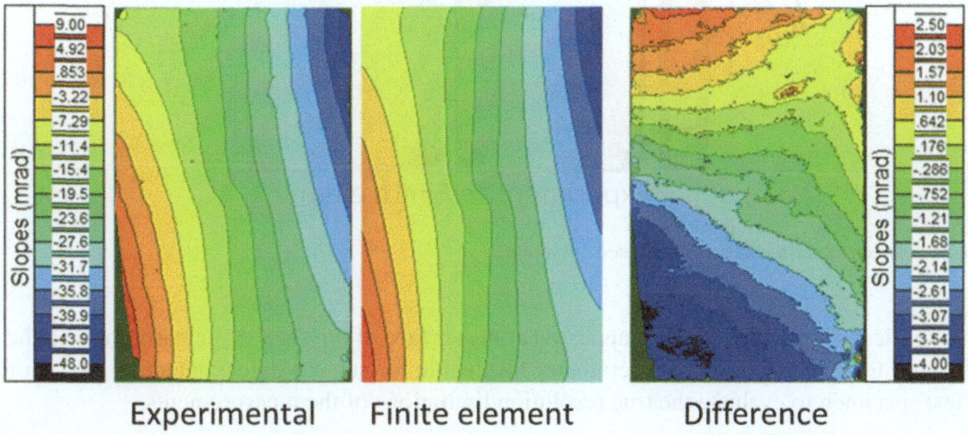

Fig. 29.4 Comparison between experimental and simulated $\frac{dw}{dz}$ slopes

of approximately 300 μm were recorded when the shape was measured. Moreover, a second likely source of discrepancy is the fact that Poisson's ratio was fixed at 0.3 in the model while from the literature, it is likely to be somewhat higher, 0.33. Finally, real boundary conditions are always difficult to reproduce in a model and some slippage may have occurs at the pinned corners. Overall, provided these, the comparison is acceptable. Future work will focus on performing sensitivity analyses to investigate the sources of discrepancy. It should also be noticed that the curvature maps are quite noisy. The reason for a large part of this is the poor quality of the specimen surface. As the specimen was bent from a flat front face

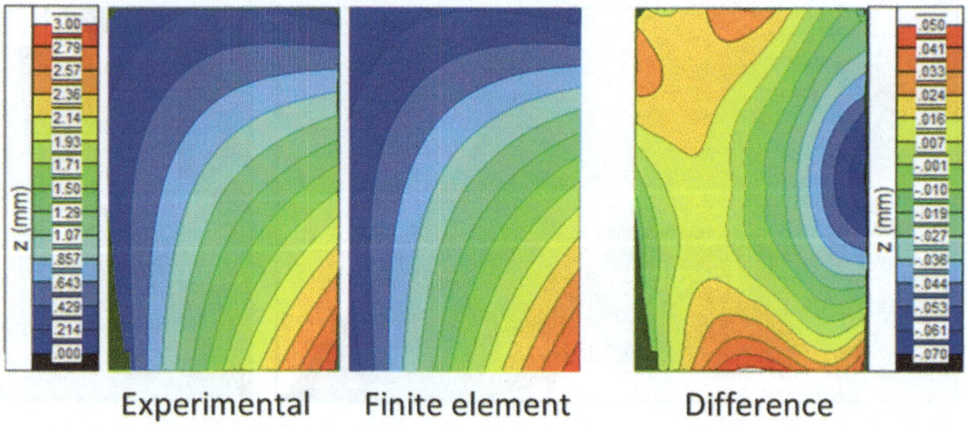

Fig. 29.5 Comparison between experimental and simulated out-of-plane displacements

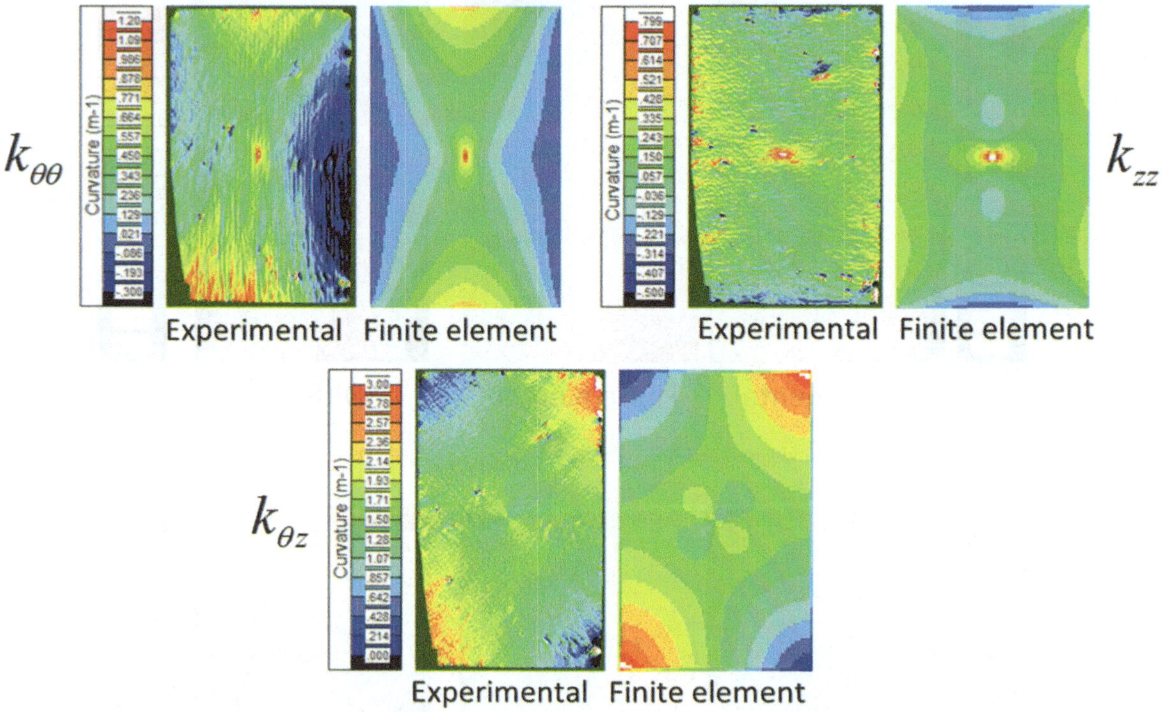

Fig. 29.6 Comparison between experimental and simulated curvatures

mirror to a near-cylindrical shape, some creases appeared at the surface of the shell that clearly show in the derivatives of the slopes. This is the very feature that makes deflectometry invaluable for surface defect detection [13]. In the future, we will look for a better test specimen to evaluate the true resolution limitations of the measurements.

29.4 Conclusion and Future Work

This is work under progress but the perspective are promising. The main difficulty encountered here was the rather imperfect quality of the cylindrical shell, as it was obtained from a flat aluminium front-face mirror and the deformation caused creases that polluted the curvature maps. Future work will concentrate on extending the experiment to more complex surfaces, but the main goal for the near future is to explore applications to dynamics, either vibrations or, more challenging, transient dynamics. In this case, spatial phase shifting is required and therefore, the grid will have to be 'reversed' so that the grid image is regular. Finally, application of the Virtual Fields Method will be explored for thin curved shells either to identify material properties [14] or to identify pressure distributions [15] in aero-acoustics.

Acknowledgments The authors gratefully acknowledge funding from EPSRC, grant EP/L026910/1.

References

1. Zhang, Z., et al.: Three-dimensional shape measurements of specular objects using phase-measuring deflectometry. Sensors. **17**(12), 2835 (2017)
2. Petz, M., Ritter, R.: Reflection grating method for 3D measurement of reflecting surfaces. In: Proceedings of SPIE—The International Society for Optical Engineering (2001)
3. Ligtenberg, F.K.: A new experimental method for the determination of moments in small slab models. In: SESA. (ed.) Proceedings of SESA XII, pp. 83–98 (1954)
4. Ritter, R.: Reflection moiré methods for plate bending studies. Opt. Eng. **21**(4), 663–671 (1982)
5. Kao, T.Y., Chiang, F.P.: Family of grating techniques of slope and curvature measurements for static and dynamic flexure of plates. Opt. Eng. **21**(4), 214721–214721 (1982)
6. Surrel, Y., et al.: Phase-stepped deflectometry applied to shape measurement of bent plates. Exp. Mech. **39**(1), 66–70 (1999)
7. Devivier, C., Pierron, F., Wisnom, M.: Damage detection in composite materials using full-field slope measurements. Compos. Part A Appl. Sci. Manuf. **43**(10), 1650–1666 (2012)
8. Devivier, C., et al., Time-resolved full-field imaging of ultrasonic Lamb waves using deflectometry. Exp. Mech., 2016. 56(3): p. 345–357
9. Periasamy, C., Tippur, H.V.: A full-field reflection-mode digital gradient sensing method for measuring orthogonal slopes and curvatures of thin structures. Meas. Sci. Technol. **24**(2), (2013)
10. Sciammarella, C.A., Trentadue, B., Sciammarella, F.M.: Measurement of bending stresses in shells of arbitrary shape using the reflection moire method. Exp. Mech. **40**(3), 282–288 (2000)
11. Bruning, J.H.: Fringe scanning interferometers. In: Optical Shop Testing. Wiley, Hoboken, NJ (1978)
12. Surrel, Y.: Phase-locked optical metrology I: identification of intrinsic camera parameters from multiple grid views. Appl. Opt. **57**(1), A39–A49 (2018)
13. Kammel, S., Leon, F.P.: Deflectometric measurement of specular surfaces. In: 2005 IEEE Instrumentationand Measurement Technology Conference Proceedings (2005)
14. Syed-Muhammad, K., et al.: Characterization of composite plates using the virtual fields method with optimized loading conditions. Compos. Struct. **85**(1), 70–82 (2008)
15. Berry, A., Robin, O., Pierron, F.: Identification of dynamic loading on a bending plate using the Virtual Fields Method. J. Sound Vib. **333**(26), 7151–7164 (2014)

Chapter 30
Measurement on a Sample of Fuel Cell at High Temperature

Ning Li, NanSheng Xu, Michael A. Sutton, and Kevin Huang

Abstract Solid Oxide Fuel Cells (SOFCs) can be designed in either a planar or non-planar configuration. Since modern fuel cell assemblies are constructed from several different materials and operate at relatively high temperatures, 550–750 °C, the experimental characterization of the in-situ response is of interest for model validation. This study employs stereo-vision and stereo-DIC to measure both full-field surface strains and the out-of-plane displacement on the surface of a heterogeneous cell during operation.

To obtain in-situ images of fuel cell materials, several technical issues were successfully addressed to perform stereo-imaging of a fuel cell at elevated temperature. These included (a) modifications of the lens system to develop a far-field microscope arrangement, acquiring images of a small region imaging with a large stand-off; (b) application of a speckle pattern capable of withstanding elevated temperatures; (c) construction of the fuel cell to allow imaging through an open hole of the furnace; (d) development of a camera support system to ensure stability for both calibration and experiment imaging and (e) ensuring continuous air circulation around the furnace and camera-lens system.

Since full field deformation are obtained using the measurement method, viewing the spatial variations in strain that occur during heating is necessary to identify conditions which induce localization in the deformations, including increases in out-of-plane motion in specific regions of the fuel cell at selected temperatures. The strain measurements are also used to determine the average thermal coefficients of expansion. Taken together, the measurements provide baseline data to assist material developers and fuel cell manufacturers.

Keywords High temperature · Strain measurement · Stereo-DIC · Speckle pattern · SOFCs

30.1 Motivation

Solid Oxide Fuel Cells (SOFCs) are in development rapidly because they have many benefits to energy industry, not only working in higher efficiency quietly but also producing much less pollution to environment. They are made entirely of solid materials and can be designed other than a flat plane configuration. Since these modern fuel cell assemblies are constructed from several different materials and operate at relatively high temperatures, 550–750 °C, the experimental characterization of the response of a fuel cell is needed to assess performance. The enclosed effort employs a non-contacting method using stereo-vision imaging that is known as three-dimensional digital image correlation (3D-DIC) to measure both the full-field strains and the out-of-plane displacements on the surface of a heterogeneous cell during operation.

30.2 Specimen

As shown in Fig. 30.1a, the specimen used in this study is the top cell in a fuel cell stack. The outer cell construction consists of three layers: (a) 20 mm diameter, 0.5 mm thick copper sheet the edges appear golden in Fig. 30.1a as the initial layer that acts as anode; (b) a thin layer of electrolyte (brown) is adhered to the top of the copper anode; (c) a final thin layer (black) is adhered to the electrolyte and will be coated as cathode.

N. Li (✉) · N. Xu · M. A. Sutton · K. Huang
Department of Mechanical Engineering, University of South Carolina, Columbia, SC, USA
e-mail: lini@cec.sc.edu

L. Lamberti et al. (eds.), *Advancement of Optical Methods & Digital Image Correlation in Experimental Mechanics, Volume 3*,
Conference Proceedings of the Society for Experimental Mechanics Series, https://doi.org/10.1007/978-3-319-97481-1_30

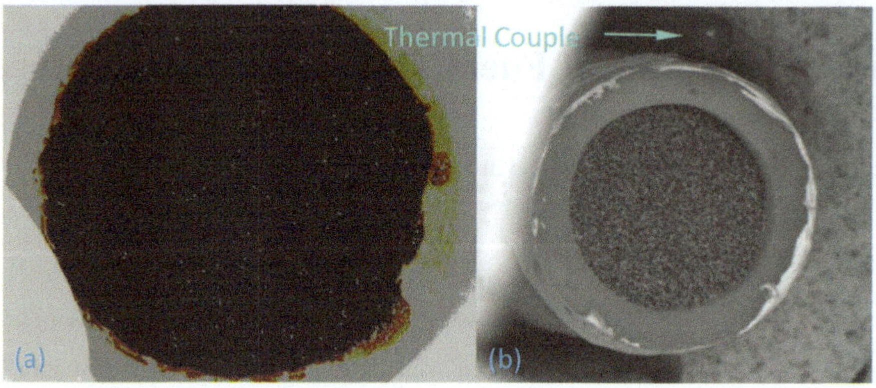

Fig. 30.1 (**a**) Image of outer layer of the tri-layer cell; (**b**) image of the patterned outer layer of the cell and the surrounding ceramic tube structure to which the outer edge of the cell is lightly adhered

Since the 3D-DIC method requires a random pattern with high image contrast, as discussed in [1], the black outer surface is sprayed with white flameproof paint to obtain a black-and-white random pattern known as a speckle pattern. The pattern is shown in Fig. 30.1b. The average size of each white dot is approximately 0.3 mm and the contrast in the image is very good, providing confidence that the measured deformations will be accurate. A separate thermocouple is bonded to the outer surface of the ceramic tube structure.

The specimen shown in Fig. 30.1a is lightly adhered to the surrounding ceramic tube and the tube-cell system is placed in the furnace. Given the construction of the tri-layer cell and the weak adherence provided, it is hypothesized that the specimen will expand "freely" during the heating process. In this regard, it is noted that several heating-cooling experiments were performed with the tube-specimen assembly between 25 and 850 °C without any obvious defects.

Prior to performing the heating-cooling process, it is noted that the furnace is placed horizontally (not vertically) on a rigid platform, where the platform is nominally isolated from large scale vibrations. The horizontal orientation is used primarily to simplify the use of a stereo-vision system for imaging. The horizontal orientation did cause some issues, of which the most important was the ability to reach high temperatures. Because heat is generated near the middle part of the furnace, and the specimen is located (a) away from the furnace center and (b) a few inches from an open end so that the specimen could be illuminated and viewed by the stereo system, the thermocouple temperature consistently lower than expected. To overcome this problem, the investigators made the following changes to the setup:

- The other end of the ceramic tube is "partially sealed" using heat resistant cotton so that only a small amount of air can flow through;
- A fan is placed outside the open end and placed on low speed to achieve relatively constant flow of heated air from the center of the tube to the specimen, improving heating of the specimen and mixing the air in the region around the end of the tube to minimize refraction effects on imaging [2–4].

30.3 Imaging System

To obtain stereo images of a specimen and use these images to obtain the complete surface strain field and the 3D motions of the specimen, a complete system such as shown in Fig. 30.2 is required. As shown in Fig. 30.2, two cameras rotated relative to each other are mounted to a rigid bar and are oriented to view the same region on a specimen of interest. The cameras are connected to a control and image storage device (typically a PC). The PC controller ensures that synchronized images of the specimen are recorded simultaneously and stored digitally.

In the experiments, two 9 megapixel cameras with relatively long focus lenses are oriented as shown in Fig. 30.2 and used to acquire stereo images of the specimen; long lenses are necessary due to the requirement for both high magnification and large stand-off distance between camera and specimen. Since the angle between cameras must be large enough to reliably measure the out of plane displacement of the specimen during heating, the angle between the cameras is approximately 10° in these studies.

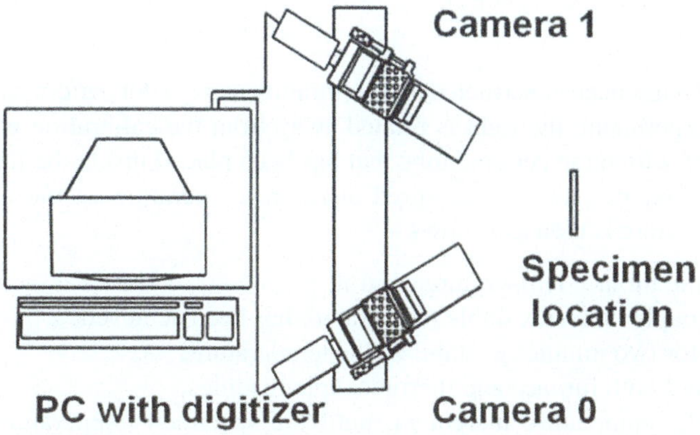

Fig. 30.2 Schematic of a typical stereo-vision system for image acquisition and storage

Fig. 30.3 Experimental setup includes (**a**) both cameras with lenses mounted to a rigid bar that is attached to a tripod; (**b**) Light to illuminate specimen; (**c**) furnace with temperature controller; (**d**) calibration target that is used for initial calibration of stereo-vision system; (**e**) computer for data acquisition control and monitor for viewing specimen images during experiment

The actual experimental setup used in this study is shown in Fig. 30.3. For stability during the experiment, each camera-lenses combination is placed in a specially designed aluminum mount. Each camera mount is firmly attached to a rigid bar and the bar is mounted to a sturdy tripod. Prior to making measurements during heating, the stereo-vision system must be calibrated. The procedures are described in detail in a recent book written by one of the investigators [1]. In our studies, the image acquisition software VIC-Snap [5] is installed on the computer and used to control the acquisition of images via a USB3 port connection. To complete calibration of the stereo-vision system shown in Fig. 30.3, a dot pattern target with the proper size was manufactured by printing a suitable design and bonding the design to a rigid plate. Following established procedures [1], images of the calibration grid were acquired in several orientations. All calibration images were analyzed with VIC-3D software [5] to obtain the complete set of calibration parameters for the stereo-vision system.

30.4 Experiment

Once the setup shown in Fig. 30.3 has been constructed and calibration using a dot pattern has been completed, the experiment is performed. To perform the experiment, the light is rotated away from the calibration grid and oriented to illuminate the fuel cell specimen that is located within the ceramic tube that has been placed inside the furnace.

Then, with the fan running, image pairs are acquired at room temperature as the reference images for later image correlation processing. The procedure is then as follows:

1. set a higher temperature on the furnace temperature control;
2. heat specimen until thermocouple indicates stable temperature has been achieved;
3. hold furnace at temperature for two minutes to stabilize the temperature;
4. take a pair of image and record both furnace and thermocouple readings;
5. repeat steps 1–4 with 50 °C temperature increment, until the specimen temperature reaches the desired maximum temperature of 750 °C.

30.5 Results

After processing 3D image pairs using the VIC-3D software [5], displacement and strain fields are obtained. By computing all the strains throughout the region being viewed, and averaging the strains obtained at each point, Fig. 30.4 shows the average of each strain component and a measure of the variability in the strain on the surface of the specimen at each temperature. As shown in Fig. 30.4, the mean values in the two normal strains shows the expected increase in positive strain with increasing temperature. If the mean value is representative of the strain field, then an estimate for the thermal coefficients of expansion in the x and y directions are $\alpha_x = 10$ με/°C and $\alpha_y = 8.7$ με/°C, respectively.

Since the measured variability in the normal strains is on the order of 0.002, which is much less than the measured values for ε_{xx} and ε_{yy} at higher temperatures, indicating that the trends are nominally reasonable. Finally, the shear strain is essentially zero everywhere, indicating that the thermal deformations are nominally isotropic, as would be expected.

Since full field deformation are obtained using the measurement method, viewing the spatial variations in strain that occur during heating is necessary to determine whether there is some form of localization in the deformations that is occurring. Figure 30.5 shows how the out-of-plane displacement field changes as the temperature changes from 562°C to 593°C to 627 °C, respectively. As shown in Fig. 30.5, around 600 °C the specimen has a slight upward warpage of 40 μm along a vertical centerline and a downward motion of nearly –50 μm near the left and right edges. These trends are not present at the other two temperatures.

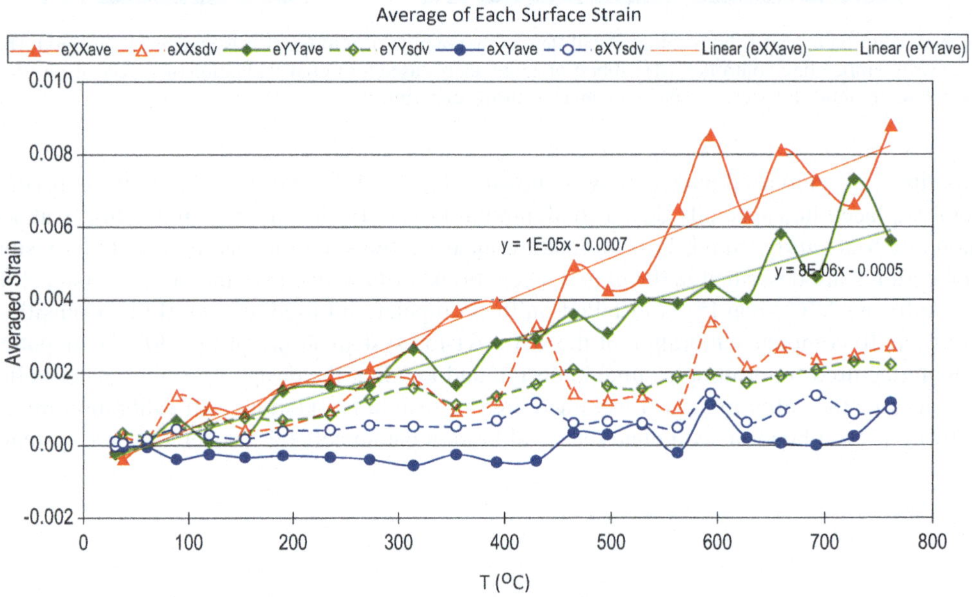

Fig. 30.4 Average of each surface strains ε_{xx}, ε_{yy} and ε_{xy} over the entire specimen surface

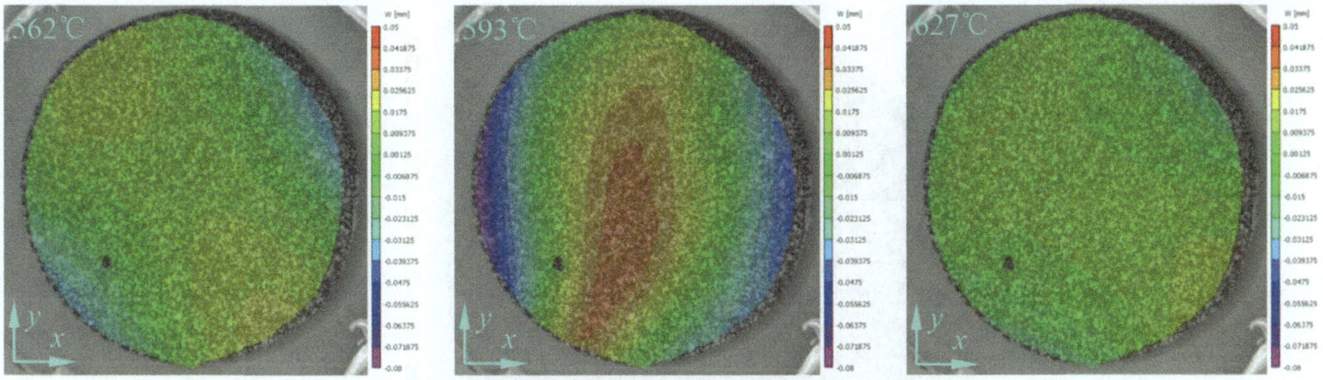

Fig. 30.5 Out of plane displacement, $w(x, y)$, at three temperatures. All units in mm

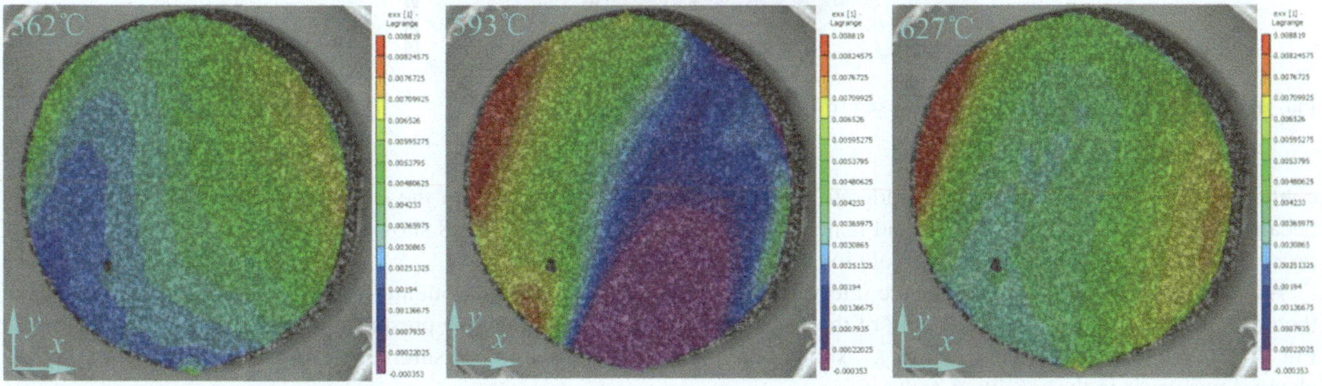

Fig. 30.6 Strain field, $\varepsilon_{xx}(x, y)$, at three temperatures

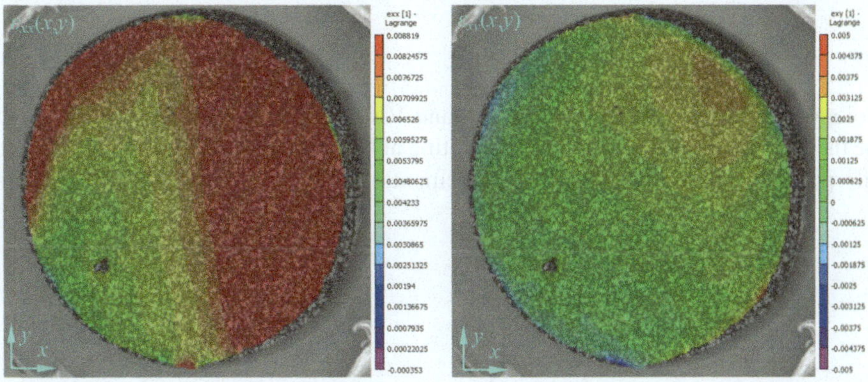

Fig. 30.7 Strain fields $\varepsilon_{xx}(x, y)$ and $\varepsilon_{xy}(x, y)$ at 761 °C

Figure 30.6 shows the horizontal strain ε_{xx} at 562 °C, 593 °C and 627 °C, respectively. As was seen for the displacement $w(x, y)$, at 593 °C the expansion strains change rapidly near the centerline where the largest $w(x, y)$ occurred, with compressive strains near zero on the right side and tensile surface strains of 8500 $\mu\varepsilon$ near the left edge. Such large variations indicate that there is some form of change in either the materials or the structure occurring at this temperature.

Figure 30.7 shows the strain fields ε_{xx} and ε_{xy} at the highest testing temperature, 761 °C. At this temperature, ε_{xx} should have reached its maximum value. Again, there is a relatively large gradient occurring as one crosses the vertical centerline, with large tensile strains on the right this time and much lower strains on the left. The shear strain ε_{xy} is generally much smaller than the normal strains, though it is somewhat elevated to nearly 3500 $\mu\varepsilon$ in a small region.

For those that are not well-versed in the robustness of the stereo-vision methodology used in this study, the presence of unusual structural response in Figs. 30.5 and 30.6 could be viewed as due to "experimental difficulties". To provide confirmation of the measurement quality at this temperature, the investigators used their previous work and developed a

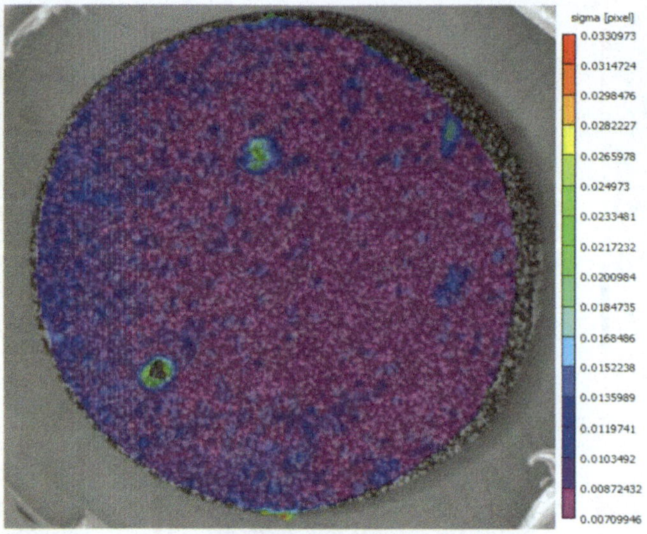

Fig. 30.8 Confidence coefficient for images at 593 °C

confidence COEFFICIENT metric to indicate when the data has small errors in the images and when it may be suspect because there are large image errors (and hence not reliable). When the Fig. 30.8 shows the confidence coefficient of DIC processing at temperature 593 °C when obvious buckling appears.

As shown in Fig. 30.8, the confidence interval is very small throughout the domain, with most image position errors on the order of 0.008–0.012 pixels (1/100th of the camera sensor element). Thus, the presence of gradients in $w(x, y)$ shown in Fig. 30.5 and the sharp changes in strain ε_{xx} shown in Fig. 30.6 are not due to measurement errors, but rather due to physical changes occurring in the specimen or structure.

30.6 Summary

The 3D-DIC methodology using stereo-vision for imaging and VIC-3D software for image analysis has been successfully applied to measure the thermal strain on fuel cell while operating at high temperature. Results indicate that the measurements were successful, while providing unique full field measurements for a model fuel cell material subjected to operational conditions.

References

1. Sutton, M.A., Orteu, J.J., Schreier, H.W.: Image correlation for shape, motion and deformation measurements. Springer, Berlin (2009). ISBN 978-0-387-78746-6
2. Lyons, J.S., Liu, J., Sutton, M.A.: High temperature deformation measurements using digital image correlation. Exp. Mech. **36**(1), 64–71 (1996)
3. Liu, J., Sutton, M.A., Lyons, J.S.: Experimental characterization of crack tip deformations in alloy 718 at high temperatures. ASME J. Eng. Mater. Tech. **20**(1), 71–78 (1998)
4. Liu, J., Sutton, M.A., Lyons, J.S., Deng, X.: Experimental investigation of near crack tip deformations in alloy 800 at 650°C. Int. J. Fract. **91**, 233–268 (1998)
5. Correlated Solutions, Incorporated; 121 Dutchman Boulevard, Irmo, SC 29063 and on the web at www.correlatedsolutions.com

Chapter 31
Simulation of 3D Reconstruction of Conical Calibration Targets

Wei-Chung Wang, Chi-Hung Hwang, and Yung-Hsiang Chen

Abstract In this study, new calibration targets based on conical 3D shapes are proposed for possible replacing the popularly used planer calibration target which has been always implement for determine the camera parameters of camera-pairs. To understand the performance of the propose calibration targets. Three different conical calibrations targets (CCTs), hemispherical, cylindrical and the conical calibration targets, were considered for briefly simulating before manufacturing those targets for future application. In this study, the spatial data associated with three different CCTs were generated and mapping to the image planes of cameras located at three different locations of a planer circle by using pin-hole imaging model. Then the generated images were used to determine the camera parameters. The determined parameters of different camera-pairs were used to reconstruct the spatial data of the generated CCTs. All reconstructed spatial data were then merged into a single data set, the departures of the reconstructed spatial data with respect to the generated ones were considered as the index for evaluating performance of the CCTs. In this study, the simulated results showed the departure of the spatial data associated to all three CCTs are small which indicates that all three proposed CCTs can be used for calibrating the camera-pairs.

Keywords Conic calibration targets · Camera calibration · Reconstruction · Simulation · Stereo-camera-pair

31.1 Introduction

Digital image correlation (DIC) method is first proposed to determine the deformation of an object by comparing two scanned supersonic images of different statuses in 1982 [1]; afterwards, camera system replaced the supersonics scanning system for generating digital images for displacement/deformation measurements. The method has been significantly improved by implementing sub-pixel registration for high accurate full-field deformation measurement. 2D-DIC method was then well developed to extract the deformation filed of objects from a series of images [1, 2] and works well for determining in-plane deformation of a flat object. To extend the deformation measurement DIC method, based on photogrammetry, 3D-DIC method was proposed to determine both in-plane and out-of-plane displacements [3]. Different from 2D DIC method, the imaging system of a 3D DIC system is consist of two cameras and always need to be calibrated to determine the intrinsic and extrinsic parameters which are the essential parameters for determining the relative depths of a set object characteristic data in space. Typically, the calibration result is summarized in the form of the camera projection matrix by combining the intrinsic camera parameters and extrinsic orientation parameters. The parameters obtained through the calibration process are extensively used to adjust the depth value determined by disparity of the stereo-camera-pair and the accurate distances.

By adopting the calibration procedures of photogrammetry, 3D DIC calibration is also commonly determined by a planar calibration target planar calibration target with black-white grid or circular dot patterns on one surface. In general calibration procedures, the planar calibration target is asked to rotate and move to at least 9 different positions to reach all possible positions of the testing object would move to during the measurement. However, in practice, the planar calibration targets are always rotated and moved randomly at least 30–40 times to avoid possible image overlap to ensure the images can be independent which can be used for determining the parameters by iteration. Therefore, to achieve better calibration

W.-C. Wang
Department of Power Mechanical Engineering, National Tsing Hua University, Hsinchu, Taiwan, Republic of China

C.-H. Hwang (✉) · Y.-H. Chen
ITRC, NARL, Hsinchu, Taiwan, Republic of China
e-mail: chhwang@itrc.narl.org.tw

L. Lamberti et al. (eds.), *Advancement of Optical Methods & Digital Image Correlation in Experimental Mechanics, Volume 3*, Conference Proceedings of the Society for Experimental Mechanics Series, https://doi.org/10.1007/978-3-319-97481-1_31

results, a skillful person is required for processing the calibration procedure, and the calibration procedures always cost time. In this paper, conical calibration targets were proposed as possible alternative solutions for the determination of camera parameters. To evaluate the performance of proposed conical calibration targets, three types of conical calibration targets were numerically generated. They are conical, cylindrical and hemispherical calibration objects. In this study, three cameras are assumed to be posited along a circular to provide different view angles. The generated surface points on the surfaces of three different calibration targets were mapping to image planes of three cameras with pin-hole camera model was adopted. Then the intrinsic and extrinsic camera parameters were determined by object-camera geometrical relations. Again, based on the determined camera parameters, the calibration target were reconstructed by re-mapping all three image data sets of three cameras to the object space. By comparing the difference of the overlapped area of the reconstructed calibration targets, the performances of three calibration targets can be evaluated for future real applications.

31.2 Calibration Targets and Simulations

To perform reliable calibration of a stereo-camera-pair, the calibration targets need to be able to provide enough knowing distances in space which can always be done by rotating and moving a planer calibration target randomly. Dots of equal space on the planer calibration target can generate different spatial data for calibration. Although varied distance can be successfully generated, the rotation angles and the relative moving distance need to be determined for camera parameter determination. As shown in Fig. 31.1, there are three proposed CCTs [4, 5] all of them are axisymmetric and the associated data point on the object surface can be described by either cylindrical or spherical coordinate system, the projection distance of equal-dot-distance along the circumference varies with respect to symmetrical axis of the CCTs and the view-angle of the camera, if the camera line-of-sight is well aligned with the symmetrical axis of CCTs. Then the distance change would be a function of camera view-angle and the angles defined by line-of-sight and conical symmetrical axis. Because of the special geometrical characteristics of CCTs, the intrinsic and extrinsic camera parameters might can be determined directly by a camera. To verify the idea, the mapped image and the corresponding spatial data are adopted for determining camera parameters. In this study, a conical of 45° top angle, a semi-circular calibration object, and a cylindrical calibration object are adopted for discussion. The intrinsic and extrinsic camera parameters were then determined by three selected conical targets with assuming that three cameras are located at three pre-defined positions constrained by a semi-circular by this way three different camera-pairs with three different viewing angles and two baselines can be generated; for discussion convenience, the cameras were named from left to right as C1, C2 and C3; and the camera-pairs were named with corresponding cameras as C_iC_j. Based on the geometrical conditions given previously, camera parameters and CCTs spatial data pointed are generated for verification. The simulations are performed by following steps:

1. Generating the spatial data sets corresponding to three conical calibration targets;
2. Mapping the generated spatial data set onto three cameras image planes by pin-hole camera model; in this study, two adjacent cameras are placed with 45° as shown in Fig. 31.2;
3. Calculating the camera parameters by using the data of image planes;
4. Calculating the difference of camera parameters between the generated and calculated ones;
5. Reconstructing the spatial data points of different camera-pairs;
6. Combining all spatial data sets determined by three camera-pairs into single object;
7. Checking the total departure of the integrated spatial data set departures from the generated ones.

31.3 Results

Figure 31.2 is the geometric layout for simulation study which is based on a semi-circular frame used for multi-camera DIC system [6]. The geometrical position of the cameras are defined to be R_i: $[\theta_x, \theta_y, \theta_z]$ in degree and T_i: $[x, y, z]$ in mm, where $i = 1, 2, 3$, all the angles and distance are defined with respect to the center of the semi-circular frame. In this study, the cameras were located at 500 mm away from the CCTs and then the layouts of three cameras can be defined by R_i and T_i as the following;

Camera 1, R_1: [0°, 45°, 0°], T_1: [−354, −354, 0];
Camera 2, R_2: [0°, 0°, 0°], T_2: [0, −500, 0];
Camera 3, R_3: [0°, −45°, 0°], T_3: [354, −354, 0].

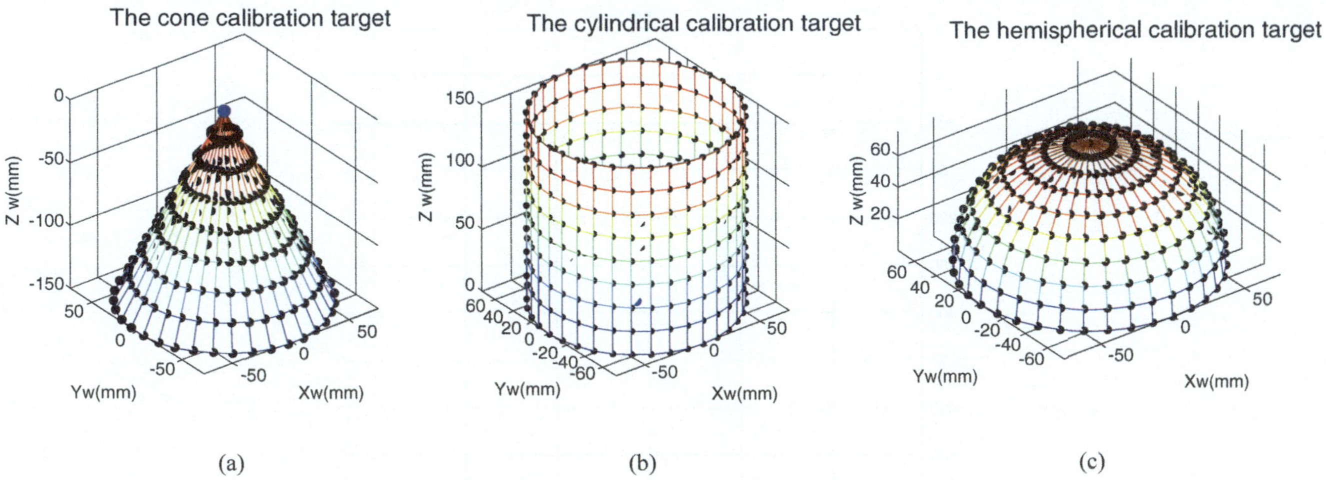

Fig. 31.1 Three types of conic calibration targets. (**a**) Conical, (**b**) cylindrical, (**c**) hemispherical

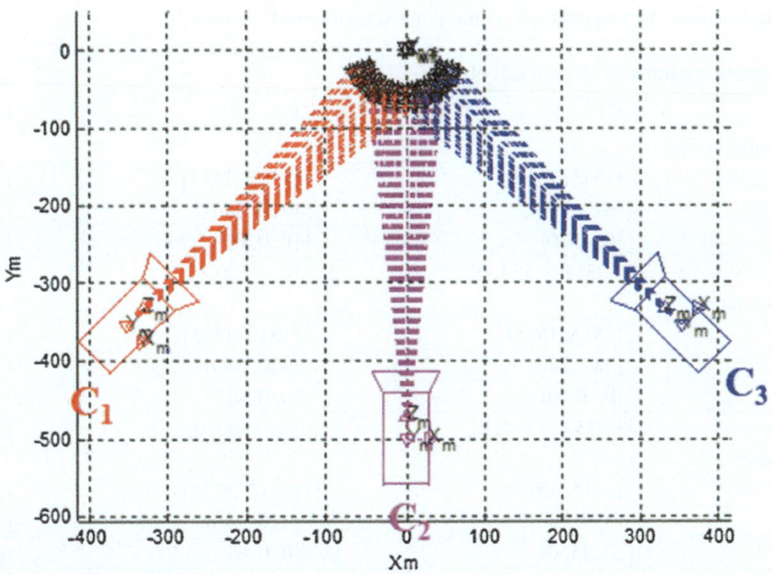

Fig. 31.2 The geometrical layout for simulation

The spatial data of marked points on the surface of CCTs, as shown in Fig. 31.1, were mapped into image planes of each camera by using pin-hole model, a typical image is shown in Fig. 31.3. In principle, the distance change should be symmetric with respect to the intersection of line-of-sight and the symmetrical axis of the CCTs. Then, by relating 2D image coordinates (u, v) with the 3D conic calibration targets (X, Y, Z) and camera calibration matrix can be determined. The results of calibrated camera parameters by using conical objects are shown as Table 31.1, obviously, the results are almost identical to the input ones. Figure 31.4a–c show the reconstructed hemispherical calibration target spatial data sets for different camera-pairs by using the calibrated data. The reconstructed 3D data sets of Fig. 31.4a, c which were determined by camera-pair C1C2 and C2C3 are the same but rotating the spatial data 90° with respect to the symmetrical axis of the hemispherical calibration target. However, as for Fig. 31.4b, the data set determined by C1C3 is related few that because the overlapped filed of views are related narrow than C1C2 and C2C3 cases. However, considering the determined camera parameters are almost the same as the data implemented for simulation, therefore, it is no surprise that the reconstructed spatial data set are also well matched with the input ones. The determined spatial data sets were then stitch together into a single spatial data set, the result is shown in Fig. 31.4d, where the data from different camera-pairs were labeled with different color, again, the data were matched well, no discontinuous or gap can be seen from the resulted data.

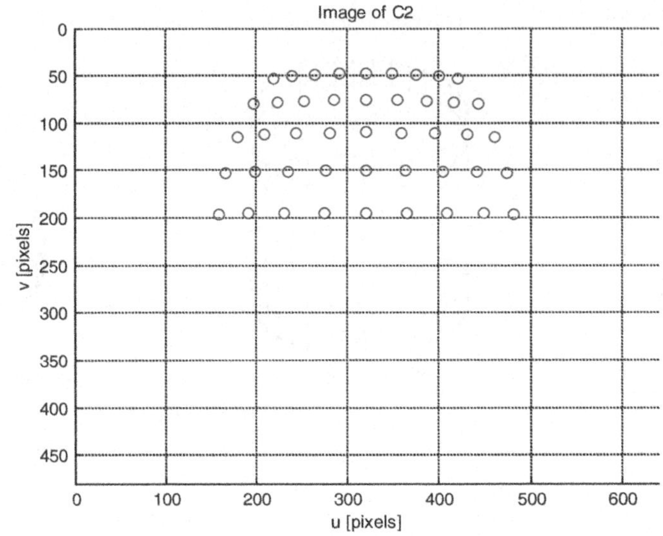

Fig. 31.3 A typical image data generated by mapping the spatial point set with pin-hole model

Table 31.1 The simulated calibration parameters by conical objects

Parameters	C1	C2	C3
(a) The hemispherical calibration target			
Focal length (F) pixels	[1515, 1515]	[1515, 1515]	[1515, 1515]
Principal point (C) pixels	[320, 240]	[320, 240]	[320, 240]
Rotation vector (R) °	[0, 45, 0]	[0, 0, 0]	[0, −45, 0]
Translation vector (T): mm	[−354, −354, 0]	[0, −500, 0]	[354, −354, 0]
(b) The cylindrical calibration target			
Focal length (F) pixels	[1515, 1515]	[1515, 1515]	[1515, 1515]
Principal point (C) pixels	[320, 240]	[320, 240]	[320, 240]
Rotation vector (R) °	[0, 45, 0]	[0, 0, 0]	[0, −45, 0]
Translation vector (T): mm	[−354, −354, 0]	[0, −500, 0]	[354, −354, 0]
(c) The conical calibration target			
Focal length (F) pixels	[1515, 1515]	[1515, 1515]	[1515, 1515]
Principal point (C) pixels	[320, 240]	[320, 240]	[320, 240]
Rotation vector (R) °	[0, 45, 0]	[0, 0, 0]	[0, −45, 0]
Translation vector (T): mm	[−354, −354, 0]	[0, −500, 0]	[354, −354, 0]

31.4 Conclusions

According to the simulated results, three proposed CCTs for determining intrinsic and extrinsic calibration parameters of camera for stereo-vision application such as 3D DIC can work well. By adopting new proposed conical calibration target, the drawbacks and limitations of traditional calibration targets can be removed because the proposed conical calibration target possess characteristics of axisymmetric and can provide different mapping distance onto the image plane with respect to different viewing angles in a single shot; obviously, the conical calibration target, in principle, can provide various geometrical data directly instead of rotating or moving the calibration targets. This simulation study shows the proposed hemispherical, cylindrical and conical targets can be used to determine the camera parameters well, and the results has been evaluated by reconstructing the spatial data sets of CCTs with the determined camera parameters together with the images.

Acknowledgements The partial financial support provided by the Office of Naval Research (ONR), United States (Award Number: N62909-16-1-2064) is greatly appreciated.

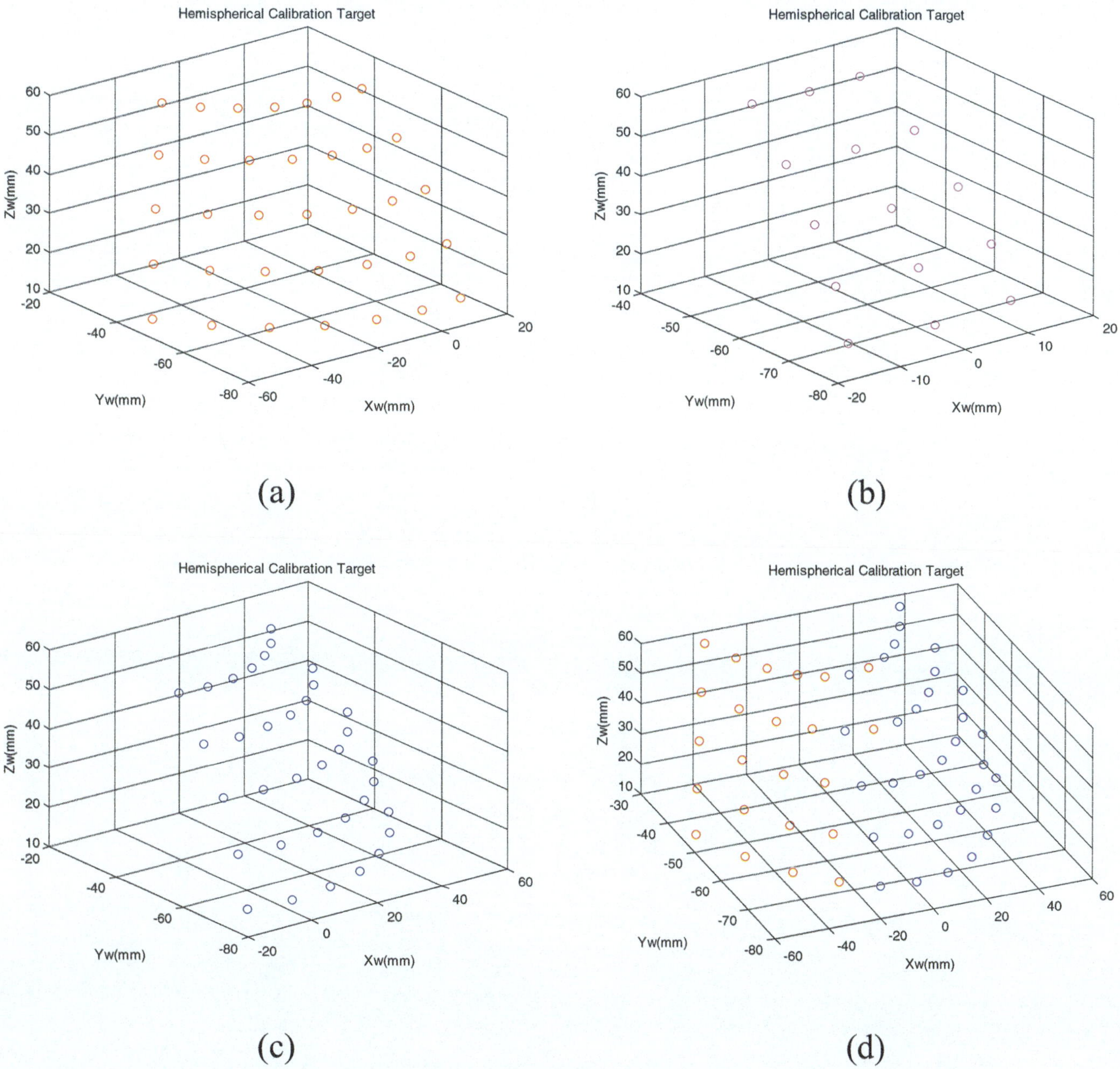

Fig. 31.4 The simulated 3D reconstruction of hemispherical calibration target. (**a**) C1C2, (**b**) C1C3, (**c**) C2C3, and (**d**) C1C2C3

Reference

1. Peters, W., Ranson, W.: Digital imaging techniques in experimental stress analysis. Opt. Eng. **21**(3), 427–431 (1982)
2. Sutton, M.A., Wolters, M.J., Peters, W.H., Ranson, W.F., McNeill, S.R.: Determination of displacements using an improved digital correlation method. Image Vis. Comput. **1**(3), 133–139 (1983)
3. Luo, P.F., Chao, Y.J., Sutton, M.A., Peters, W.H.: Accurate measurement of 3D deformations in deformable and rigid bodies using computer vision. Exp. Mech. **33**(2), 123–132 (1993)
4. Huang, C.H., Chen, Y.H., Wang, W.C.: Conical calibration target used for calibrating image acquisition device. TW patent, Publication number: I576652. Publication date: 2017.04.01, Original Assignee: National Applied Research Laboratories
5. Huang, C.H., Chen, Y.H., Wang, W.C.: Conic section calibration object for calibrating image capturing device. US patent, Publication number: US9762897 B2. Publication date: 2017.09.12, Original Assignee: National Applied Research Laboratories
6. Hwang, C.H., Wang, W.C., Chen, Y.H.: Camera calibration and 3D surface reconstruction for circular DIC multi-camera system. In: 2013 International Conference on Optics in Precision Engineering and Nanotechnology (icOPEN2013), Paper No. 876913, 8p, Singapore (2013)

Chapter 32
Recent Advancements and Perspective About Digital Holography: A Super-Tool in Biomedical and Bioengineering Fields

F. Merola, B. Mandracchia, L. Miccio, P. Memmolo, V. Bianco, M. Mugnano, P. L. Maffettone, M. Villone, E. Di Maio, V. Ferraro, Z. Wang, V. Pagliarulo, S. Grilli, and P. Ferraro

Abstract Digital holographic microscopy (DHM) has become a technique utilized widely for sample inspection, having many applications in different fields of science and technology. The capability for recovering the complex amplitude distribution scattered by the sample permits numerical refocus after acquisition and quantitative phase imaging. These are two of the features that make DHM a very versatile microscopy technique. The standard DHM system is based on a Mach–Zehnder interferometer that can be configured for operating in transmission or reflection modes, working in either the in-line or off-axis architecture. With the benefit of such special characteristics, DHM is used in basic research as much in the industry. Here we review some of the recent advancements for the label-free inspection of biological samples and the study of thin films.

Keywords Digital holography · Microscopy · Tomography · 3D tracking · Microfluidics

Digital holographic microscopy (DHM) has become a technique utilized widely for sample inspection, having many applications in different fields of science and technology. The capability for recovering the complex amplitude distribution scattered by the sample permits *a posteriori* numerical refocus and the quantitative phase imaging of the sample. These are two of the features that make DHM a very versatile microscopy technique. The standard DHM system is based on a Mach–Zehnder interferometer that can be configured for operating in transmission or reflection modes, working in either the in-line or off-axis architecture.

One of the most powerful imaging tools for analyzing biological samples is, without a doubt, optical tomography, able to furnish complete mapping of the refractive index of an object in 3D [1]. Various classes of tomographic concepts exist, such as X-ray computed tomography, optical coherence tomography [2, 3] and tomographic phase microscopy (TPM) [4–18]. TPM exploits quantitative phase imaging (QPI) techniques to map the 3D refractive index (RI) of cells, by adopting laser beam deflection, direct mechanical rotation or holographic optical tweezers to probe the sample along a number of controlled directions. At present, three are the main experimental methods adopted for image recording in order to obtain tomography of microscopic objects, i.e. 3D imaging of cells: (1) the illumination (laser beam) angle variation, (2) the sample mechanical/optical rotation, (3) the in-flow sample rotation in microfluidic environment.

For high-throughput 3D cell measurements, microfluidic devices have been integrated into TPM [10, 19–21]. For example, the 3D RI mapping of suspended live cells has been achieved, where the cells flow in a microfluidic channel, are trapped, and then rotated by dielectrophoretic forces [22].

In general, all TPM set-ups require the sample to be observed along different directions with respect to the probing beam. Many drawbacks arise for each of the aforementioned techniques. For example, in recording by adopting beam deflection [5, 14, 21] the angles are limited to 150°, thus affecting the accuracy of the tomographic reconstruction. Alternatively, direct rotation of the sample is problematic because mechanically manipulating the biological specimen [6] introduces the risk

F. Merola · B. Mandracchia · L. Miccio · P. Memmolo · M. Mugnano · Z. Wang · V. Pagliarulo · S. Grilli · P. Ferraro (✉)
CNR-ISASI, Istituto di Scienze Applicate e Sistemi Intelligenti "E. Caianiello", Pozzuoli NA, Italy

NEAPoLIS Numerical and Experimental Advanced Program on Liquids and Interface Systems, Naples, Italy
e-mail: p.ferraro@isasi.cnr.it

V. Bianco · P. L. Maffettone · M. Villone · E. Di Maio · V. Ferraro
NEAPoLIS Numerical and Experimental Advanced Program on Liquids and Interface Systems, Naples, Italy

Dipartimento di Ingegneria Chimica, dei Materiali e della Produzione Industriale—DICMaPI, University of Naples Federico II, Naples, Italy

L. Lamberti et al. (eds.), *Advancement of Optical Methods & Digital Image Correlation in Experimental Mechanics, Volume 3*, Conference Proceedings of the Society for Experimental Mechanics Series, https://doi.org/10.1007/978-3-319-97481-1_32

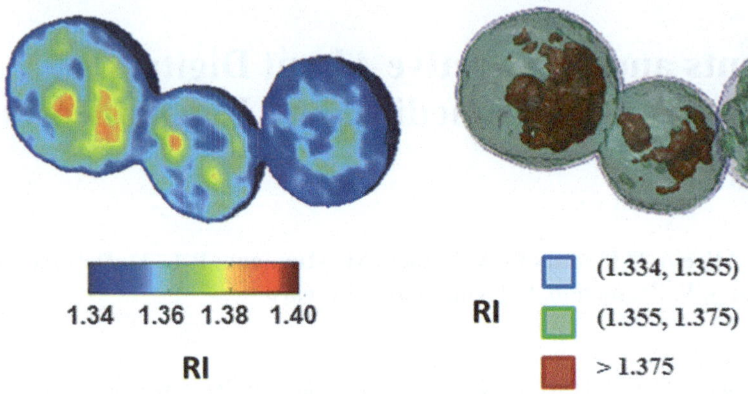

Fig. 32.1 Refractive index maps of MCF-7 cells cluster, obtained by FA-TPM

of altering the sample, while holographic optical tweezers [23–26] do not allow high-throughput tomography. In summary, to date, all tomographic methods require a high-precision, opto-mechanical and/or opto-electronic device to acquire a set of many images by probing the sample along a large number of controlled directions. Moreover, to successfully apply the tomography algorithms, each recorded image has to be tagged with the correct cell's rotation angle with respect to the probing beam, thus requiring robust and effective control of the angular positions. Overcoming such difficulties represents the main critical issue in all tomographic approaches and, in practice, limits the implementation of tomography, especially at the microscopic scale, for widespread use in the life sciences.

A step forward has been achieved in [20], where an optofluidic platform has been proposed to provide high-resolution images of both flowing and rotating samples, composed by a holographic microscope (HM) and a fluorescence module used as a control. The HM is essentially based on a Mach-Zehnder interferometer, a high numerical aperture microscope objective and a fast CCD camera, providing a label-free and non-invasive tool for cells imaging flowing in microfluidic channels. Thanks to the TPM performed in flow, hundreds of cells per minute have been analyzed; for each one, the hologram's sequence is recorded, and the corresponding quantitative phase map (QPM) is retrieved. In particular, thanks to the ability of HM to retrieve a posteriori the focal plane of imaged objects by a full numerical processing [27, 28], the accurate recovery of the 3D positions of multiple cells during their motion is performed by implementing suitable techniques of holographic 3D particles tracking [28–30]. Thus, each QPMs sequence is realigned and its 3D reconstructions (tomograms) been obtained, with a high-throughput furnished by the continuous flow of the samples. The results of in-flow TPM have been successfully applied to Red Blood Cells (RBCs) and diatom algae. By this technique, it has been possible to measure important parameters such as volume, corpuscular hemoglobin and refractive index of two common anemias, i.e. iron deficiency anemia and thalassemia, obtaining results at least comparable with biochemical techniques, but in a label-free modality [20]. For what concerns diatoms, as a test case in [20] the *skeletonema marinoi* and the *thalassiosira rotula* have been analyzed. The tomographic reconstruction permits to exactly identify chloroplasts location and dimensions, in perfect agreement with the corresponding fluorescence images. This achievement is a key point as diatom chloroplasts are the main targets of some water contaminants, such as fluoranthene, which is one of the principal constituent of PAH-contaminated aquatic systems, and copper. It has been demonstrated that in the presence of these elements, chloroplasts show signs of structural rupture or even disintegration [31].

But the tomographic algorithm used in [20] is based on the assumption that samples should have mainly asymmetric shape [32]. This aspect is a clear limitation in the case of cells having an almost spherical morphology. In order to overcome these limitations, in [33] a robust and general framework of rolling angle recovery in tomographic reconstructions has been conceived and implemented, allowing full angle TPM (FA-TPM) of quasi-spherical cells. As proof of concept, FA-TPM has been applied to circulating metastatic breast carcinoma MCF-7 and normal human mammary epithelial MCF-10A cells in microfluidic flow [33]. Moreover, the analysis has been extended to circulating MCF-7 clusters by simultaneously mapping the RI structures of all cells in the cluster, as shown in Fig. 32.1. This approach could open a realistic route for the detection and sorting of circulating tumor cells (CTCs) in the peripheral blood flow.

The impressive growth of microfluidics over the last decade is mainly due to the capability to fabricate chips equipped with devices for the accurate control and manipulation of fluids. These LoC platforms are used for accurate biomedical research studies in the developed world, and point-of-care diagnostics in low-resource settings, where adequate instruments, costly facilities, and clinical laboratories for accurate analysis are still missing. It is clear that the possibility to observe the samples inside LoCs is crucial for a deep understanding of the processes occurring onboard such platforms. Hence, great

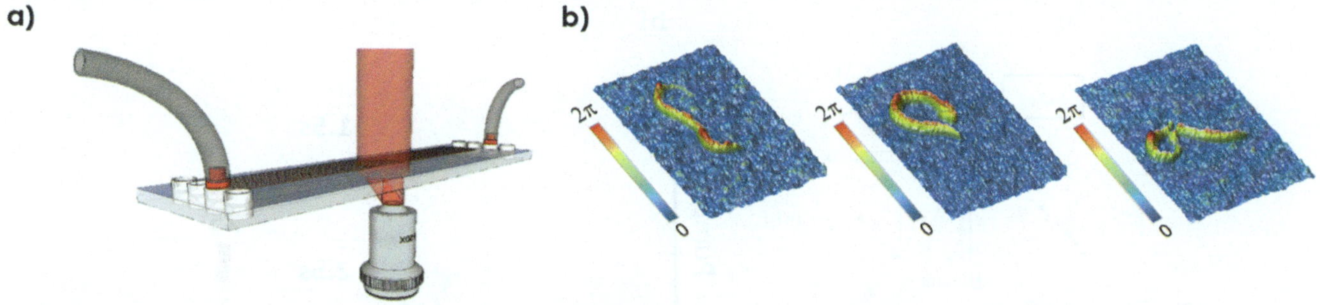

Fig. 32.2 (**a**) Working principle of the Holographic Microscope Slide. (**b**) Time-lapse phase imaging of a *C. elegans* worm

effort has been spent to efficiently apply well-known diagnostic techniques to the LoC investigation of biological specimens. The development of new imaging strategies to study biological samples in microfluidic channels is basically guided by four major needs. First, quantitative information is required for a deeper understanding of cell structure and behaviors. On the other hand, the demand for compactness needs to be satisfied in order to promote LoC portability for point-of-care diagnostic purposes. In order to obtain statistically relevant information, high-throughput imaging systems need to be developed that allow analyzing multiple specimens in reduced time. Besides, non-invasive techniques are preferred to avoid the decreases in throughput due to sample pre-treatment, and to prevent undesired alterations of the specimens.

The Optofluidic microscope introduced in 2006 represented a novelty in this direction, overcoming the resolution limits imposed by the finite size of the CCD recording elements [34]. However, phase-retrieval was not allowed, thus impairing the access to quantitative information. On-chip flow cytometry of leukocyte cells was demonstrated with the design of a compact Digital In-Line Holography (DILH) microscope using a pulsed, fiber-coupled laser source [35], and since then a number of approaches based on DILH have been proposed, but DILH-based techniques turn out not to be quantitative in a number of practical situations [36–38]. Furthermore, the lensless approach constitutes an inner limitation to the available resolution.

To get rid of the shortages deriving from the twin image superposition, the use of off-axis Digital Holography in place of the in-line configuration has been proposed in [39]. Here, instead of avoiding the use of lenses, an alternative lens-based approach that integrates the imaging functions onboard chip is adopted. In particular, a microfluidic chip was functionalized adding a photoresist grating and polymer lenses. This avoids the need for a reference arm of the interferometer as well as external optical components, see Fig. 32.2a. Label-free imaging and quantitative phase contrast mapping of live samples is demonstrated, along with flexible refocusing capabilities, see Fig. 32.2b. In addition, thanks to the single beam scheme, the system is completely integrated and robust against vibrations, sharing the useful features of any common path interferometer.

In addition, [40] shows the potentialities of the combination with spatio-temporal scanning acquisition modality for high-throughput counting and 3D tracking of RBCs. Space-Time Digital Holography (STDH) exploits the sample motion inside microfluidic channels to obtain a synthetic hologram, with unlimited FoV along the scanning direction and reduced noise. STDH is demonstrated to exhibit just the same properties of common DHM when mapped in the 2D space domain. The possibility to inspect an entire liquid volume by flexible refocusing, combined with the unlimited FoV along the flow direction allows to dramatically increase the throughput of the imaging system.

Thin liquid films flows are of fundamental importance in numerous areas of engineering, geophysics and biophysics [41–43]. Nevertheless, there are not appropriate quantitative experimental tools for an adequate study of film evolution in case of not-ideal conditions. Recently, a custom holographic microscope combining quantitative phase imaging with materials engineering [44]. This results in a combination of a full-field thickness estimation, three-dimensional imaging and fast acquisition time. This technique offers an unprecedented level of details that will promote a deeper understanding of the underlying physics of thin film dynamics.

The setup consists in an off-axis Mach-Zhender interferometer and engineered for in order to allow the formation of thin liquid films and to control the bubble growth, see Fig. 32.3. To form the films, it has been used a metal pipe with a side inlet connected to e syringe pump.

Given the pseudo-3D nature of the DHM acquisition the measured thickness profile, s, is a projection on the image plane of the three-dimensional one, see Fig. 32.1. However, the thickness parallel to the bubble surface normal, $\bar{s}$ can be found thanks to geometrical calculus. Notice that on the top of the bubble the values are almost the same ($\Delta s < 1\%$). Furthermore, DHM does not need multiple exposures and can be matched with high-speed cameras to measure rapidly changing features, such as the rim of the hole formed by the rupture of the liquid film.

Last but not least, the combination of evanescent wave microscopy and DHM has proven itself as a good tool for the characterization of thin films, cellular focal adhesions and biomolecular binding events [45–47]. Despite the high sensitivities

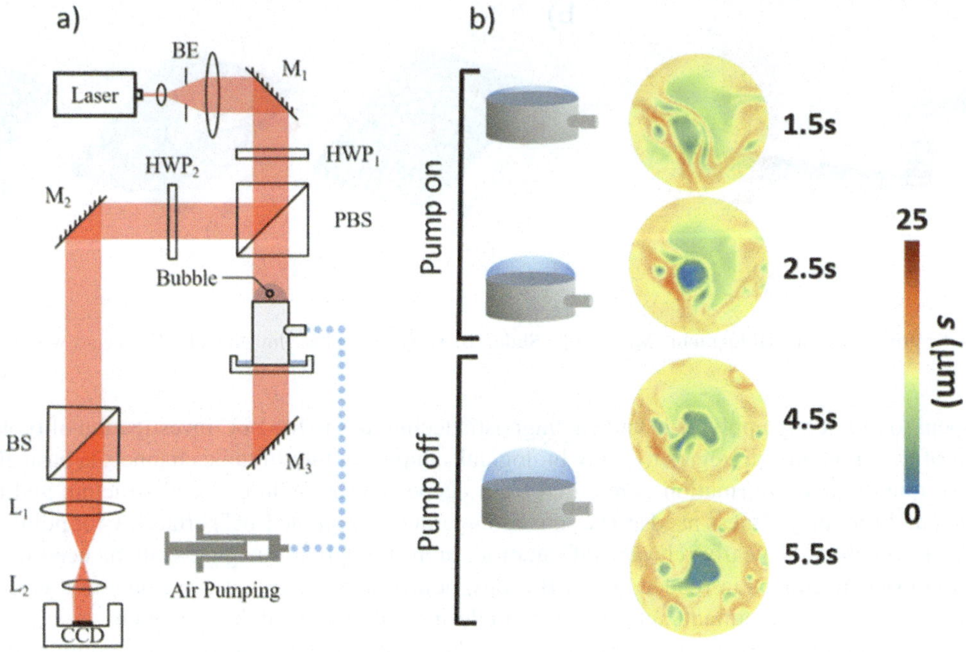

Fig. 32.3 (**a**) Sketch of the DH Setup. (**b**) Some frames of the thickness, $\bar{s}$, profile over time

achieved so far, each technique alone has a limited detection range with optimal sensitivity while their combination paves the way for a new field of application of DHM. An improvement in this direction has been marked by the development of innovative and compact systems for quantitative phase imaging and, in particular, the implementation of through-the-objective configurations for Holographic Total Internal Reflection (HoloTIR) microscopy and Holographic Surface Plasmon Resonance imaging (HoloSPRi) [48, 49].

The study of cell adhesion on different substrates evidenced an actual sensitivity enhancement for HoloTIR microscopy, Fig. 32.4a, b. Experimental data show a more accurate detection of focal contacts in comparison with conventional TIR microscopy [48, 50]. HoloTIR microscopy has the advantages of being relatively easy to implement on commercial microscopes, and generally cheaper than its fluorescence counterpart is. Also sample preparation time is reduced with a resulting improvement of data throughput and the absence of labels avoids the photobleaching issues for time-lapse imaging. Furthermore, illuminating light is mostly recovered by the camera, which allows for low-light imaging with subsequent reduction of phototoxicity damage for long time exposures of biological samples.

Objective-based HoloSPRi technique, instead, offers the possibility of simultaneous detection of reflectivity and phase changes in SPR images [49]. It extends the range of detectable samples and yields a higher throughput capacity in respect to both phase-contrast and amplitude-contrast imaging methods. This is because SPRi methods usually must scan the sample in order to retrieve quantitative information. Using HoloSPRi is possible to perform quantitative measurements without any angular or planar scanning, thanks to the fact that the whole complex reflection coefficient is measured, see Fig. 32.4c, d. In comparison with previous attempts of Holographic SPR imaging, it avoids geometrical aberrations and limitations due to the use of a prism. Finally, the employment of a configuration similar to the widely used objective-type TIR, and the modular design keep it compatible with conventional microscopes, paving the way for an easy implementation with other optical techniques, from standard bright field microscopy to fluorescence.

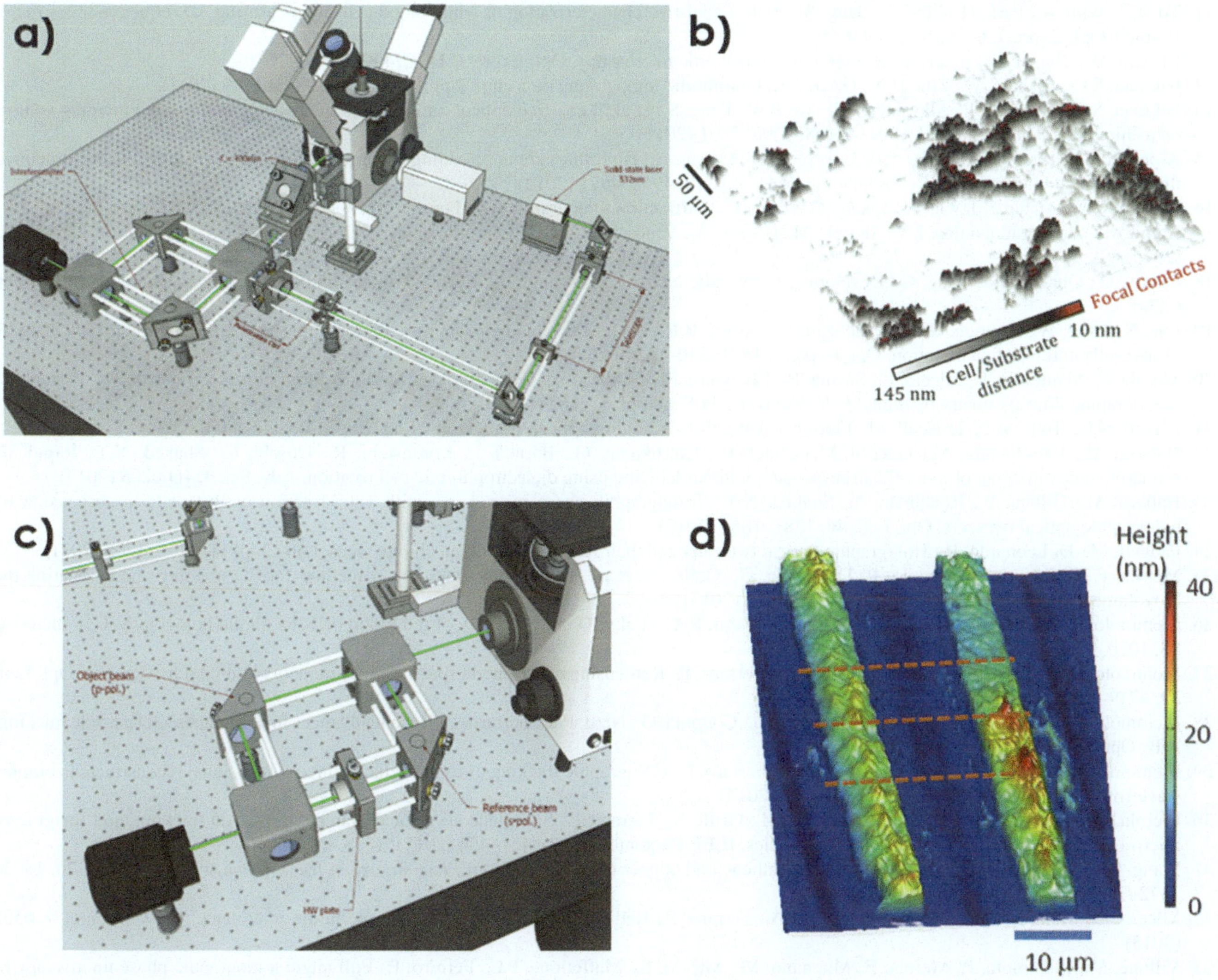

Fig. 32.4 (**a**) Holographic Total Internal Reflection setup. (**b**) Mosaic of phase images showing the focal contacts of fibroblasts adhering on a slab of Lithium Niobate. (**c**) Modular interferometer for HoloSPRi. (**d**) Quantitative reconstruction of a pattern of photoresist. This image was obtained mixing the information from amplitude and phase-contrast images

References

1. Mertz, J.: Optical sectioning microscopy with planar or structured illumination. Nat. Methods. **8**, 811–819 (2011)
2. Huang, D., Swanson, E.A., Lin, C.P., Schuman, J.S., Stinson, W.G., et al.: Optical coherence tomography. Science. **254**, 1178–1181 (1991)
3. Yi, J., Liu, W.Z., Chen, S.Y., Backman, V., Sheibani, N., et al.: Visible light optical coherence tomography measures retinal oxygen metabolic response to systemic oxygenation. Light Sci Appl. **4**, e334 (2015)
4. Choi, W., Fang-Yen, C., Badizadegan, K., Oh, S., Lue, N., et al.: Tomographic phase microscopy. Nat. Methods. **4**, 717–719 (2007)
5. Kim, K., Kim, K.S., Park, H., Ye, J.C., Park, Y.K.: Real-time visualization of 3-D dynamic microscopic objects using optical diffraction tomography. Opt. Express. **21**, 32269–32278 (2013)
6. Charrière, F., Marian, A., Montfort, F., Kuehn, J., Colomb, T., et al.: Cell refractive index tomography by digital holographic microscopy. Opt. Lett. **31**, 178–180 (2006)
7. Kim, T., Zhou, R.J., Mir, M., Babacan, S.D., Carney, P.S., et al.: White-light diffraction tomography of unlabelled live cells. Nat. Photonics. **8**, 256–263 (2014)
8. Mir, M., Tangella, K., Popescu, G.: Blood testing at the single cell level using quantitative phase and amplitude microscopy. Biomed. Opt. Express. **2**, 3259–3266 (2011)
9. Kim, Y., Shim, H., Kim, K., Park, H., Jang, S., et al.: Profiling individual human red blood cells using common-path diffraction optical tomography. Sci. Rep. **4**, 6659 (2014)
10. Sung, Y., Lue, N., Hamza, B., Martel, J., Irimia, D., et al.: Three-dimensional holographic refractive-index measurement of continuously flowing cells in a microfluidic channel. Phys Rev Appl. **1**, 014002 (2014)

11. Yoon, J., Kim, K., Park, H., Choi, C., Jang, S., et al.: Label-free characterization of white blood cells by measuring 3D refractive index maps. Biomed. Opt. Express. **6**, 3865–3875 (2015)
12. Bishara, W., Zhu, H.Y., Ozcan, A.: Holographic opto-fluidic microscopy. Opt. Express. **18**, 27499–27510 (2010)
13. Isikman, S.O., Bishara, W., Zhu, H.Y., Ozcan, A.: Optofluidic tomography on a chip. Appl. Phys. Lett. **98**, 161109 (2011)
14. Isikman, S.O., Bishara, W., Mavandadi, S., Yu, F.W., Feng, S., et al.: Lens-free optical tomographic microscope with a large imaging volume on a chip. Proc. Natl. Acad. Sci. U. S. A. **108**, 7296–7301 (2011)
15. Arpali, S.A., Arpali, C., Coskun, A.F., Chiang, H.H., Ozcan, A.: High-throughput screening of large volumes of whole blood using structured illumination and fluorescent on-chip imaging. Lab Chip. **12**, 4968–4971 (2012)
16. Horstmeyer, R., Chung, J., Ou, X., Zheng, G., Yang, C.: Diffraction tomography with Fourier ptychography. Optica. **3**, 827–835 (2016)
17. Kamilov, U.S., Papadopoulos, I.N., Shoreh, M.H., Goy, A., Vonesch, C., et al.: Learning approach to optical tomography. Optica. **2**, 517–522 (2015)
18. Psaltis, D., Quake, S.R., Yang, C.: Developing optofluidic technology through the fusion of microfluidics and optics. Nature. **442**, 381–386 (2006)
19. Lue, N., Choi, W., Popescu, G., Badizadegan, K., Dasari, R.R., Feld, M.S.: Synthetic aperture tomographic phase microscopy for 3D imaging of live cells in translational motion. Opt. Express. **16**, 16240–16246 (2008)
20. Merola, F., Memmolo, P., Miccio, L., Savoia, R., Mugnano, M., Fontana, A., D'Ippolito, G., Sardo, A., Iolascon, A., Gambale, A., Ferraro, P.: Tomographic flow cytometry by digital holography. Light Sci Appl. **5**, e16241 (2017)
21. Pégard, N.C., Toth, M.L., Driscoll, M., Fleischer, J.W.: Flow scanning optical tomography. Lab Chip. **14**, 4447–4450 (2014)
22. Habaza, M., Kirschbaum, M., Guernth-Marschner, C., Dardikman, G., Barnea, I., Korenstein, R., Duschl, C., Shaked, N.T.: Rapid 3D refractive-index imaging of live cells in suspension without labeling using dielectrophoretic cell rotation. Adv. Sci. **4**, 1600205 (2017)
23. Habaza, M., Gilboa, B., Roichman, Y., Shaked, N.T.: Tomographic phase microscopy with 180° rotation of live cells in suspension by holographic optical tweezers. Opt. Lett. **40**, 1881–1884 (2015)
24. Padgett, M., Di Leonardo, R.: Holographic optical tweezers and their relevance to lab on chip devices. Lab Chip. **11**, 1196–1205 (2011)
25. Merola, F., Miccio, L., Memmolo, P., Di Caprio, G., Galli, A., et al.: Digital holography as a method for 3D imaging and estimating the biovolume of motile cells. Lab Chip. **13**, 4512–4516 (2013)
26. Memmolo, P., Miccio, L., Merola, F., Gennari, O., Netti, P.A., et al.: 3D morphometry of red blood cells by digital holography. Cytometry A. **85**, 1030–1036 (2014)
27. Memmolo, P., Paturzo, M., Javidi, B., Netti, P.A., Ferraro, P.: Refocusing criterion via sparsity measurements in digital holography. Opt. Lett. **39**, 4719–4722 (2014)
28. Memmolo, P., Miccio, L., Paturzo, M., Di Caprio, G., Coppola, G., Netti, P.A., Ferraro, P.: Recent advances in holographic 3D particle tracking. Adv. Opt. Photon. **7**, 713–755 (2015)
29. Memmolo, P., Miccio, L., Finizio, A., Netti, P.A., Ferraro, P.: Holographic tracking of living cells by three-dimensional reconstructed complex wavefronts alignment. Opt. Lett. **39**, 2759–2762 (2014)
30. Merola, F., Miccio, L., Memmolo, P., Paturzo, M., Grilli, S., Ferraro, P.: Simultaneous optical manipulation, 3-D tracking, and imaging of micro-objects by digital holography in microfluidics. IEEE Photonics J. **4**(2), 451–454 (2012)
31. Wang, L.P., Zheng, B.H.: Toxic effects of fluoranthene and copper on marine diatom *Phaeodactylum tricornutum*. J. Environ. Sci. **20**, 1363–1372 (2008)
32. Miccio, L., Memmolo, P., Merola, F., Netti, P.A., Ferraro, P.: Red blood cell as an adaptive optofluidic microlens. Nat. Commun. **6**, 6502 (2015)
33. Villone, M., Memmolo, P., Merola, F., Mugnano, M., Miccio, L., Maffettone, P.L., Ferraro, P.: Full-angle tomographic phase microscopy of flowing quasi-spherical cells. Lab Chip. **18**, 126 (2018)
34. Heng, X., et al.: Optofluidic microscopy-a method for implementing a high resolution optical microscope on a chip. Lab Chip. **6**, 1274–1276 (2006)
35. Vercruysse, D., Dusa, A., Stahl, R., Vanmeerbeeck, G., de Wijs, K., Liu, C., Prodanov, D., Peumans, P., Lagae, L.: Three-part differential of unlabeled leukocytes with a compact lens-free imaging flow cytometer. Lab Chip. **15**(4), 1123–1132 (2015)
36. Wu, J., et al.: Optical imaging techniques in microfluidics and their applications. Lab Chip. **12**, 3566–3575 (2012)
37. Ferraro, P., Wax, A., Zalevsky, Z. (eds.): Coherent Light Microscopy: Imaging and Quantitative Phase Analysis, Springer Science & Business Media, vol. 46, Berlin (2011)
38. Latychevskaia, T., Fink, H.W.: Solution to the twin image problem in holography. Phys. Rev. Lett. **98**, 233901 (2007)
39. Bianco, V., Mandracchia, B., Marchesano, V., Pagliarulo, V., Olivieri, F., Coppola, S., et al.: Endowing a plain fluidic chip with micro-optics: a holographic microscope slide. Light Sci Appl. **6**(9), e17055 (2017)
40. Mandracchia, B., Bianco, V., Wang, Z., Mugnano, M., Bramanti, A., Paturzo, M., Ferraro, P.: Holographic microscope slide in a spatio-temporal imaging modality for reliable 3D cell counting. Lab Chip. **17**(16), 2831–2838 (2017)
41. Marmottant, P., Hilgenfeldt, S.: Controlled vesicle deformation and lysis by single oscillating bubbles. Nature. **423**, 153–156 (2003)
42. Massol, H., Koyaguchi, T.: The effect of magma flow on nucleation of gas bubbles in a volcanic conduit. J. Volcanol. Geotherm. Res. **143**, 69–88 (2005)
43. Salerno, A., Di Maio, E., Iannace, S., Netti, P.A.: Tailoring the pore structure of PCL scaffolds for tissue engineering prepared via gas foaming of multi-phase blends. J. Porous. Mater. **19**, 181–188 (2012)
44. Wang, Z., Mandracchia, B., Ferraro, V., Tammaro, D., Di Maio, E., Maffettone, P.L., Ferraro, P.: Interferometric measurement of film thickness during bubble blowing. Proc. SPIE Int. Soc. Opt. Eng. **10333**, 6936 (2017)
45. Calabuig, A., Mugnano, M., Miccio, L., Grilli, S., Ferraro, P.: Investigating fibroblast cells under "safe" and "injurious" blue-light exposure by holographic microscopy. J. Biophotonics. **9**, 1–9 (2016)
46. Ash, W., Kim, M.: Digital holography of total internal reflection. Opt. Express. **16**, 9811–9820 (2008)
47. Li, S., Zhong, J.: Simultaneous amplitude-contrast and phase-contrast surface plasmon resonance imaging by use of digital holography. Biomed. Opt. Express. **3**, 3190–3202 (2012)

48. Mandracchia, B., Gennari, O., Marchesano, V., Paturzo, M., Ferraro, P.: Label free imaging of cell-substrate contacts by holographic total internal reflection microscopy. J. Biophotonics. **10**(9), 1163–1170 (2017)
49. Mandracchia, B., Gennari, O., Bramanti, A., Grilli, S., Ferraro, P.: Label-free quantification of the effects of lithium niobate polarization on cell adhesion via holographic microscopy. J. Biophotonics. (2018). https://doi.org/10.1002/jbio.201700332
50. Mandracchia, B., Pagliarulo, V., Paturzo, M., Ferraro, P.: Surface plasmon resonance imaging by holographic enhanced mapping. Anal. Chem. **87**(8), 4124–4128 (2015)

Chapter 33
High-Speed Shape and Transient Response Measurements of Tympanic Membrane

Payam Razavi, Haimi Tang, Nima Maftoon, John J. Rosowski, Cosme Furlong, and Jeffrey Tao Cheng

Abstract We are developing a High-speed Digital Holographic (HDH) system to measure acoustically induced transient displacements of live mammalian Tympanic Membranes (TM) for research and clinical applications. To date, the HDH can measure one-dimensional displacements along a single 1-D sensitivity vector. However, because of the TM's tent-like shape and angled orientation inside the ear canal, 1-D measurements need to be combined with measurements of the shape and orientation of the TM to determine the true surface normal (out-of-plane) displacements. Furthermore, TM shape also provides invaluable information for better diagnostic and modelling of the TM. To introduce shape measurements capabilities into our HDH, a tunable laser (line width <300 kHz, 770.2–790 nm, <8 nm/s) is incorporated in tandem with the single frequency (line width <100 MHz, 532.3 nm) laser used for displacement measurements in a manner that provides a common sensitivity vector for the two measurements. Interferograms gathered with continuous high-speed optical phase sampling and wavelength tuning allow the reconstruction of the TM shape with <50 μm measuring resolution and <120 μm repeatability. With our modified HDH, shape measurements immediately follow transient displacement measurements (>67 kHz temporal and <15 nm displacement resolutions) in response to broadband acoustic click excitations (50 μs duration). Both shape and displacement can be measured in less than 150 ms, which avoids slow disturbances introduced by breathing and heartbeat, with the promise of future measurements in vivo. Representative shape and displacement measurements capabilities are demonstrated on cadaveric human temporal bones.

Keywords Holographic interferometry · Multiple wavelength · Shape measurement · Transient response · Tympanic membrane

33.1 Introduction

The displacements at each point of the moving surface in a holographic measurement of deformation, give rise to the optical path length differences essential for interferometric measurements. The optical path length in an off-axis configuration is shown in Fig. 33.1a, as the summation of the magnitudes of the blue *path-vectors* $\boldsymbol{r_1}$, $\boldsymbol{r_2}$. The *propagation-vectors* ($\boldsymbol{k_1}$, $\boldsymbol{k_2}$) along each path-vector show the direction of the light array with an amplitude of $2\pi/\lambda$ (wave number, k) such that their dot product with the corresponding path-vector (e.g., $\boldsymbol{r_i}\cdot\boldsymbol{k_i}$) provides the optical path length in phase cycles of light [1].

P. Razavi (✉) · H. Tang
Center for Holographic Studies and Laser micro-mechaTronics (CHSLT), Worcester Polytechnic Institute, Worcester, MA, USA

Mechanical Engineering Department, Worcester Polytechnic Institute, Worcester, MA, USA
e-mail: prazavi@wpi.edu

N. Maftoon · J. J. Rosowski · J. T. Cheng
Eaton-Peabody Laboratory, Massachusetts Eye and Ear (MEE), Boston, MA, USA

Department of Otology and Laryngology, Harvard Medical School, Boston, MA, USA

C. Furlong
Center for Holographic Studies and Laser micro-mechaTronics (CHSLT), Worcester Polytechnic Institute, Worcester, MA, USA

Mechanical Engineering Department, Worcester Polytechnic Institute, Worcester, MA, USA

Eaton-Peabody Laboratory, Massachusetts Eye and Ear (MEE), Boston, MA, USA

Department of Otology and Laryngology, Harvard Medical School, Boston, MA, USA

L. Lamberti et al. (eds.), *Advancement of Optical Methods & Digital Image Correlation in Experimental Mechanics, Volume 3*,
Conference Proceedings of the Society for Experimental Mechanics Series, https://doi.org/10.1007/978-3-319-97481-1_33

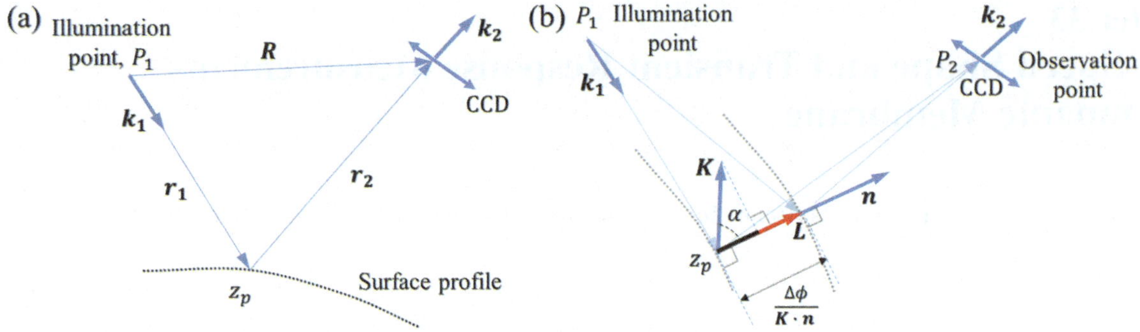

Fig. 33.1 Nomenclature for fringe analysis: (**a**) schematic representation of optical path length from the observation point P_1 to a point on the sample surface z_p to a point of observation on the CCD P_2; and (**b**) schematics of optical path length difference of a thin-shell having dominant out-of-plane displacements. $\Delta\phi$ is the interference phase, $\boldsymbol{K} = \boldsymbol{k}_2 - \boldsymbol{k}_1$ is the sensitivity vector, $\boldsymbol{L}$ is the displacements vector, and $\boldsymbol{n}$ is the surface normal at the point z_p

In most holographic applications, the lengths of path vectors are on the order of centimeters to meters while the displacements are on the order of nanometers to micrometers, and the optical path length differences are related to the displacements by,

$$\Delta\phi = \boldsymbol{K} \cdot \boldsymbol{L}, \quad \boldsymbol{K} = \boldsymbol{k}_2 - \boldsymbol{k}_1, \tag{33.1}$$

where $\boldsymbol{K}$ is the sensitivity vector, $\boldsymbol{L}$ is the displacement vector, and $\Delta\phi$ is the interference phase. Since the displacement vector has three unknowns, at least three independent sensitivity vectors (i.e., 3D holography) are needed to determine displacements. Another way to retrieve the displacements is by determining the direction of the displacements vector $\boldsymbol{L}$. Based on previous studies, the TM has a thin-shell structure with the dominant displacement component being the out-of-plane component of the displacements along the surface normal [2]. With knowledge of the surface normal direction, the measured displacement along the sensitivity vector can be decomposed to define the magnitude of the surface normal motion component (i.e., $\boldsymbol{L} = |\boldsymbol{L}| \cdot \boldsymbol{n}$), and we can rewrite Eq. (33.1) as,

$$|\boldsymbol{L}| = \frac{\Delta\phi}{\boldsymbol{K} \cdot \boldsymbol{n}} = \frac{\Delta\phi}{|\boldsymbol{K}| \cos(\alpha)}, \tag{33.2}$$

where the surface normal $\boldsymbol{n}$, sensitivity vector $\boldsymbol{K}$, and the angle α, which is the angle between the two vectors, are shown schematically in Fig. 33.1b. The holographic measurements of the Tympanic Membrane (TM, eardrum) inside the ear canal require about 10–15 mm working distance, which is large compared to the millimeter scale spatial variations of the optical path length and surface normals due to shape and orientation of the TM. This phenomenon produces a spatially varying sensitivity of the measurements that can generate errors in displacements as large as 50%.In this paper, we present a novel high-speed digital holographic shape measurement method incorporated into our high-speed holographic displacement measurement system to measure the TM shape and orientation in-situ. Knowledge of shape helps to account for the spatial variations of the measuring sensitivity to determine surface-normal displacements from 1-D holographic measurements.

33.2 Methods

33.2.1 High-Speed Digital Holographic Interferometry

We have previously developed and implemented a High-speed Digital Holographic (HDH) method that 'instantaneously' measures the full-field (>200,000 points at 67,200 Frames Per Seconds (FPS)) transient displacements of the TM in response to pulsed acoustical and mechanical excitations [3–5]. To measure the transient response of the TM in-vivo, an adaptive phase sampling holographic method was developed to compensate for the effects of large in-vivo physiological motions using high-speed continuous phase sampling [1].

33.2.2 Shape Measurements Requirements

The development of the shape measurement system was performed with the aim of satisfying all of the following functional requirements:

1. Utilizing the imaging optics of the High-speed Digital Holographic Displacement measurement system,
2. Measurements from a close millimeter-scale distance and within small triangulation volumes of the ear canal,
3. Adjustable shape resolution to account for local variations in TM shape and orientation,
4. Uniform and sufficient axial and lateral shape resolutions to accurately determine surface normals,
5. Short intervals (<100 msec) between displacement and the shape measurements to maintain the same coordinate system for both measurements,
6. High-speed shape measurements to reduce the effects of in-vivo physiological motions.

Meeting these functional requirements was not a trivial task. Many of the methods that have been previously applied to TM shape measurements are inappropriate. For example, structured light projection [6, 7] requires large triangulation angles compared to the ear canal's confined volume while others that are capable of single shot in-vivo endoscopic measurements [8] require high-resolution sensors (>10 Megapixels) and specialized optics that are incompatible with the HDH system.

33.2.3 Multiple Wavelengths Holographic Interferometry (MWHI)

The MWHI shape measurement method uses a tunable laser source that can vary the illumination wavelength during repeated sets of phase-shifted holograms. While the Optical Path Length (OPL) is constant (with a stationary sample), the variations in wavelength generate contours of constant height due to sample shape. The relation between the interference phase $\Delta\gamma$, the constant OPL, and the tuned set of wavelengths at each hologram exposures (i.e., $\lambda_1 - \lambda_2$) is described by,

$$\Delta\gamma = \frac{2\pi}{\lambda_1} OPL - \frac{2\pi}{\lambda_2} OPL = \frac{2\pi}{\Lambda} OPL, \tag{33.3}$$

where Λ is the synthetic or equivalent wavelength determined by,

$$\Lambda = \frac{\lambda_1 \lambda_2}{\mid \lambda_1 - \lambda_2 \mid}. \tag{33.4}$$

To determine the shape, additional calibration efforts are required to transform the interference phase $\Delta\phi$ onto a coordinate system where variations in $\Delta\gamma$ define the shape of the sample [9, 10] along the camera axis (z in Fig. 33.2) by,

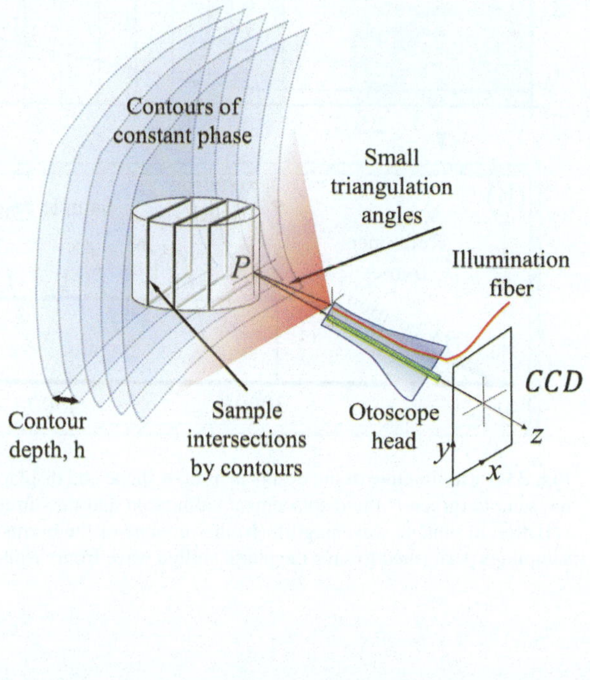

Fig. 33.2 Schematic representation of contours of constant phase intersecting the sample with contour depth (h) in the MWHI shape measurements method intersecting the sample shape

$$z_p(x, y) = \frac{\Lambda}{4\pi} \Delta\gamma_c(x, y), \tag{33.5}$$

where $\Delta\gamma_c$ is the calibrated phase and z_p is the z component of point P corresponding to the shape of the sample.

33.2.4 High-Speed Multi-Resolution MWHI Method

We used a new technique to increase the speed of acquisition without using multiple lasers for shape measurements. By performing phase samplings at a temporal resolution much higher than the phase shifts introduced by sample or wavelength variations, we were able to quantify the interference phase (i.e., $\Delta\gamma$) during a laser tune. This achievement significantly increased the speed of acquisition by removing the delays required for laser wavelength stability. Furthermore, this technique provided multiple measurements of the shape over a continuous range of contour depths, and enabled selection of an optimum contour depth (a synthetic wavelength that provides the shape with maximum quantifiable fringe density) in less than 0.2 s. As a result, the shape of the sample can be extracted robustly at the highest possible spatial resolution of the sensor.

33.2.5 Timeline of High-Speed Shape and Displacement Measurements

Figure 33.3 illustrates the timeline of the shape and displacement measurement with a total duration of about 200 ms. The displacement measurement (the green illuminated section at the start of Fig. 33.3a that is detailed in Fig. 33.3b) is performed in less than 25 ms. As shown in Fig. 33.3b, the displacement measurements start by capturing a set of reference holograms during a continuous piezo-controlled phase shift (within 3 ms) followed by a broadband acoustic click (a 50 μs square pulse). After 25 ms, the piezo moves back to its original position, and the shutters are set to switch to the variable wavelength laser. During the later 175 ms, the illumination is shifted to the variable wavelength laser as it undergoes a change in wavelength. During that transition 17 rapid (500 μs each) phase samplings are performed; one is shown in Fig. 33.3c. Both shape and displacement measurements are performed at 67,200 frames per second temporal resolution. Each of the 17 shape phase sampling triggers is synchronized with a continuous ramp of the piezo positioner to allow for quantitative phase imaging while the tunable laser slowly varies the illumination wavelength. The last phase sampling is performed after the wavelength reaches its final value. Both shape and displacement measurements are repeated three times.

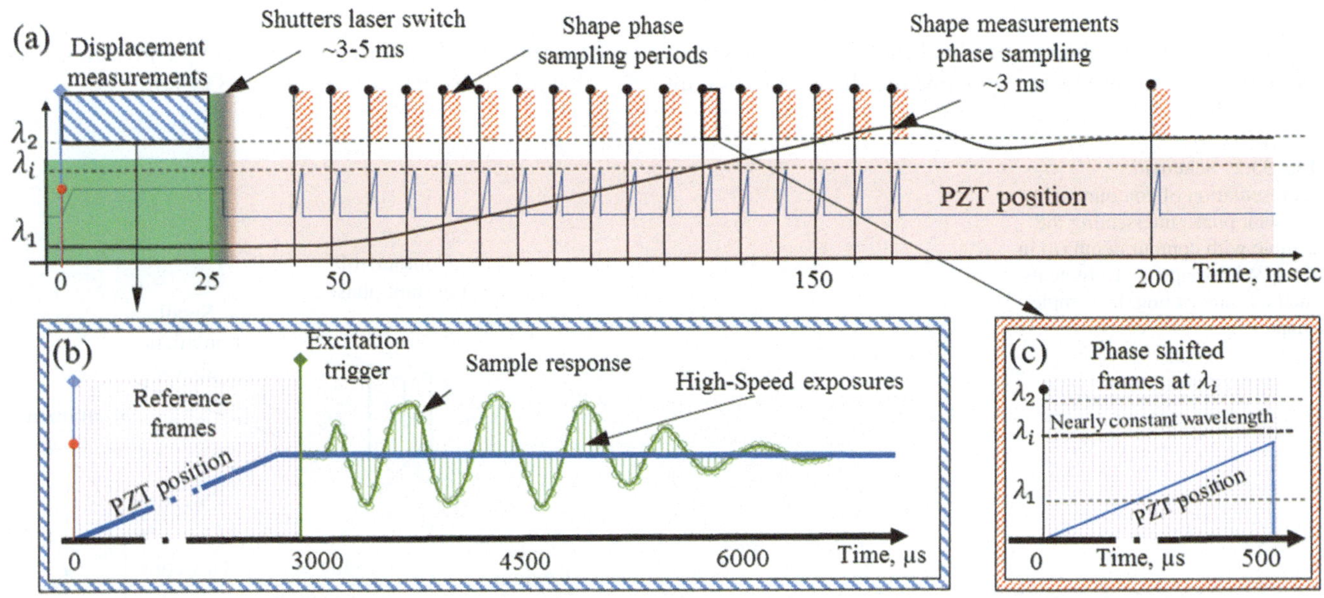

Fig. 33.3 (**a**) timeline of the events in a set of shape and displacements measurements. The color of the background shows which laser illuminates the sample (green is the displacement calibration and measurement, pink is the region of repeated shape measurements during a relatively slow variation in tunable wavelength); (**b**) the timeline of the events during the displacements measurements; and (**c**) one of the 17 high-speed phase samplings performed to save the phase shifted wave fronts with nearly constant wavelength during the slow wavelength change

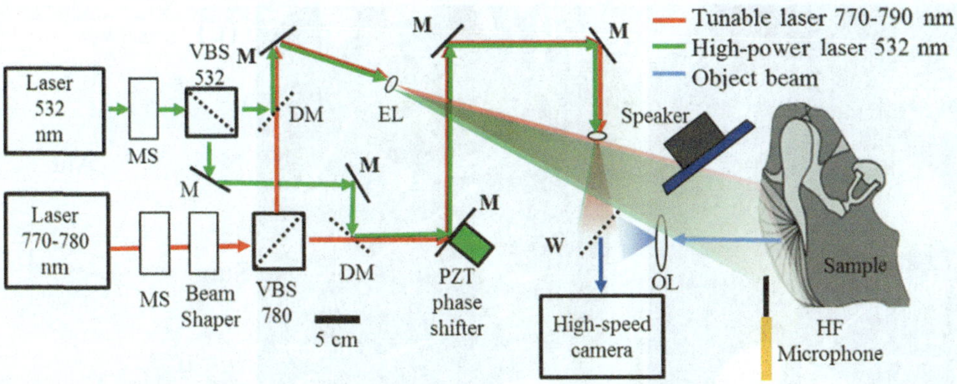

Fig. 33.4 Schematic representation of the shape and displacements measurements setup. Variable Beam Splitters (VBS) are to balance the intensity ratio between the reference and object at different wavelengths. *MS* mechanical shutters, *DM* dichroic Mirror, *W* wedge, *EL* expansion lens, *OL* objective lens

33.2.6 Experimental Setup

Different lasers were used for shape and displacement measurements: A 780 ± 10 nm tunable laser for shape and a highly stable CW 532 nm laser for displacements. As shown in Fig. 33.4, using variable beam splitters (VBS), each laser is divided into reference and illumination beams. To maintain the same observation vector for both measurements, the lasers are guided onto the same illumination and reference axis using dichroic mirrors (DM). The beam reflected from the sample (object beam) is combined with the reference using a wideband wedge to form an interference pattern on the sensor of the high-speed camera. Each laser is equipped with an automated mechanical shutter (MS) to switch rapidly between the lasers (<5 ms).

33.3 Representative Results

33.3.1 Sample Preparation

A fresh (non-fixed) cadaveric human ear was used for these measurements (Fig. 33.5). The cartilaginous and boney parts of the ear canal were removed to provide a larger view of the TM. The middle-ear cavity was opened to check the middle ear ossicles healthiness and then it was sealed using silicone sealant. A small vent equilibrated the static pressure within the middle ear. The sample was sprayed with a thin layer of white face paint using a modified airbrush. To ensure the paint layer was sufficiently thin and uniform the airbrush was set to generate the finest particles while the sample was placed >10 cm away from the airbrush. After painting, parts of ossicles were still visible under white light illumination (Fig. 33.5c).

33.3.2 Representative Shape and Displacement Measurements

The shape and acoustically induced transient displacements of the human post-mortem TM were measured by the combined multi-resolution high-speed MWHI method and 2 + N high-speed correlation interferometry. Using an optimization approach based on phase quality, an optimum resolution phase-map (i.e., having the maximum phase quality and fringe density) was chosen from tens of shape measurements at varied resolutions. The optimum map was used for determination of surface-normal and sensitivity vectors. Figure 33.6a shows the optimum modulo 2π phase map (in this case with a synthetic wavelength of $\Lambda \cong 1.2$ mm). Figure 33.6b shows the wrapped phase after the calibration procedure (i.e., using a flat surface calibration method), and Fig. 33.6c illustrates the shape after spatial filtering and masking are performed to remove any surface discontinuities that might interfere with the surface normal determination. The colorbarin 6-c quantifies the depth of the sample along the z-axis.

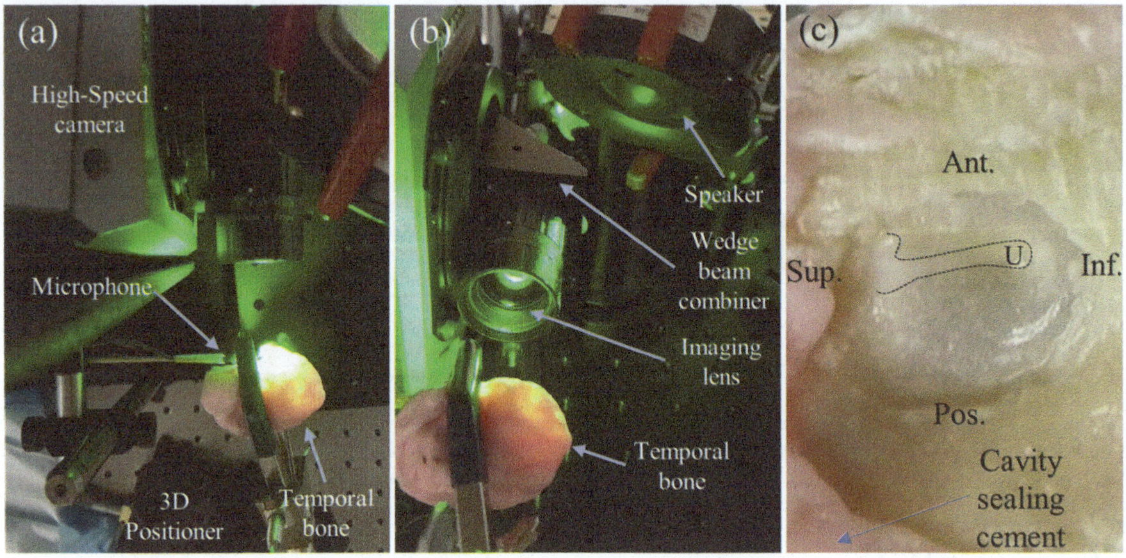

Fig. 33.5 The setup, measuring instrumentations and the sample: (**a**) top view; (**b**) back view; and (**c**) photograph of the human cadaveric TM. The black contour shows the outline of the manubrium and umbo (shown with U symbol)

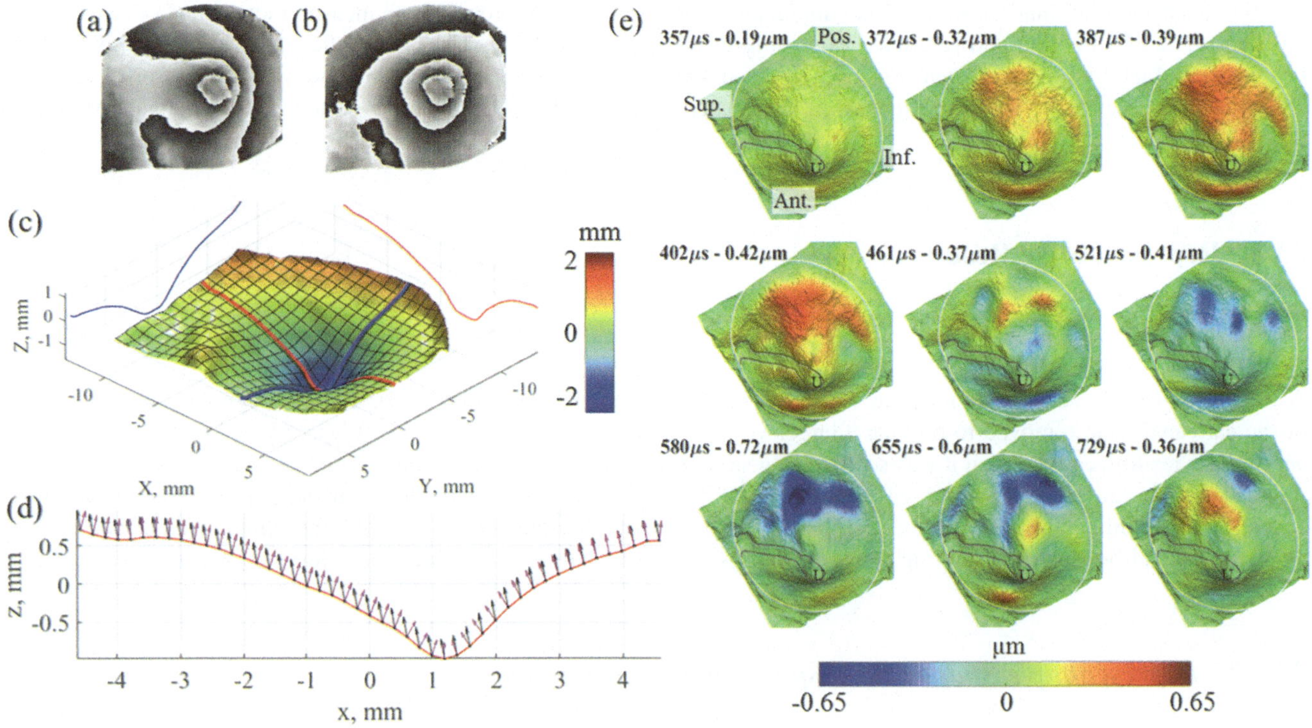

Fig. 33.6 Results of shape and transient response measurements of a human post-mortem sample: (**a**) wrapped interference phase (i.e., $\Delta\gamma$); (**b**) wrapped interference phase after flat surface calibration (i.e., $\Delta\gamma_c$); (**c**) 3D shape and orientation of the TM after it was unwrapped, scaled, spatially filtered, and masked. The shape of the red and blue lines are shown in the *y-z* and *x-z* planes and the colorbar represent the depth of the sample along z-axis; (**d**) computed surface normals (magenta) and sensitivity vectors (black) along the red line shown in panel (**c**); and (**e**) 3D representation of the TM transient response surface normal displacements due to an acoustic click (50 μs square pulse). Timing and the maximum displacement at each time instance is given in each panel in (**e**). The nanometer-scale displacements are 1000 times exaggerated and shown on top of millimeter-scale shape. The manubrium of the malleus is shown with a black line and U represents the umbo

Knowing the shape and orientation of the TM together with the coordinates of the observation and the illumination points, the surface normal and sensitivity vectors are computed at each location on the TM surface. Figure 33.6d illustrates the surface normals (magenta arrows) and sensitivity vectors (black arrows) distribution along a single contour of the TM surface that passes through the umbo (i.e., the red line shown in Fig. 33.6c). The angle α, between the sensitivity and normal vectors, varies across the field of view. Using α and Eq. 33.2, a sensitivity multiplier map is computed from the surface-normal and sensitivity vectors at each point, where, the sensitivity multiplier is a 2D scalar map that converts the measured interference phase to the corresponding surface-normal displacements.

With the sensitivity map, the acoustically induced TM transient response surface-normal displacements are determined from the interference phase measured by the 2 + N high-speed local phase correlation method [3]. Figure 33.6e illustrates 3D maps representing the shape and normal-displacements at nine selected temporal instances out of 1700 consecutive measurements (25 ms, at 67,200 FPS) during the TM transient response due to a broadband acoustic click.

The results show the TM response from the very first moments of acoustic to solid interaction, the initiation of complex traveling and standing surface waves, and the complete dissipation of the vibrations across the entire surface of the TM. The first four consecutive frames of the measurements in Fig. 33.6e shows the initial response of the membrane as the sound waves start to move the entire surface of the TM with essentially the same motion phase. We call this stage of the response "initiation" and this early response can be used to determine TM stiffness, overall volume change, etc. After initiation, surface waves are seen to reflect from both the manubrium and a complex superposition of traveling and standing waves occur across the TM ("Surface wave superposition stage"). The TM response at this stage is temporally separated from the excitation, making it suitable for methods such as experimental modal analysis to determine mechanical properties, damping etc. The last stage of the response includes the TM vibration at one of the primary mode shapes until the motions are completely damped out (dissipation stage).

33.3.3 Significance of Surface-Normal Determination

During the short period of the displacements measurements, the temporal variations of the measuring sensitivity (i.e., spatial distribution of the angle α) are minimal. In order to demonstrate the significance of the surface-normal displacements, the results are compared against the maximum spatially uniform sensitivity of the holographic system; under ***ideal sensitivity*** when α is zero and spatially uniform (i.e., $\alpha\ (x, y) = 0$ or $\mid \boldsymbol{L} \mid = \frac{\lambda \Delta \phi}{4\pi}$). Figure 33.7a, shows the spatial distribution of the sensitivity of the holographic system compared to the ideal case determined by the setup geometries, the shape, and orientation of the TM. It is visible that at regions near umbo the sensitivity is less than half of the ideal sensitivity.

Figure 33.7b show the error maps determined by taking the absolute errors between the surface-normal displacements and the ones obtained from the maximum sensitivity condition. Comparison of the spatial distribution of errors, Fig. 33.7b, and the actual displacements, Fig. 33.6e, show different spatiotemporally patterns. The results also show the initiation of

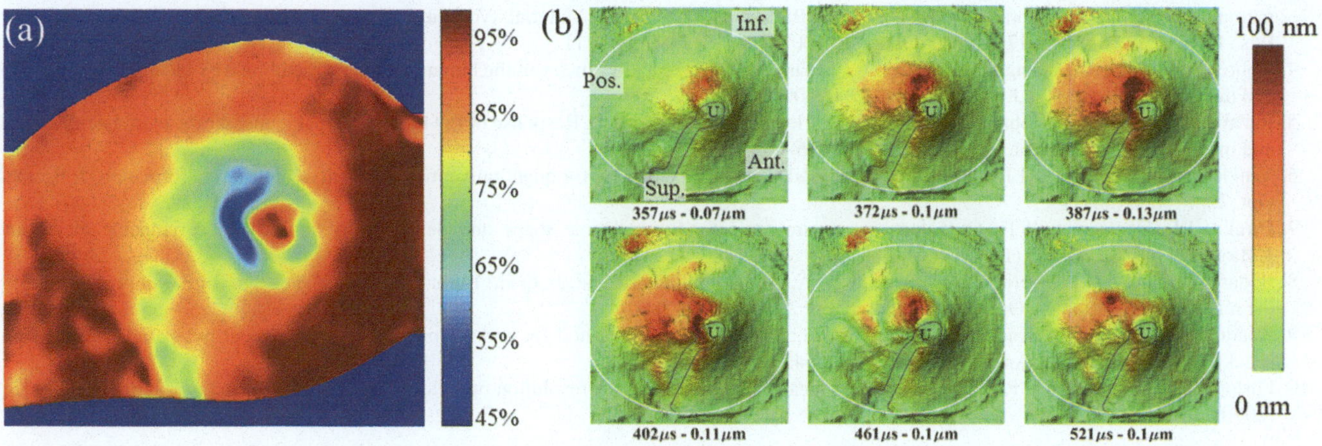

Fig. 33.7 (**a**) sensitivity map distribution percentile compared to the maximum spatially uniform sensitivity (i.e., $\alpha\ (x, y) = 0$); and (**b**) the spatiotemporal absolute differences between surface-normal displacements and the displacements determined under ideal sensitivity condition. The maximum differences occur at 387 μs with an amplitude of 138 nm

the motions are accompanied by about 40% maximum error with a peak of 52% error at 387 μs (top right panel of Fig. 33.7b). The maximum error amplitude is about 138 nm and occurred at 595 μs. The spatiotemporal distribution of the errors together with the regions that the errors occur show the importance of the shape measurements in parallel with displacement measurements to account for these variations. It should be noted again that the measurements presented in this paper were performed by using nearly collimated wave fronts with the TM annulus surface normal nearly parallel to camera axis. For an endoscopic configuration to observe the TM inside ear canal, TM orientation as well as larger spatial variations of the sensitivity vector generate significantly larger errors.

33.3.4 Conclusions and Future Work

A high-speed holographic shape measurement method is developed to measure TM shape parameters (i.e., curvature, dimensions, etc.) as well as to compensate for geometrically induced spatial variations of the sensitivity vector. These improvements provide measuring capabilities to evaluate in-situ and near instantaneous (<200 ms) TM displacements and shape with near six orders of magnitude different resolution ranges (from millimeters to nanometers). The new system enables the rapid sequential activation of a high-speed multi-resolution MWHI shape measurement and high-speed holographic displacement measurements, enabling shape sand displacement within a fraction of a second to help to reduce the effects of in-vivo physiological motions. Furthermore, the knowledge of shape and orientation makes the displacements measurements independent of direction of observation, which facilitates comparison of measurements after the specimen is moved. The results also help better interpret the TM transient response either by accurate measurements of speed and trajectory of the travelling waves or by use of finite element or experimental modal analysis. Finally, the TM shape parameters such as curvature, area, etc. have direct clinical diagnostic values that can be used for rehabilitation surgeries as well as development of personalized grafts and artificial TMs. Future work includes the analysis of the data gathered from several human samples and development of miniaturized optical head to enable in-vivo measurements inside the ear canal.

Acknowledgments This work has been funded by the National Institute on Deafness and Other Communication Disorders (NIDCD), the National Institute of Health (NIH), and the Massachusetts Eye and Ear (MEE). The authors would like to acknowledge contributions from other members of the CHSLT at the Worcester Polytechnic Institute.

References

1. Vest, C.M.: Holographic interferometry, vol. 1, p. 476. John Wiley and Sons, Inc., New York (1979)
2. Khaleghi, M., Cheng, J.T., Furlong, C., Rosowski, J.J.: In-plane and out-of-plane motions of the human tympanic membrane. J. Acoust. Soc. Am. **139**, 104–117 (2016)
3. Razavi, P., Cheng, J.T., Furlong, C., Rosowski, J.J.: High-speed holography for in-vivo measurement of acoustically induced motions of mammalian tympanic membrane. In: Mechanics of Biological Systems and Materials Volume 6, C.S. Korach, S.A. Tekalur, and P. Zavattieri, Eds., Proc. SEM 75–81 (2017). https://doi.org/10.1007/978-3-319-41351-8_11
4. Dobrev, I., Furlong, C., Cheng, J.T., Rosowski, J.J.: Full-field transient vibrometry of the human tympanic membrane by local phase correlation and high-speed holography. J. Biomed. Opt. **19**, 096001 (2014)
5. Razavi, P., Ravicz, M.E., Dobrev, I., Cheng, J.T., Furlong, C., Rosowski, J.J.: Response of the human tympanic membrane to transient acoustic and mechanical stimuli. Hear. Res. **340**, 15–24 (2016)
6. Van der Jeught, S., Dirckx, J.J.J.: Real-time structured light-based otoscopy for quantitative measurement of eardrum deformation. J. Biomed. Opt. **22**, 016008 (2017)
7. Dirckx, J.J., Decraemer, W.F.: Optoelectronic moire projector for real-time shape and deformation studies of the tympanic membrane. J. Biomed. Opt. **2**, 176–185 (1997)
8. Bedard, N., Shope, T., Hoberman, A., Haralam, M.A., Shaikh, N., Kovačević, J., et al.: Light field otoscope design for 3D in vivo imaging of the middle ear. In: Biomed. Opt. Express, vol. 8, pp. 260–272 (2017)
9. Kandulla, J., Kemper, B., Knoche, S., von Bally, G.: Two-wavelength method for endoscopic shape measurement by spatial phase-shifting speckle-interferometry. In: Appl. Opt., vol. 43, pp. 5429–5437 (2004)
10. Furlong, C., Pryputniewicz, R.J.: Absolute shape measurements using high-resolution optoelectronic holography methods. Opt. Eng. **39**, 216 (2000)

Chapter 34
High-Speed Digital Image Correlation for Endoscopy: A Feasibility Study

H. Tang, K. Pooladvand, P. Razavi, J. J. Rosowski, J. T. Cheng, and C. Furlong

Abstract Endoscopes are imaging tools utilized to observe objects in confined volumes. However, endoscopic imaging is qualitative, and only specific methods allow corresponding quantitative imaging under such conditions. This paper reports our progress in incorporating endoscopic imaging with digital image correlation (DIC) to perform high-speed quantitative measurements of shapes and displacements. A high-speed camera with a maximum framerate of two million per second is set up in an endoscopic configuration for 2D DIC measurements. Measurements are first verified against the displacements of a manual translation with a micrometer resolution. Then the system is applied in order to study the transient response of a piezoelectric actuator via step response. These results allow for the characterization and identification of the measuring challenges observed when performing high-speed endoscopic DIC. In addition, this study explores the potential of incorporating endoscopes into DIC imaging systems and providing the preliminary testing parameters of future optical designs for both biological and industrial applications.

Keywords Endoscopy · High-speed digital image correlation · PZT · Optical design

34.1 Introduction

Digital Image Correlation (DIC) has been utilized for the measurement of shapes and deformations of different applications, ranging from aerospace and automation to health and energy [1, 2]. With regard to imaging of confined volumes or regions that pose heightened risks for the optical system [3], the utility of endoscopes (borescopes, otoscope etc.) remains inevitable. However, large distortions (i.e., barrel, pincushion, coma, etc.) are known to adversely affect the DIC measurement results, implying that large spatially non-uniform correlation errors will disturb the measurements. In order to enable spatially uniform correlation coefficients within an endoscopic configuration, it is paramount to eliminate these distortions. Despite the significance of this topic, there are very few published works specifically focused on the utility of endoscopes in DIC as well as the methods to remove these distortion effects [4]. In this paper, we report our progress in incorporating endoscopic optics, high-speed digital image correlation (HS-DIC) techniques, and camera calibration methods to minimize the effects of

H. Tang (✉) · K. Pooladvand · P. Razavi
Center for Holographic Studies and Laser micro-mechaTronics (CHSLT), Worcester, MA, USA
Mechanical Engineering Department, Worcester Polytechnic Institute, Worcester, MA, USA
e-mail: HTANG3@WPI.EDU

J. J. Rosowski · J. T. Cheng
Eaton-Peabody Laboratory, Massachusetts Eye and Ear (MEE), Boston, MA, USA
Department of Otology and Laryngology, Harvard Medical School, Boston, MA, USA

C. Furlong
Center for Holographic Studies and Laser micro-mechaTronics (CHSLT), Worcester Polytechnic Institute, Worcester, MA, USA
Mechanical Engineering Department, Worcester Polytechnic Institute, Worcester, MA, USA
Eaton-Peabody Laboratory, Massachusetts Eye and Ear (MEE), Boston, MA, USA
Department of Otology and Laryngology, Harvard Medical School, Boston, MA, USA

L. Lamberti et al. (eds.), *Advancement of Optical Methods & Digital Image Correlation in Experimental Mechanics, Volume 3*, Conference Proceedings of the Society for Experimental Mechanics Series, https://doi.org/10.1007/978-3-319-97481-1_34

optical distortions. We present our preliminary results of performing HS-DIC measurements through an endoscopic imaging configuration. This study is our first step in building an endoscope based HS-DIC system in order to study high-speed dynamic events in confined volumes.

34.2 Method

We utilize 2D DIC with a high-speed camera to quantitatively measure displacements of moving objects through an endoscope. First, translational motions of an aluminum plate under steady-state conditions are captured by a high-speed camera through an endoscope to characterize and compensate for optical distortions. Camera calibration is applied to remove distortions by specific procedures using a planar checkerboard target. In order to determine the effectiveness of the calibration, known displacements (i.e., pure translations along *x*, or *y*) are applied, and these translations estimated by an open source DIC algorithm are compared before and after calibration [5]. Furthermore, the step response of a piezoelectric actuator is measured with the calibrated system. The preliminary experimental data provides essential guidelines for the future design and manufacturing of a miniaturized optical sensor for imaging and illumination under confined volumes.

34.2.1 Digital Image Correlation (DIC)

Measurements of displacements by DIC is based on the assumption that the sample's random surface feature (painted with speckle, if necessary) undergoes small displacements between two adjacent images such that the relativity among adjacent surface features does not change during the displacement. The images are divided into small subsets that are assumed to have a unique grayscale matrix. Tracking these matrices from the reference image through subsequent images enables the determination of displacements and strain of the sample's surface [6]. A typical correlation function of a point (x, y) is [7],

$$S\left(x, y, u, v, \frac{\partial u}{\partial x}, \frac{\partial u}{\partial y}, \frac{\partial v}{\partial x}, \frac{\partial v}{\partial y}\right) = 1 - \frac{\sum\left[F(x, y) * G(x^*, y^*)\right]}{\left[\sum\left(F(x, y)^2\right) * \sum\left(G(x^*, y^*)^2\right)\right]^{1/2}}, \tag{34.1}$$

where u and v refer to pixel displacements parallel to the image plane, $F(x, y)$ and $G(x^*, y^*)$ are the image's grayscale value before and after motions. Values of u and v are found by minimizing the correlation function.

In this paper, planar checkerboard targets with known dimensions are used to calibrate the imaging system in order to determine the intrinsic and extrinsic parameters of the camera in addition to radial and tangential distortion coefficients (Fig. 34.1). To find these parameters, the target checkerboard is placed at the imaging plane of the endoscope to capture different views after applying multiple rotations and translations to the target [8]. These recorded images are used to solve an overdetermined system of equation based on a pinhole camera model [9].

34.2.2 Experiment Setup

This study uses an ultra-high-speed camera (Photron SAZ) with settings for high-frequency acquisition that include a spatial resolution of 512 × 512 pixel, framerate of 67,200 fps, and exposure time of 15 μs [11]. A rigid body endoscope (KARL STORZ 26033 AP Endoscope) with a relay lens (f = 35 mm) attached to the eyepiece is utilized; the endoscope assembly is attached to the camera using a C-mount adapter. The flange distance is adjusted by C-mount spacers to utilize the maximum number of pixels of the camera.

An aluminum plate painted with black and white speckles is mounted atop a 10 μm (displacement resolution) manual *x-y* translation stage (Fig. 34.2a). The stage applies controlled discrete single axis translational motions, while the camera captures the motion at steady state. Even though the motion of this sample does not require high-frequency acquisition, the camera settings remain at low exposure time (15 μs) to mimic its performance for high-speed measurements. After measurement of discrete motions, more than 30 calibration images are recorded by manually rotating the calibration target at random angles while ensuring that the center of the target is at the same plane as the sample. Additionally, a step response of a piezoelectric (PZT) tilt stage is measured by the HS-DIC system (Fig. 34.2b, c) with the same camera configuration.

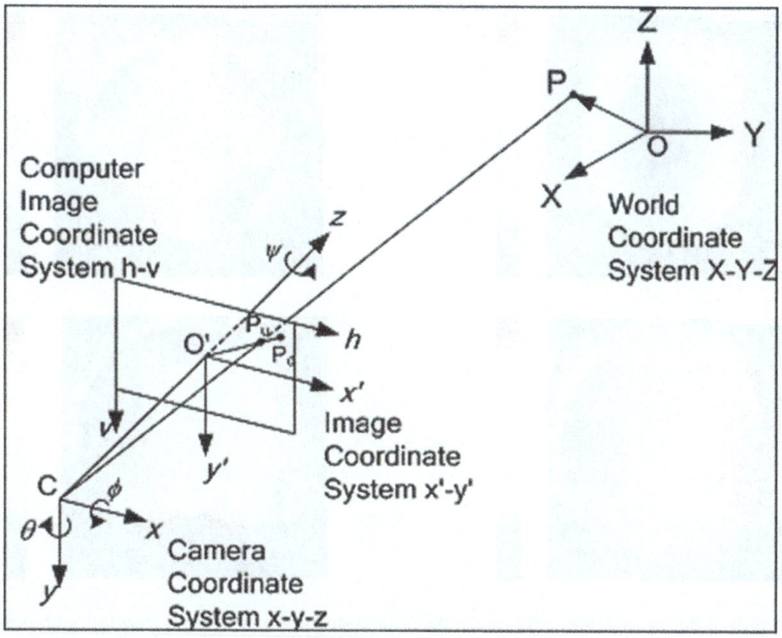

Fig. 34.1 The camera, image, and world coordinates in a pinhole model [10]

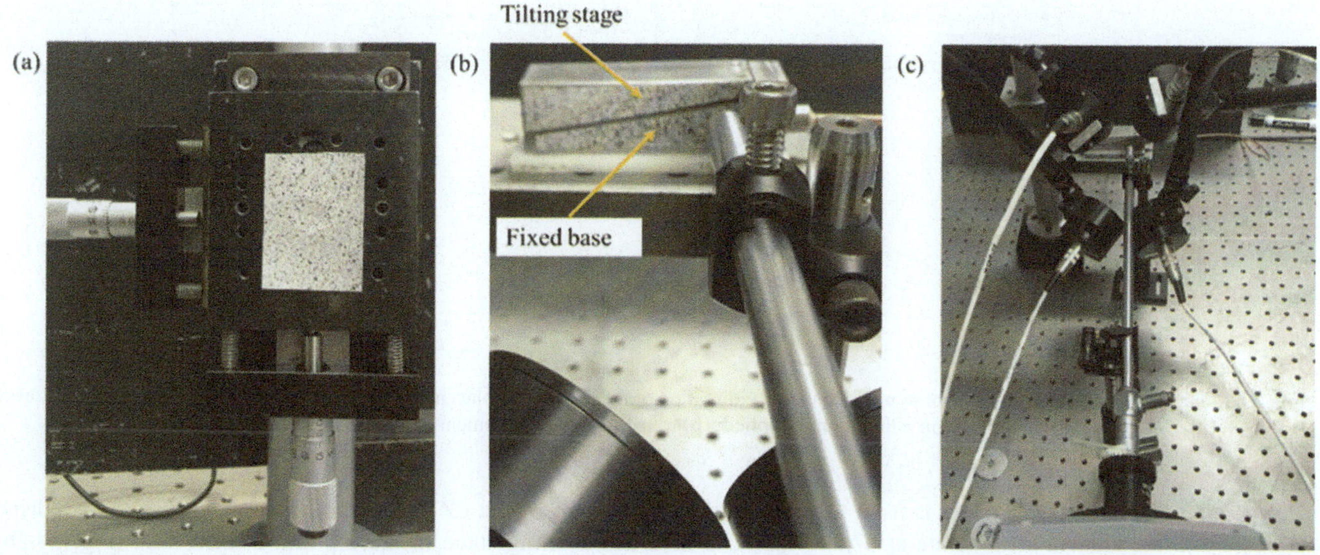

Fig. 34.2 Experiment setup. (**a**) Aluminum plate mounted on a manual *x*-*y* translation stage used for discrete displacement measurement; (**b**) piezoelectric tilt stage is observed by the endoscope; (**c**) experiment setup showing the sample, endoscope and camera assembly and the illumination lamps

34.3 Representative Results

A discrete translational displacement in the *x*-direction of 1 mm is measured by the endoscopic HS-DIC. The *x*-direction displacement error (e_U) and $y-$ direction displacement error (e_V) without applying the calibration process and those of the same data set calibrated for optical distortions are shown in Fig. 34.3. The results demonstrate that through the calibration, the errors in *x* and *y*-directions are reduced from ±24 to ±8%, and ±12 to ±2.5%, respectively.

In addition, a series of single direction translational displacements are measured by the same DIC configuration, and the measurement results after calibration are depicted in Fig. 34.4. Figure 34.4a shows the calibrated DIC measurements of *x*-direction translational displacement of 1 mm at different step size. Blue line is the actual displacement applied by the stage, and orange and yellow lines are the averaged DIC results of *x*-direction displacement (*U*) and *y*-direction displacement (*V*), respectively. Similar translational displacement along y-direction is also measured shown in Fig. 34.4b with green line being the actual displacement.

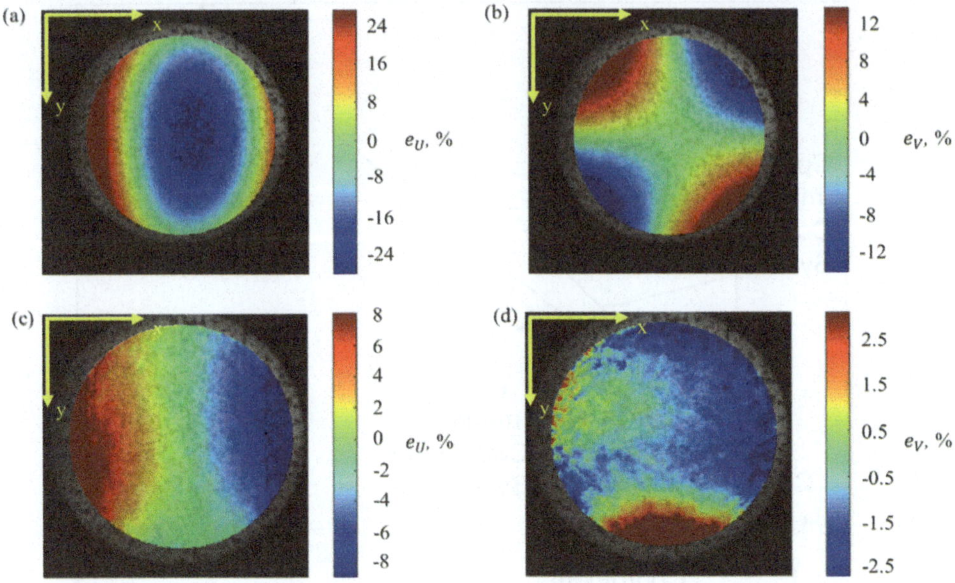

Fig. 34.3 DIC results comparison of before and after calibration for optical distortion of a 1 mm translational motion in x-direction. (**a**) The error distribution in x-direction e_U without calibration; (**b**) the error distribution in $y-$ direction e_v without calibration shows a fake motion; (**c**) the error distribution in $x-$ direction e_U with calibration; (**d**) the error distribution in $y-$ direction e_v with calibration

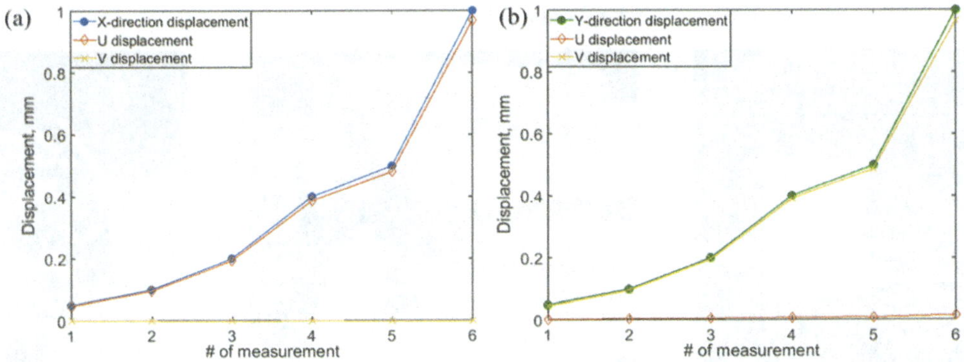

Fig. 34.4 The displacement measurement in x-direction and y-direction of the discrete displacement measurement with a calibration for distortion is shown. (**a**) Translational displacement in x-direction is applied; (**b**) translational displacement in y-direction is applied

The calibrated HS-DIC system is used to measure the step response of a PZT with the input signal to the PZT driver at 10 V and the camera's framerate at 67,200 fps. The PZT has a tilting stage in the top and a fixed base (Fig. 34.2b); the endoscope is focused on the tip of the tilting stage which has the most displacement during the excitation. Figure 34.5 illustrates the displacement distribution of the tilting stage of the PZT in x-and y-directions, respectively when the PZT reaches its maximum displacements. Figure 34.6 depicts the 100 times downsampled averaged displacement of the tip of the PZT's tilting stage during the step response. The x-direction displacement U and y-direction displacement V indicate that the PZT is an overdamped system with a time constant of 78 ms.

34.4 Conclusion and Future Work

The results reveal that by performing camera calibrations before applying DIC algorithm, it is possible to remove a significant amount of optical distortions, thereby making HS-DIC feasible in the context of endoscopic configuration. The future work will focus on performing validation measurements for the HS-DIC system in 3D. Measurements with a calibrated high-speed DIC using conventional optics and Laser Doppler Vibrometry (LDV) will be used to compare those with high-speed endoscopic DIC configuration in order to investigate the accuracy of these measurements.

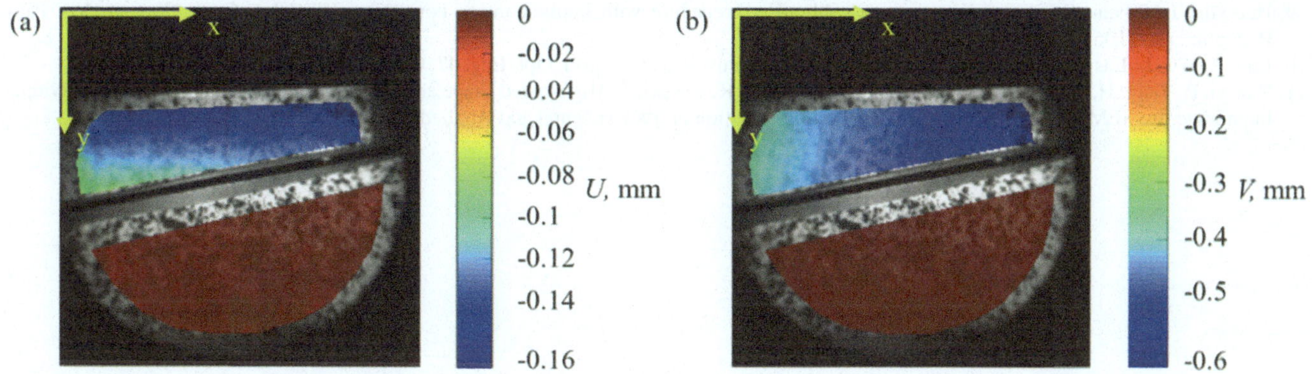

Fig. 34.5 The step response of the PZT showing the displacement distribution of the tilting stage and 0 displacements of the fixed bottom. (**a**) x-direction displacement distribution; (**b**) y-direction displacement distribution

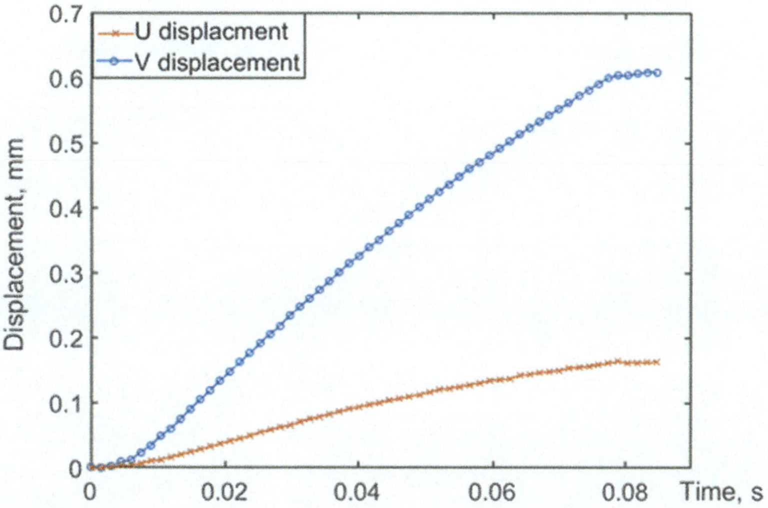

Fig. 34.6 Averaged tip displacement of the step response of the PZT. The data is downsampled 100 times due the long response time of the PZT

Acknowledgments This work has been supported by National Institute on Deafness and other Communications Disorders (NIDCD) and has been partially supported by NSF award CMMI-1428921. We would also like to gratefully acknowledge the support of the Mechanical Engineering Department at the Worcester Polytechnic Institute (WPI) and the contributions by other members of the CHSLT and MEE.

References

1. Sutton, M.: Digital image correlation for shape and deformation measurements. In: Sharpe, W.N. (ed.) Springer Handbook of Experimental Solid Mechanics, pp. 565–600. Springer, Boston, MA (2008)
2. Chu, T., Ranson, W., Sutton, M.: Applications of digital-image-correlation techniques to experimental mechanics. Exp. Mech. **25**, 232–244 (1985)
3. Razavi, P., Tang, H., Pooladvand, K., Furlong, C., Cheng, J.T., Ravicz, M.E., Remenschneider, A., Rosowski, J.J., Furlong, C.: Application of high-speed DIC to study damage of thin membranes under blast. In: Submitted to SEM 75th, Greenville, SC (2018)
4. Reu, P.L.: Digital image correlation through a rigid borescope. In: Optical Measurements, Modeling, and Metrology, vol. 5, pp. 141–145. Springer, New York, NY (2011)
5. Blaber, J., Adair, B., Antoniou, A.: Ncorr: open-source 2D digital image correlation matlab software. Exp. Mech. **55**, 1105–1122 (2015)
6. Bing, P., Kemao, Q., Huimin, X., Anand, A.: Two-dimensional digital image correlation for in-plane displacement and strain measurement: a review. Meas. Sci. Technol. **20**, 062001 (2009)
7. Bruck, H., McNeill, S., Sutton, M., Peters III, W.: Digital image correlation using Newton-Raphson method of partial differential correction. Exp. Mech. **3**, 261–267 (1989)
8. Zhang, Z.: A flexible new technique for camera calibration. IEEE Trans. Pattern Anal. Mach. Intell. **22**(11), 1330–1334 (2000)

9. Heikkila, J., Silven, O.: A four-step camera calibration procedure with implicit image correction. IEEE Int. Conf. Comput. Vis. Pattern Recognit., pp. 1106–1112 (1997)
10. Luo, P., Wu, J.: Easy calibration technique for stereo vision using a circle grid. Opt. Eng. **47**, 033607–1-033607-10 (2008)
11. Razavi, P., Tang, H, Maftoon, N., Furlong, C., Cheng, J.T., Rosowski, J.: High-speed shape and transient response measurements of tympanic membrane. In: SEM XV International Congress and Exposition on Experimental and Applied Mechanics, Greenville SC (2018)

Chapter 35
Holographic Interferometry: Then and Now

Karl A. Stetson

Abstract In the autumn of 1964, Robert L. Powell and I discovered holographic interferometry while working at the Radar Laboratory of the University of Michigan's Institute of Science and Technology. I have worked in this field ever since, and I have watched it grow from an unexplored technology to a widespread industrial testing method, and I have contributed to these developments. In this paper, I will trace my history in this field—our discovery and my involvement in its theory and applications. I will conclude with a discussion of digital holography, which is currently replacing photographic holography for most research and industrial applications.

Keywords Holography · Interferometry · Holographic interferometry · Metrology · Digital holography

35.1 Discovery

35.1.1 Off-Axis Holography

Off-axis holography was invented by Emmett Leith at the Optical Processing Group of the Radar Laboratory of the Institute of Science and Technology at the University of Michigan in the early 1960s. The main function of that group was optical processing of synthetic aperture radar, a new technology that had been invented in the 1950s at that laboratory. Data from a coherent side-looking airborne radar system was recorded on photographic film and optically processed to obtain images. At that time, photographic film was the only material capable of storing massive amounts of data, and the Fourier transform properties of lenses made optical processing of this data very convenient. Leith visualized the process as recording data at radar wavelengths and converting it to optical wavelengths for processing. By this insight, the processing was reformulated as an imaging process, and Emmett Leith found a precedent for this in Denis Gabor's hologram process which proposed recording electron microscope data as an optical pattern and thus converting it from electron wavelengths to optical wavelengths. Gabor's application was never put into practice, but synthetic aperture radar was, and it was quickly learned that an angle between the object and reference fields was necessary to avoid the twin image problem that plagued Gabor's in-line reference process.

I joined the group in September of 1962, just before they bought their first lasers, and I was in the laboratory on Dec. 24th when Emmett and Juris Upatnieks recorded and reconstructed their first laser hologram of a continuous tone optical transparency. Work progressed at a modest pace, and by 1964, laser holograms had progressed from continuous tone transparencies to transparencies illuminated by diffuse light, to three-dimensional objects. In September of that year, Emmett assigned me to work on the problem of the holograms. The diffraction efficiency of holograms with diffused light illuminating a transparency was about half what you got for a transparency without the diffuser. The corresponding diffraction efficiency for a hologram of a three-dimensional object was maybe one tenth of that. They had one hologram, made of a model train, that was far brighter than any of the others, and they were unable to duplicate it. In my memory, that hologram had been recorded during the demonstration of an experimental laser from Spectra Physics that was much more powerful than the 1 mW lasers we were using. That experimental laser eventually became the model 125 that was rated for 50 mW, but normally put out about 90. Around that time, Robert Powell, who had been one of my professors at undergraduate school, joined our group

---

K. A. Stetson (✉)
Karl Stetson Associates, LLC, Coventry, CT, USA
e-mail: kastetson@holofringe.com

L. Lamberti et al. (eds.), *Advancement of Optical Methods & Digital Image Correlation in Experimental Mechanics, Volume 3*, Conference Proceedings of the Society for Experimental Mechanics Series, https://doi.org/10.1007/978-3-319-97481-1_35

and we began working together. This laboratory had two large granite tables about a foot and a half thick on rubber isolation pads. The photographic material was Kodak 649F, a spectrographic emulsion of very high resolution that was extremely insensitive to light. With the mere 1 mW of power available, our exposures routinely took over a minute, and this, together with the fact that the tables were not on air-suspension vibration isolators and our room was next to the air conditioners, was the real problem of the holograms. We eventually solved this problem, and could consistently make very bright holograms of the model train.

35.1.2 Holographic Interferometry

It can be argued that holographic interferometry is the most wide-spread and successful application of holograms, and it is the only application that makes full use of a hologram's unique ability to reproduce the optical field reflected or transmitted by an object. It uses two unforeseen properties of off-axis laser holograms. First, a hologram can record fields that are incoherent and reconstruct them coherently so that they interfere. Those fields may come from separate laser modes, they may exist at different times, or they may be incoherent due to motion such as vibration. Second, a hologram can reconstruct the field from an object so precisely that it can interfere with the field from the object itself. Robert L. Powell and I discovered the first of these properties between late October and early December 1964, and the second in April of 1965.

It is documented [1] that the first observations of phenomena due to holographic interferometry were by Emmett Leith and Juris Upatnieks dating back to December 1963. The first case involved holograms recorded of paper cemented to an aluminum block with a hole in it that showed a black spot in the unsupported area. In another, a sheet of cardboard overlapped the block to which it was cemented and in that area the hologram reconstruction showed what looked like fringes. Upatnieks estimates the latter hologram as recorded in February or March of 1964. These observations, however, were not followed up to determine exactly what was happening. As stated by Upatnieks in an email to me [2], "We observed the effect and noted it, but did not get into any detail. Our main interest was image quality." The first observation Powell and I made of interference in a hologram reconstruction was the result of an unstable mounting of a holographic plate, which I pressed into a U-shaped frame against some rubber. It moved slightly during the hologram recording, and we saw a set of vague diagonal dark bands in the reconstruction. We abandoned this hologram and the mount and had a more stable one made with which we were able to record good quality holograms.

The Spectra-Physics representative had told me that these lasers emitted typically 3 longitudinal modes, and thus their longitudinal coherence was periodic in twice the cavity length, which was 60 cm. In response to concerns by others about the laser's coherence, I decided we should demonstrate that, and we set up an experiment where the unexpanded object beam was allowed to strike an object set between two mirrors. We positioned our hologram plate so that we could look through it over one mirror into the other and see a sequence of images, each 10 cm further in path length. We recorded several holograms with the zero path-length point lying at different points in that sequence of dots, and all of these holograms exhibited a periodic variation in image brightness as a function of path length just as we expected.

We noticed in one of these holograms what looked like a horizontal black fringe in the nearest image of the object beam spot. More important, this fringe moved as we raised or lowered our view through the hologram, so it was clearly not something on the object itself. We immediately set up an experiment to contain the entire object beam on a white surface, and capture the entire reference beam on the hologram plate. This recording gave us fringes that moved depending on the vertical position on the hologram plate through which we looked. I checked the laser and found that it was operating in more than one transverse mode. It was a Spectra-Physics/Perkin-Elmer model 110 laser that had a confocal mirror configuration, a spherical output mirror with a flat high reflectance mirror located at its center of curvature. There was a longitudinal adjustment for the flat mirror, and a single TEMoo mode was only obtained when the distance was set correctly. We initially had a combination of a 00 mode and a 20 mode and, after adjusting the laser, we recorded a hologram with a 00 mode and a 10 mode.

The question was: What gave us these fringes? I suggested to Bob that the photographic plate was recording separate holograms for each object and reference beam mode combination. When the hologram was reconstructed, however, both recordings would reconstruct their respective object beams, but since both fields were now being generated by the same reconstruction beam, the two reconstructed fields were coherent and could interfere. The shifting of the fringe positions would be due to phase variations between the two transverse mode recordings. Bob found this very disturbing, because it implied that fields that were incoherent during the recording of a hologram were being made coherent in its reconstruction. At this point that I referred back to the hologram we had made with the plate wedged against the rubber pad as proof of what I was suggesting. Bob was still unconvinced, and I said something like, "look, if we record a double-exposure hologram and give the object a small rotation between the exposures, wouldn't we expect to see fringes in the reconstruction?" That question spurred immediate action. It took us some time to learn how much rotation to give the object, hologram plate, or reference beam, but we got this to work and showed double-exposure interference patterns in hologram reconstructions.

As with Leith and Upatnieks, our object was a sheet of paper cemented to an aluminum plate, and in one of the reconstructions, we noticed a perturbation of the fringes in a region where there was a bolt hole. In retrospect, I'm sure the paper had statically deformed between the two exposures, but at the time we wondered if the effect could be vibration. We set up an empty can from a reel of 35 mm film with a solenoid mounted under it, and this generated the remarkable set of J_0 fringe patterns which we presented at the 1965 spring meeting of the Optical Society of America and published in the December 1965 issue of JOSA [3]. My record books are long gone, but to the best of my memory, we did the coherence function experiments in October 1964, the double-exposure holograms in November 1964, and the vibration holograms in early December 1964. In contrast to Leith and Upatnieks, Robert Powell and I observed a phenomenon, studied it in detail, proposed an explanation, and then tested and confirmed that explanation.

As we submitted our paper to the OSA for the spring meeting of the Optical Society of America at Dallas, TX, in April 1965, I proposed to Emmett that we should also present a paper on the entire set of experiments we had done defining holographic interferometry. He vetoed the idea saying that one paper was adequate, an unfortunate decision, and the reason for it became clear to me in 1970. He and Juris had filed a number of patent disclosures on holographic applications, including one on coherence measurement and one on vibration analysis, and Bob and I had unwittingly walked into these areas. We presented our additional material at the OSA annual meeting in Philadelphia in October 1965, and published it in JOSA in September 1966 [4].

Bob and I searched for prior work using interferometry for vibration analysis and found a paper by Harold Osterberg [5], which identified the fringes we saw in our reconstructions as a zero-order Bessel function of the first kind. It also suggested a method for real-time vibration analysis. Just after the spring meeting of the OSA, it occurred to me that if we could replace a hologram in its holder accurately enough, its reconstruction ought to interfere with the field from the original object. We tried this and it worked, giving us a real-time vibration measurement technique, a method of evaluating the quality of a hologram reconstruction, and measurement of object deformation in real-time. We submitted a letter to the editor in JOSA, which appeared in the same issue as our vibration analysis paper [6].

Kenneth Haines and Percy Hildebrand, who worked in the Willow Run Laboratory, were also thinking about holography in April 1965. They borrowed our real-time holographic interference setup to record the holograms showing deformation of a plate due to bolt tightening. In April 1996, Haines and Hildebrand published a paper [7] presenting real-time holographic interferometry as a "new method" solely attributable to them. The prior work by Robert Powell and me is described by the sentence "A related technique for vibration measurement was used by Powell and Stetson." Their paper presented an analysis of the fringes and fringe localization, but this paper and a nearly identical one following it [8] were so mathematically dense that neither has been used for that purpose. That year, they also published a theoretical paper [9] describing how object contours could be obtained by changing the laser wavelength. In the following years, I and many others developed comprehensive descriptions of the relationship between object deformations and observed fringes and fringe localization.

A possible contender for discovery prior to ours was Melvin H. Horman [10]. He presented a paper at the same OSA meeting in Dallas where we presented our vibration work, and published in Applied Optics in March of 1965, with a received date of October 1964. His paper was essentially theoretical with no experimental work presented, and it proposed that a hologram could be used to replace an object in an otherwise conventional Mach-Zehnder interferometer. To perceive that theoretically was impressive, but there is no mention in the paper of the key concepts of holographic interferometry—the ability of a hologram to make fields that are incoherent during recording coherent during reconstruction, nor was a hologram proposed as a beamsplitter between the reference and object beams used in its recording.

The year 1965 saw a number of independent discoveries of holographic interferometry. One of the first was by Collier, Doherty, and Pennington [11], who described interference effects within a hologram reconstruction as a moiré effect. Pulsed laser holographic interferometry was reported by Brooks, Heflinger, and Wuerker [12]. Their discovery was a fortunate accident that occurred while they were recording pulsed laser holograms of a bullet in flight. The laser fired two pulses during one of these recordings, one of which occurred with the bullet in the field of view and one without it. The shock waves in the air were clearly visible as an interference pattern. In England, J. M. Burch published a paper in which he described experiments done at the National Physical Laboratory in Teddington that showed both double-exposure and real-time holographic interferometry [13]. In the Soviet Union, holographic interferometry was discovered about this same time by Yu I. Ostrovsky.

35.1.3 Characteristic Fringe Functions and Separable Object Motions

The function that describes the fringes observed in a time-average hologram reconstruction is the time average of the phase generated by the object motion. This can be expressed as

$$M(\Omega) = (1/T)\int_0^T \exp(i\Omega\ f(t))\ dt, \tag{35.1}$$

where M is the fringe function, Ω is the fringe locus function, T is the exposure time, f(t) is the function of time describing the object motion, and t is time. The fringe locus function can be described in terms of the object displacement vector, **L**, and propagation vectors describing light propagation from the object to the hologram, $\mathbf{K_2}$, and from the illumination to the object, $\mathbf{K_1}$, as

$$\Omega = (\mathbf{K}_2 - \mathbf{K}_1) \cdot \mathbf{L} = \mathbf{K} \cdot \mathbf{L}, \tag{35.2}$$

where **K** is called the sensitivity vector. Implicit in this formula is the idea that the object motion is separable, i.e., that it can be expressed as the product of a single object deformation vector, **L**, times a single function of time, f(t). This means that all points on the object move with the same time function. With vibrating objects, this is not always the case.

Adam Kozma, one of the major figures at the Optical Processing Group, made a significant contribution to the theory of holographic interferometry very shortly after our demonstration of holograms of vibrating objects. He sought to describe hologram reconstructions for random object motions. He observed that for object motions are stationary and ergodic in a statistical sense, the values it would assume could be described by a probability density function, p(f), describing the percentage of the exposure time at which it assumed each value. Thus, we could rewrite Eq. (35.1) as

$$M(\Omega) = \int_{-\infty}^{+\infty} p(f)\ \exp(i\Omega\ f)\ df = \mathcal{F}_\Omega\left[p(f)\right], \tag{35.3}$$

where $\mathcal{F}_\Omega$ [p(f)] indicates the Fourier transform of p(f) with respect to the variable Ω. This formulation makes it quite easy to tabulate fringe functions, and, for Gaussian motion in particular, the fringe function becomes itself a Gaussian function. This formulation also stresses the fact that the fringe function is not determined by the exact motion of the object but rather by how much of the exposure time it spends in each position.

35.2 The Search for Applications

Robert Powell and I were not mechanical engineers. For example, one of the vibration patterns we recorded of our film can bottom showed a five-lobed pattern that was an impossible vibration mode for a circular vibrating structure. In retrospect, it was a combination of a five-diameter mode and a two-ring circumferential mode and not a mode by itself, but we did not understand that at the time. I communicated information about our discovery quite freely, and one of the people I told was Ralph Grant, a professor of acoustics at the University. I also came in contact with Donald Gillespie, who was very interested in finding a way to exploit holograms financially. Don worked in a laser laboratory and was able to make his own lasers. I helped him set up a holography laboratory in his basement in which he made holograms for demonstration purposes. Eventually he and his brother John Gillespie formed the first company to make equipment specifically designed for holography laboratories, Jodon Engineering. Ralph Grant had teamed up with a lawyer, Joseph Crafton to form Grant-Crafton Optronics, aka GCO. They were fortunate to find that this new interferometry could make bonding flaws visible on laminated structures, such as pneumatic tires, when subjected to changes in pressure, and this became a major business. Around 1973 or so, GCO ended and was replaced by Industrial Holographics, Inc., which, in turn, was replaced by Grant Engineering, Inc. around 1986. Approximately 1997, this business was obtained by another company who sold systems under the trade name, L-Ray.

The aircraft industry became a major user of holographic interferometry. Holographic testing required a significant initial investment which added considerable cost to products, and this was practical for items like airplanes which cost a lot to make and for which failures are very serious. In jet engines, high cycle fatigue due to vibration is a major source of failure for the blades, and they are designed to have resonances that are not excited at the rotation speeds of operation. Holography offered a way to visualize the mode shapes and confirm mathematical analyses. This was used very early on by Pratt & Whitney,

GE, and Rolls-Royce, and also by Westinghouse for analysis of vibration modes for steam turbine blades. Aircraft structures made of honeycomb cores and metal or composite skins were also inspected by holography for delaminations. Another major area of application has been analysis of air flow and analysis of plasmas, which generally required use of pulsed lasers. A wide survey of techniques can be found in the book, *Holographic Nondestructive Testing*, edited by R. K. Erf [14], and a detailed discussion of holographic interferometry can be obtained in the book *Holographic Interferometry*, by C. M. Vest [15].

35.3 Recording Materials

35.3.1 Photographic Film

In 1967, I joined the Institute for Optical Research at the Royal Institute of Technology in Stockholm to work on a doctoral degree. They were already working with holograms, and to my surprise and delight, they had new photographic materials from Agfa Gaevert for recording holograms that were much faster than the Kodak 649F. As I recall, there were a total of four materials: 8E70, 8E75, 10E70, and 10E75. The 8E series were higher resolution and slower than the 10E series, and were suitable for recording standing waves such as in Lippmann holograms. The 70 series were red sensitive and had a band of low sensitivity to blue-green light, which allowed them to be used under a blue-green safe light. That series was eventually eliminated. When I moved to England in 1969, to join James Burch's group at the National Physical Laboratory, I found that Ilford had also brought out a high speed holographic material. I returned to the United States in 1971, and within a couple of years Kodak came out with a material they called SO-253 which was the fastest holographic material made. All of these materials were available as either photographic emulsions on glass plates or as reels of film at typically 35 mm width. Jodon Engineering eventually came out with a unit for recording holograms on photographic film. It held a supply reel and a take-up reel, and the film threaded through an open area that was slightly curved for higher stability. A matching fixture was supplied for holding the film in the same position for reconstruction. It was typical to record image-plane holograms on photographic film because warping of the film would affect the quality of reconstructions if the light had to propagate much distance to form an image.

35.3.2 Thermoplastic Holograms

The first thermoplastic recording of holograms was reported in 1966 by Urbach and Meier [16]. A thin photo-conducting layer is covered by a transparent layer of thermoplastic material. The thermoplastic material was charged, the photoconductor was exposed, the layer was charged a second time, and the outer layer heated. The charge pattern caused the thermoplastic to deform, and after it is cooled it diffracted light to form a hologram reconstruction. The advantage of this was that the hologram was exposed and developed without being moved. In-situ processing of photographic plates was developed in 1970 [17], but the process involved holding a plate within a liquid gate. Typically, the exposure would be with water in the gate, followed by developer, followed by a fixing bath, and often returned to water for the reconstruction phase. By comparison, thermoplastic holograms were completely dry, and required less time between exposure and reconstruction. Both of these techniques were a boon to real-time holography, because nearly perfect, fringe-free, interferograms could be obtained. Hans Rottenkolber, of Rottenkolber Holo-Systems, marketed a thermoplastic recording system in the 1970s and eventually Newport Research Corp. produced a system of this sort as well. The Rottenkolber system used a movable strip of thermoplastic which was advanced for each hologram recording whereas the Newport system reused the same thermoplastic layer which was bonded to the photoconductor.

35.4 The Speckle Interferometer

One result of world-wide interest in holographic interferometry was that it prompted people to think about laser speckles. One of the first innovations was the speckle interferometer [18], a device invented at the National Physical Laboratory (NPL) in 1969. This was an optical instrument that allowed a person to look at an object through an aperture small enough that speckles were clearly visible on its surface. This image was combined with a smooth reference beam to create speckles that would blur if the object vibrated but which would remain of high contrast at the vibration nodes. I was fortunate to spend 2 years at NPL starting in September 1969, and I had the opportunity to make substantial improvements to this instrument [19]. Television cameras were eventually used to observe and process these images, and Electronic Speckle Pattern Interferometry (ESPI) was born. About the same time, John Leendertz at Loughborough University published work demonstrating the use of speckles for measuring displacements and strains of objects [20], and this gave rise to an independent field of measurement called speckle metrology [21].

35.5 Heterodyne Interferometry

Heterodyne readout of interference fringes in holographic interferometry was introduced by Dändliker, et al., in 1973 [22], and this introduced a new level of precision and flexibility. By recording two exposures of a double-exposure hologram with separate reference beams, it became possible to introduce a frequency shift between the two reconstructing beams and convert the fringes into sinusoidal irradiance fluctuations. These could be converted to electrical signals by a photodetector, and these could be analyzed by an electronic phase meter with a phase resolution of 0.1°. The fringes could be evaluated anywhere within the interferogram and to a much higher level of accuracy than by locating fringe centers.

35.6 Phase Stepping and Digital Holography

An alternative to heterodyne readout of interference fringes is to step the phase of a fringe pattern through a set of discrete values and record the irradiance values at each step. If the phase steps are known, and three or more steps are used, it is possible to calculate the phase from the irradiance values. This method is well suited to array detectors such as solid-state TV cameras which can measure irradiance values for an array of picture elements, or pixels, and this technique was introduced into holographic interferometry by P. Hariharan, et al. [23] in 1982 and further developed by K. Creath [24]. Solid-state cameras are an essential enabling factor for this, for unlike the vidicons and orthicons that proceeded them, solid-state cameras provide accurate irradiance measurements and are much more tolerant of overexposure. In addition, their fabrication via photolithography results in a pixel arrays that have essentially no distortion.

In the late 1980s, I combined phase stepping with the speckle pattern interferometer to create the first system to record and display in real time what are now called digital holograms [25]. The origin of digital holography is generally taken to be the work of Schnars and Jüptner [26] where a charge-coupled device (CCD) TV camera was used directly to replace film in an off-axis hologram recording system. The image of the object was obtained from the recorded data by a numerical calculation that duplicates optical propagation. In our system, the image of the object was focused on the TV camera, so there was no incentive to explore the possibility of focal plane shift via numerical calculation. The data recorded by our system, however, always allowed for this possibility, as was eventually shown in 2009 [27].

The main advantage digital holography has over photographic holograms is that it is possible to calculate the phase of the object image, and, correspondingly, the phase of holographic interference fringes. With double-exposure holograms, this is most easily done by calculating the random image phase of the object before and after deformation, and subtracting these two random functions. The phase calculations generally result in eight-bit pixel values, and subtraction of two such numbers creates a nine-bit number, which also exhibits a phenomenon I call random wrapping. For example, if a larger number is subtracted from a smaller one, the result is a negative number, which will be represented as a positive number in the upper range of the nine-bit numbers. This problem can be eliminated by wrapping the nine-bit numbers into eight bits. The result of this is a wrapped phase map of the object deformation which then has to be unwrapped to obtain a continuous map of the object deformation. Numerous methods for phase unwrapping are discussed in a book by Ghiglia and Pritt [28]. I introduced a method not included in their book, which makes use of calculated phase unwrap regions [29].

The ability to obtain phase maps via phase stepping suggested a way to apply this method to vibration fringes. Because the zero-order Bessel function is nearly periodic, it is possible, approximately, to phase step J_0 fringes by introducing a bias vibration into the holographic reference beam. The bias vibration has to be at the same frequency as the object vibration and has to match its phase. Under these conditions, the bias vibration simply adds a scalar number to the argument of the fringe function, and by setting its amplitude correctly, it can be set to yield the equivalent of a 120° phase shift. Data recorded with

±120 and 0° can be used to calculate the resulting argument of the J_0 function under the assumption that it were actually a cosine function. The error resulting from this calculation can be precalculated and a table constructed to correct the results to obtain the actual argument of the J_0 function [30].

Deformation and vibration analyses of objects important in engineering applications, such as turbine blades and structural components, are probably most easily analyzed by the phase-stepped, image-plane digital holography system we developed 25 years ago. Digital holography, generally, has found extensive applications in microscopy where the focusing capabilities are more important. Details of these applications are found in numerous books, such as those by Schnars and Jüptner [31] and Kim [32].

35.7 Conclusion

Looking back over the past 54 years is an interesting experience. It has given me the opportunity to recall the state of science and technology as it was back then and compare it to what we have available now. Of course, a major factor has been the remarkable rise of personal computers, one of which I am using to write this document and which I am using to listen, simultaneously, to my local classical music station. Fifty years ago, I would have given nearly anything to have the technical capabilities I have now, and had we had them back then, history would have been very different. Optical processing of synthetic radar data would probably not have been developed. Correspondingly, the entire subject of Fourier Optics might not have been pursued either. The development of lasers back in the 1960s, however, would still have galvanized the scientific and engineering community, and many of the things that they have brought to light would still have occurred.

I would like to close with my favorite "what if" story. While I was at NPL, we were visited by Samuel Tolansky a couple of years before his death. He was well known for his work in multiple-beam interferometry and was a major figure in the optics community. He told of having done experiments back in the 1930s with a Fabry-Perot interferometer, an instrument that consists of two parallel mirrors between which light reflects back and forth many times as it does in a laser cavity. He was investigating the phenomenon, predicted by Einstein, of stimulated emission using a neon discharge, and I remember his comment to this day. "I wouldn't have felt half so bad if I hadn't had helium on the bench, but I never thought of mixing the two." Had he decided to see what might happen with a mixture of helium and neon, he might have happened upon the He-Ne laser 30 years before it was discovered. Who can say how much history would have been changed had this happened?

References

1. Johnston, S.F.: Holographic Visions, pp. 191–200. Oxford University Press, London (2006)
2. Upatnieks, J.: Email to Stetson, K.A., 10 Dec 2013
3. Powell, R.L., Stetson, K.A.: Interferometric vibration analysis by wave front reconstruction. J. Opt. Soc. Am. **55**(12), 1593–1598 (1965)
4. Stetson, K.A., Powell, R.L.: Hologram interferometry. J. Opt. Soc. Am. **56**(9), 1161–1166 (1966)
5. Osterberg, H.: An interferometer method of studying the vibrations of an oscillating quartz plate. J. Opt. Soc. Am. **22**(1), 19–35 (1932)
6. Stetson, K.A., Powell, R.L.: Interferometric hologram evaluation and real-time vibration analysis of diffuse object. J. Opt. Soc. Am. **55**(12), 1694–1695 (1965)
7. Haines, K.A., Hildebrand, B.P.: Surface-deformation measurement using the wavefront reconstruction technique. Appl. Opt. **5**(4), 595–602 (1966)
8. Haines, K.A., Hildebrand, B.P.: Interferometric measurements on diffuse surfaces by holographic techniques. IEEE Trans. Inst. Meas. **IM-15**(4), 149–161 (1966)
9. Haines, K., Hildebrand, B.P.: Contour generation by wavefront reconstruction. Phys. Lett. **19**(1), 10–11 (1965)
10. Horman, M.H.: An application of wavefront reconstruction to interferometry. Appl. Opt. **4**(3), 333–336 (1965)
11. Collier, R.J., Doherty, E.T., Pennington, K.S.: Application of moiré techniques to holography. Appl. Phys. Lett. **7**(8), 223–225 (1965)
12. Brooks, R.E., Heflinger, L.O., Wuerker, R.F.: Interferometry with a holographically reconstructed comparison beam. Appl. Phys. Lett. **7**(9), 248–249 (1965)
13. Burch, J.M.: The application of lasers in production engineering. Prod. Eng. **44**(9), 431–442 (1965). (The 1965 Viscount Nuffield Memorial Paper)
14. Erf, R.K. (ed.): Holographic Nondestructive Testing. Academic, New York (1974)
15. Vest, C.M.: Holographic Interferometry. Wiley, New York (1979)
16. Urbach, J.C., Meier, R.W.: Thermoplastic xerographic holography. Appl. Opt. **5**, 666–667 (1966)
17. Biedermann, K., Molin, N.-E.: Combining hypersensitization and rapid *in-situ* processing for time-average observation in real-time hologram interferometry. J. Phys. E Sci. Instrum. **3**, 669–680 (1970)
18. Archbold, E., Burch, J.M., Ennos, A.E., Taylor, P.A.: Visual observation of surface vibration node patterns. Nature. **222**, 263–265 (1969)
19. Stetson, K.A.: New design for a laser image-speckle interferometer. Opt. Laser Technol. **2**(4), 179–181 (1970)

20. Leendertz, J.A.: Interferometric displacement measurement on scattering surfaces utilizing speckle effect. J. Phys. E Sci. Instrum. **3**, 214–218 (1969)
21. Erf, R.K. (ed.): Speckle Metrology. Academic, New York (1978)
22. Dändliker, R., Ineichen, B., Mottier, F.: High resolution hologram interferometry by electronic phase measurement. Opt. Commun. **9**(4), 412–416 (1973)
23. Hariharan, P., Oreb, B.F., Brown, N.: A digital phase-measurement system for real-time holographic interferometry. Opt. Commun. **41**(6), 393–396 (1982)
24. Creath, K., "Phase measurement interferometry techniques," Prog. Opt., Vol. 26, 349-393 (1988). E. Wolf, ed
25. Stetson, K.A., Brohinsky, W.R., Wahid, J., Bushman, T.: Electro-optic holography system with real-time arithmetic processing. J. Nondestruct. Eval. **8**(2), 69–76 (1989)
26. Schnars, U., Jüptner, W.: Direct recording of holograms by a CCD target and numerical reconstruction. Appl. Opt. **33**(2), 179–181 (1994)
27. Stetson, K.A.: Mathematical refocusing of images in electronic holography. Appl. Opt. **48**(19), 3565–3569 (2009)
28. Ghiglia, D.C., Pritt, M.D.: Two-dimensional Phase Unwrapping. John Wiley & Sons, New York (1998)
29. Stetson, K.A., Wahid, J., Gauthier, P.: Noise-immune phase unwrapping by use of calculated wrap regions. Appl. Opt. **36**(20), 4830–4838 (1997)
30. Stetson, K.A., Brohinsky, W.R.: A Fringe shifting technique for numerical analysis of time-average holograms of vibrating objects. J. Opt. Soc. Am. A. **5**(9), 1472–1476 (1988)
31. Schnars, U., Jüptner, W.: Digital Holography: Digital Hologram Recording, Numerical Reconstruction, and Related Techniques. Springer, Berlin (2010)
32. Kim, M.K.: Digital holographic microscopy: principles, techniques, and applications. In: Springer series in optical sciences (2011)

The manufacturer's authorised representative in the EU is Springer Nature Customer Service Centre GmbH, Europaplatz 3, 69115 Heidelberg, Germany. If you have any concerns regarding our products, please contact ProductSafety@springernature.com

Printed and bound by CPI Group (UK) Ltd, Croydon, CR0 4YY
30/09/2026
02225270-0001